Autodesk® Robot™ Structural Analysis Professional - 2016

Essentials

The essential guide to learning Autodesk Robot Structural Analysis Professional...

KEN MARSH

ISBN-13: 978-0-9915181-3-5

Marsh API
179 Albion Street • Unit 2
Somerville, MA 02144
RobotEssentials@marshapi.com

Dear Reader,

Thank you for choosing Autodesk® Robot™ Structural Analysis Professional 2016 – Essentials. Learning any new software can be a challenge. With this in mind, we have attempted to present a thorough guide to the basics of Robot Structural Analysis along with some insight into the deeper functioning of the program. We hope that you will find this text to be very helpful on your way to mastering the program. We have tried to be as accurate as possible, give you some deeper color around the features of the program, and to hopefully demystify some of the more challenging commands in the program. We are always happy to hear your feedback and look forward to the opportunity to continue improving this text and subsequent texts through your feedback. Please feel free to contact me at ken@marshapi.com with any thoughts you would like to share. I am also happy to receive requests for training or consulting in the areas of Robot Structural Analysis, Revit, Revit Structure, and AutoCAD Structural Detailing

Kind regards and best wishes on your path to mastering Robot Structural Analysis Professional,

-Ken

Owner, Marsh API, LLC

To Li, thank you for your patience, support, and encouragement: You've kept this drive alive!
—Ken

Acknowledgements

I recall being surprised to learn that my piano teacher in middle school also had a teacher. I suppose, at that point in life, I believed that it was possible to know everything or to have accomplished everything… with age comes wisdom: First, I would like to thank Autodesk for the amazing opportunity to work with an incredible team on the Revit® Product. It has been an awesome adventure and has led me to some amazing places in the world. I would also like to take a moment to thank the customers of Autodesk software: Without you, I would never have had the opportunity to be involved with an amazing revolution in our industry which we have experienced in the growth and adoption of Building Information Technology. I still believe that we have only scratched the surface and there are still many exciting things to come.

There are a few people who have been amazing inspiration along my journey, not only in writing this book, but also in my personal journey. I would like to first thank Nicolas Mangon for bringing his vision of Revit for Structural Engineers to fruition in the form of Revit Structure. Even at version 2.0, I knew it was going to be a major revolution in our industry and I'm thrilled to have been a part of it. I would like to thank Anthony Hauck for helping me learn to see the bigger picture and for his willingness to share his vision of the future. I would like to thank Pawel Piechnik for his encouragement and positive outlook. I would like to thank my first Robot teachers Tomek and Waldek. Without your help I would probably have never begun this adventure in the first place. I would like to extend a huge thank you to Pawel Pulak, an amazingly patient, brilliant, kind, and gifted product support specialist at Autodesk. His insight, tireless accurate diagnoses, and deep product and engineering knowledge have been indispensable to me throughout this process. I would also like to thank the other Robot product support specialists Autur and Rafal: you guys rock.

Contents at a Glance

Contents

Introduction

Robot Structural Analysis Professional is an incredibly powerful tool for general purpose structural engineering, analysis and design. As it has been said with regard to software: "With great power comes great complexity". While this has not always been the case for all software, Robot Structural Analysis Professional seems to fall into this category. The standard methods of interacting with typical windows style programs do not necessarily apply to Robot which can lead to some confusion while trying to learn this product. I vividly recall my personal process of learning Robot and I would say that the learning process continues on a day to day basis.

I live in Boston. I was not born in Boston. If you know anything about Boston, you know that there is nary a straight road in the town (save Back Bay: lucky us). Learning to drive around Boston is quite possibly one of the most exhausting experiences I've had. It makes it impossible to enjoy the sights, impossible to find where you want to go and infuriating after you've missed the turn and can't figure out how on earth you can get back to where you were before. I feel quite the same about Robot as a learning experience, but once you get the hang of it, you'll be off and running. Now I love driving in Boston and I also think you'll have a lot of fun with this powerful tool.

The Goal of this Book

My goal in writing this introduction to Robot is to give you the nuts and bolts of Robot: Navigating the interface, the methods and modes of interacting with Robot, the General workflows for design so that we can demystify the unique nature of the interface and allow you to begin your own investigation of the product. No practicing structural engineer I have ever known, would swap out a structural analysis package without a thorough investigation of the capabilities and not until having gained a thorough understanding of the settings and their effects on the outcome of the analysis and design. After working through this book, you should have the basic tools to quickly and efficiently set up validation problems and understand the analysis and design parameters.

Non-Goals

This book is not intended to be the ultimate Robot handbook nor a detailed, complete reference for all the advanced capabilities of Robot. The product is incredibly powerful and has a 20+ year history of highly advanced and theoretically based development. The list of capabilities and configuration options is mind-numbing. We will not cover advanced analysis types (e.g., modal, harmonic, moving load, pushover, dynamic, footfall), we will not cover selection of solver (several are available), we will not cover wall and slab reinforcing design, and we will not cover timber design. However, on the topic of timber design and slab/wall reinforcing, the topics we do cover should give you a great start on those functions as well. For the topics we cover, I will try to give a thorough explanation of the essential functionality with enough detail that you can make good progress. We will leave advanced topics for another text.

I think it's important to point out that this book is also not about Finite Element Modeling nor a course in best practices for Finite Element Modeling. The topic of FEM is deep and wide ranging and there are many books already available on the subject which would be more enlightening than I could probably ever be on the subject (much to the chagrin of my FEM professor…). Please see the section on Further Reading for suggestions on additional texts which can help you with the finer points of Finite Element Modeling.

Assumptions

This book is written by a structural engineer for structural engineers. It has been said that Engineering is the art of using materials we don't fully understand to build structures from components we cannot fully analyze which must resist forces of nature which we cannot fully quantify nor accurately predict in such a way that the public at large has no reason to suspect the extent of our ignorance. The truth is that we, as a profession, have done extensive research to understand the statistical probabilities of load variation, material variation, and to understand the limits of our ability to analyze structural components. The safe design of structures for use by the public relies on this body of work which has been codified in large part and for which tools such as Robot Structural Analysis have been designed to help us implement. This book presumes that you are already familiar with and proficient with both the process and science of building structural analysis by the finite element method. The subject of Finite Element Analysis (FEA) is extensive and is thoroughly covered in other texts. It should be said here that, while Robot Structural Analysis is a very powerful tool for building structural engineering, it is a tool which requires a strong foundation in the engineering principles behind FEA and using it without understanding the principles behind it can lead to unsafe designs and very bad consequences.

I also assume you have successfully installed and licensed the software. Autodesk technical support is highly skilled at assisting with any installation problems. I would feel remiss at spending pages here to detail the installation process.

What you will learn

Chapter 1 - Getting Started From opening your first project, starting to understand the interface and navigating a project, using workplanes, graphical views and tables. You will gain familiarity with the basic function of the interface. Most importantly you will learn all about how to manage selections in Robot.

Chapter 2 - Preferences and Project Setup Take a look at the basic program preferences and initial setup. Configure material databases, unit formats, design and combination codes, and take a look at analysis settings.

Chapter 3 - Basic Structural Modeling Learn how to layout structural grids and stories, add nodes if desired, model bars, beams, columns, and apply support definitions. You will begin to learn about Robot labels: a very important concept in Robot.

Chapter 4 - Modeling Loads Learn the ins and outs of adding loads to your structure model. Some of the loading techniques are unique and you will get insight into adding all types of loads for static analysis.

Chapter 5 - Preliminary Structural Analysis In this chapter, we will look at Robot's automatic load combinations: a very important feature of Robot analysis. We will discuss model verification and techniques for resolving issues with calculations as well as analysis types and an introduction to Robot's nonlinear solver.

Chapter 6 - Exploring Preliminary Results This chapter covers the tools Robot provides to investigate your analysis results. Robot result terminology will be explained, directions of forces and stresses are covered, as well as the use of Robot result tables for drilling down into the result information.

Chapter 7 - Printout Composition Learn the secrets of Robot's screen capture utilities, how to put your report together with impactful images and things to watch out for while prepping your presentation.

Chapter 8 - Basic Steel Design Workflow Get the hang of steel design in Robot and how to manage both deflection and strength control. Look at code parameter lables in Robot for AISC 360 – 05 and learn techniques for creating design groups to control your member design.

Chapter 9 - Basic Wood Design Workflow Learn the ins and outs of Robot's implmeentation of the AWC-NDS 2012 wood design code. Learn the right way to add materials, assign wood/timber sections, and configure design code parameters.

Chapter 10 - Structure Modification Robot has some very powerful modification tools. We'll take a look at the edit menu, how to apply model corrections correctly and how to adjust the structure when needed.

Chapter 11 - Beginning Decks and Walls Learn about how to define walls and decks, how to configure materials and thickness, understand panel calculation options, meshing, and reviewing panel results.

Chapter 12 - Basic Seismic Analysis Use the equivalent lateral force method to generate a seismic load case for your structure. Learn how to use Robot's results tools to understand what loads have been applied to your structure.

Chapter 13 - Beginning Concrete Design Learn the basics of concrete design with both the required reinforcement module for simplified concrete member design as well as an introduction to the provided reinforcement module for precise member design and concrete drawing generation.

Chapter 14 - Trouble-shooting Sometimes something goes wrong with a model. Here are a few tips and tricks to try and help smooth out some of the bumps and give you some tools for investigating and tracking down issues in models.

Using The Exercise Files

Exercise project files can be found on the companion website here:

http://marshapi.com/robot-essentials.html

You will need to download the exercise files and place them in a location on your own machine to use them.

Robot Structural Analysis Overview

History

1983 The first version of ROBOT is written by Andrew NIZNIK as part of his doctoral thesis for the French Civil Engineering Institute, INSA in Toulouse France

1985 First Commercial release of ROBOT software for DOS, called ROBOT Structures and distributed by the firm ROBOT Diffusion in Toulouse.

1988 – 1995 Creation of Robobat incorporated firm in France. With over 2,500 companies using it, ROBOT V6 is the largest of any top-end structural software packages.

1996 – 2000 ROBOT 97 is chosen the best new software in 1998 in France. Robobat Group is already represented in 50 countries. Partnership agreement with Autodesk, Inc. to become ADN Partner (Autodesk Developer Network). Robobat launches ROBOT Millennium.

2001- 2005 Robobat launches two new products detailing CAD solutions within AutoCAD: RCAD Steel and RCAD Reinforcement. The Idea of global product integration becomes the driving force of development. Robobat launches a new product - RCAD Formwork. Development of the Partnership Agreement with Autodesk, Inc. (USA) - the flagship Robobat product, ROBOT Millennium, is linked with Revit Structure.

2006 – 2007 Further development of the Partnership Agreement with Autodesk, Inc. (USA) - Robobat released a series of Extensions for Revit®, Autodesk's building information modeling (BIM) software for structural engineers. Robobat acquired by Autodesk

2008 First release of the newly branded product – Autodesk® Robot™ Structural Analysis 2009.

General Capabilities

Robot Structural Analysis is a very powerful general purpose analysis and design software primarily targeted towards the building analysis and design market. It includes several different solvers designed and optimized for quad-core and multicore computer processors to make processing times as efficient as possible and can handle the simplest to the most complex of structure models with speed. While we focus mostly on the two node bar element, multi-node floor, and wall elements here, Robot is also able to perform non-planar shell and volumetric finite element analysis using geometric construction utilities such as extrusions, revolves and Boolean geometry operations.

Robot can provide many of the most commonly used (and some uncommonly used) analyses required in the design of most buildings including static, modal, non-linear/p-delta, time history analysis, harmonic, footfall, pushover and buckling. Robot also includes load generation utilities for developing and applying seismic loading according to equivalent lateral force, spectral, or building code based lateral force based on a modal analysis of the structure. Wind loads can also be generated for simple structures according to several different building codes.

The Robot interface can be configured for a language of your choice and output configured for a different language which makes it useful for design work in many different geographies around the world. Supported languages include English, French, Romanian, Spanish, Dutch, Russian, Polish, Chinese, and Japanese. You can use imperial and/or metric units.

With over 60 section and material databases and 70 localized design codes supported by Robot, you can work with country specific member shapes, materials and design codes. Robot supports over 40 different design codes for steel and over 30 design codes for concrete/reinforcing design.

Robot is also fully integrated with Autodesk® Revit® Structure and AutoCAD® Structural Detailing allowing you to work seamlessly between Revit and Robot to perform designs in both steel, concrete, timber, and foundations and have all member size designs and rebar configurations returned to Revit for full construction documentation. Robot can also export directly to AutoCAD Structural Detailing to facilitate steel detailing, concrete reinforcing detailing, and formwork detailing. This is by far one of the most compelling stories of fully integrated analysis, design, documentation, and detailing currently available in the market.

If that wasn't enough, Robot also includes an open and flexible API (Application Program Interface) which uses the Microsoft® COM (Component Object Model). The ease of operating the Robot API through familiar interfaces like Microsoft® Excel® allows you access to analysis results, member and materials databases to extend the power of Robot to your own custom in-house design and analysis spreadsheets based on real information and data directly from your model.

The capabilities of this software definitely make learning the product a worthwhile endeavor.

General Limitations

While there are a tremendous number of capabilities of the software, there are a few items which are not currently supported by Robot Structural Analysis:

Composite design: Robot does not currently support composite (steel with integral concrete deck) design or code checks directly. There is, however, a utility integrated with Autodesk Revit Structure in the Extensions for Revit available as part of the "Subscription Advantage Pack" which provides composite floor design and code checks according to US based design code. You will also find a new extension for Robot itself which provides integrated composite design for Robot though it is not native to the software.

Pre-stress/post-tension: Designs in concrete with pre-stressing or and/or post-tensioning are not currently directly supported by Robot Structural Analysis. There are techniques for simulating this type of design with tension-only cables though the process is somewhat tedious and may not be the type of project you necessarily want to attempt with Robot.

Serviceability design iteration: A somewhat smaller issue is that in design of steel, designs can only be performed for either ultimate or service at one time which leads to a somewhat iterative process of design and checking, then redesign and re-checking. Because Robot allows several design optimization options, design currently cannot be performed for both service and ultimate at the same time.

What's New in 2016 Highlights

Robot Structural Analysis 2016 includes some really exciting enhancements over 2015 including implementation of the American Wood Council National Design Specification for Wood Structures. Here are the major enhancements to Robot 2016:

Modeling: The most visible enhancements here are around element pre-highlighting and visualization. Elements will now have a nice preview of what will be selected to assist you in selecting exactly what you want. Also several dialogs have been re-arranged to enhance usability. You'll find that the Section dialog and the Materials dialog in particular have a new and more usable look.

Simulation and Analyses: Robot 2016 has redefined its non-linear analysis. Non-linear becomes "P-Delta" and "P-Delta" becomes "Large Displacements". This is an attempt to more clearly convey the intension and function of these non-linear analyses. In Wind Load Simulation, wind load profiles may now be saved and reused in future analyses. Finally, Robot has implemented a new multi-threaded Intel Pardiso solver which is a fully thread-safe solver designed for enhanced performance for large degree of freedom systems.

Steel and Timber Modules: Most notably, Autodesk has implemented the American Wood Council National Design Specification (AWC-NDS 2012) for Timber Stuctures. Robot 2016 also includes new Tube Endplate Connections for Eurocode EC3 and PN90 codes. 2016 also includes several updates to Eurocode 3, and Eurocode 5. Review the "what's new" page of Robot help for the gory details.

Concrete Design Modules: Robot now offers access to the calculation messages dialog in the Required Reinforcement module after it has been closed. Previously you would need to run the calculations again to review the results. One of the most significant enhancements is that RC Beam verification is now based on real bars position. This allows better consideration of multi-layer reinforcement in calculating member capacity. For RC Beam calculations, multi-core machines can now take advantage of parallel processing for calculations.

Composite Beam Design Extension: A new Robot extension has been added which enables composite steel design. This extension is available for download from the Autodesk Exchange App Store (apps.exchange.autodesk.com) and will allow you to check design and behavior of composite beams and design an entire composite floor system on a floor by floor basis. All calculations are to the AISC360-10 Steel Design Code.

Results Connect: Robot 2016 also brings a Microsoft Excel extension for connecting Robot results data with Excel. This addin for Excel allows you to access a wide range of results data from your Robot model. E.g., Forces in bars, nodal displacements, section properties.

Chapter 1 - Getting Started

Starting a new project

Go ahead and open Robot Structural Analysis. After accepting the license agreement and opting in (or out) of the Autodesk Customer Involvement Program (CIP) you will see the Robot Open/New project page:

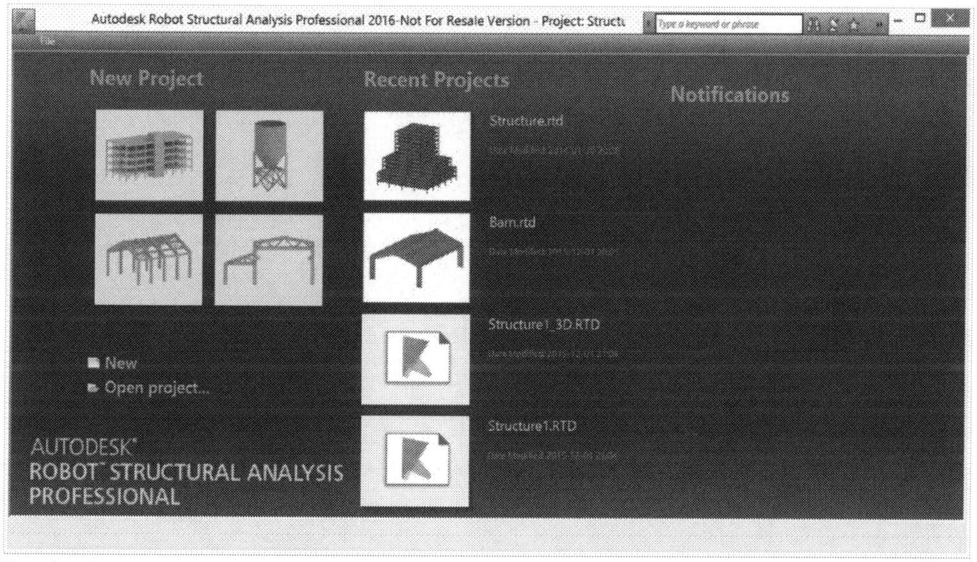

In the Projects area you will see the 4 most recent models you have worked on. Simply clicking on one will open that model. You may also choose Open Project… which will present you the standard windows file explorer so that you can locate your file to open.

NOTE: ROBOT DATA FILES ARE SAVED WITH THE FILE EXTENSION .RTD

Structure Type

In the New Project section, you will see the 4 most recently used project type templates. I use the term "template" loosely. These are, in fact, simply initial UI configurations for your new project. They have almost no significance as you may change the UI based on structure type at any point in the project with the exception of the last 4 options which are specific tools which may be accessed inside of the program at any point. Choose "More…" and you will be presented with the "Select Project" dialog:

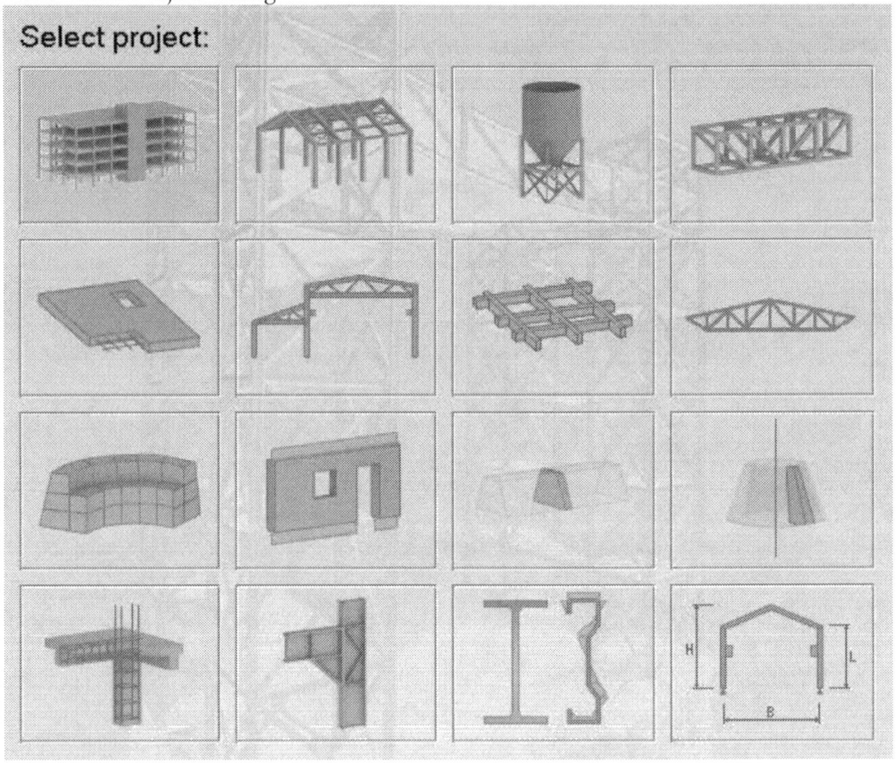

Building Design: Special tools for working in plan. I believe this was added to mimic working in Revit Structure plan views for placing columns and beams in particular.

Frame 3D Design: No tools exposed for panels or extrusions but claddings are available.

Shell Design: All tools except volumetric extrusions available.

Truss 3D Design: reduced set of tools similar to Frame 3D Design.

Plate Design: Restricted view plane, customized for single slab/floor analysis.

Frame 2D Design: Tools optimized for 2D frame work. Supports only have 3 degrees of freedom, no cladding tools, but snow and wind load calculators are available.

Grillage Design: Tools optimized for working in 2D plane with only vertical and rotational degrees of freedom about the in-plane axes.

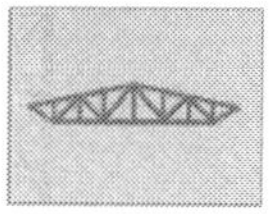

Truss 2D Design: Similar to Frame 2D Design.

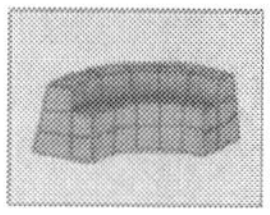

Volumetric Structure Design: Shell design tools with additional volumetric extrusion and volumetric results exploration tools.

Plane Stress Structure Design: Typically used for the design of plates where major forces act in the plane of the plate. That is, the normal and shear stresses on the surface of the plate are assumed to be zero.

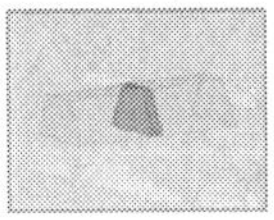

Plane Deformation Structure Design: A simplifying set of assumptions used to model elements supporting only loads perpendicular to its long axis and the long axis dimension is significantly larger than its cross sectional dimensions. The analysis of dams, tunnels and other geotechnical works are often analyzed in this manner. Strain normal to the x-y plane ε_z and shear strains γ_{xz} γ_{yz} are assumed to be zero.

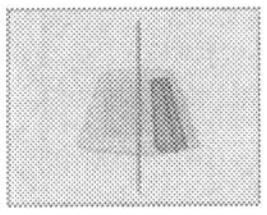

Axisymmetric Structure Design: In axisymmetric problems, the stresses are independent of the θ coordinate (in a cylindrical coordinate system) from which circumferential displacement as well as shear strains $\gamma_{r\theta}$ $\gamma_{\theta z}$ and shear stresses $\tau_{r\theta}$ $\tau_{\theta z}$ are all zero. This type of analysis can be used when both the part and the loading are revolved about the same axis. Some typical applications of this type of analysis are thick-walled cylinders or similar pressure vessels.

RC Elements Design: Independent module for reinforcement design of columns, beams, walls, floors and isolated footings.

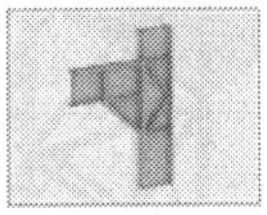

Connection Design: Independent module for designing steel connections. (Note: No connection design is available for North American Steel Codes at this time.)

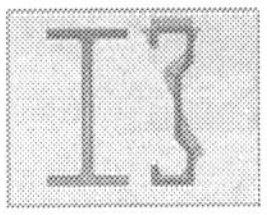

Section Definition: Custom cross section design and analysis.

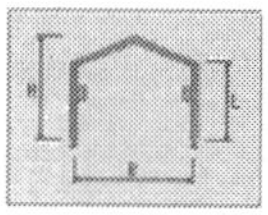

Parameterized Structure Design: Similar to Frame 2D Design but begins by inserting a parametric structure. Parametric structures may be inserted in the other standard design configurations as well.

HISTORICAL NOTE: Robot Structural Analysis, when it was Robot Millenium and produced by the Robobat company in France, was sold as a series of modules for specific purposes. The "Select Project" part of starting a new project is a hold-over from this historical modularization of the product. You will find that as you move between different structure types that certain menu items, layout selectors, and available views will change dramatically.

If you ever find yourself thinking "Gee, I could have sworn that I saw that menu item before…" and not sure what could have happened to it, it is no small likelyhood that you have a project with a different current-structure-type than the one you were using when you saw that menu item. Menu items will *literally* appear and disappear as the structure type is changed. You can customize the menus you see and add in necessary tools from **Tools>Customize>Customize Menu...** to add them in.

Exercise 1: Open an existing project.

From Robot start page, select Open Project… and use the windows file browser to navigate to example project file Exercise_01-UserInterface.rtd.

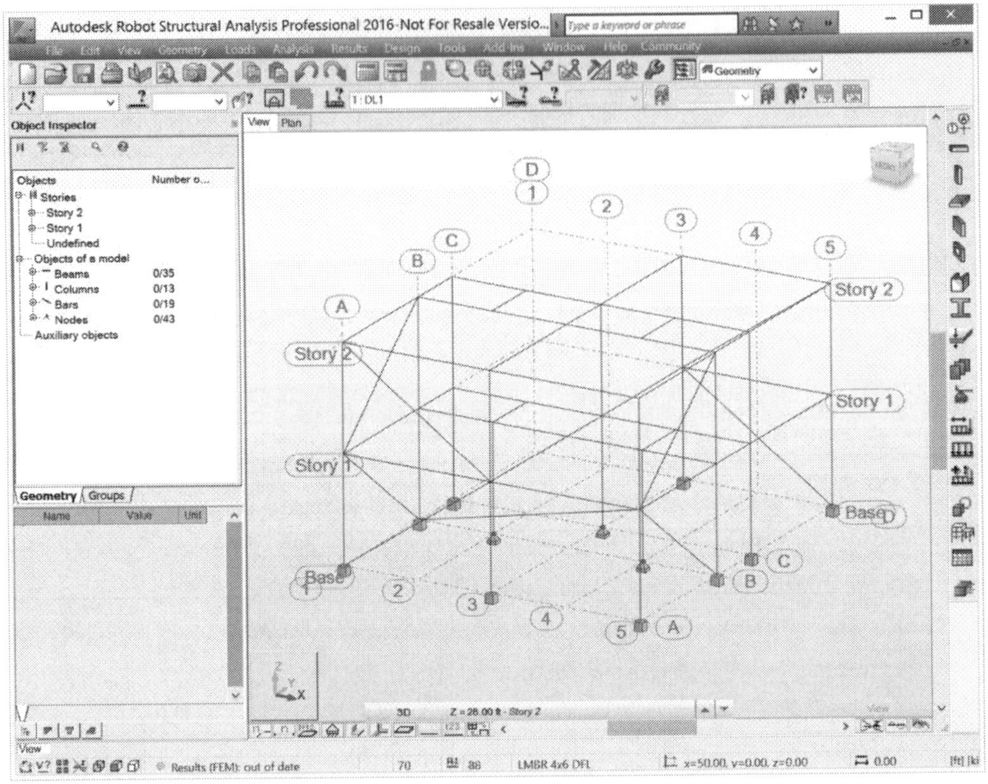

When Robot opens a project, the interface will be in the Pan/Zoom/Orbit tool. You can recognize this by moving the cursor around the model window. The cursor will change based on which quadrant of the view window you hover the mouse.

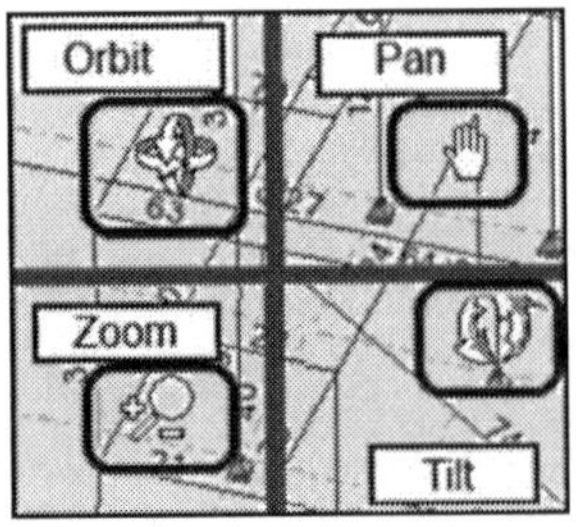

To use the Pan/Zoom/Orbit tool, holding down the left-click button move the mouse to Orbit/Pan/Zoom depending on which quadrant of the view you are currently hovering the mouse. Take a few minutes and use Pan/Zoom/Orbit tool to look at the model. The Escape key or right-click>Cancel will exit the Pan/Zoom/Orbit tool. Clicking on this button  will bring it back:

--End of Exercise--

The User Interface

When you first open Exercise_01-UserInterface.rtd you will see a simple structure model. Here are the major parts of the user interface:

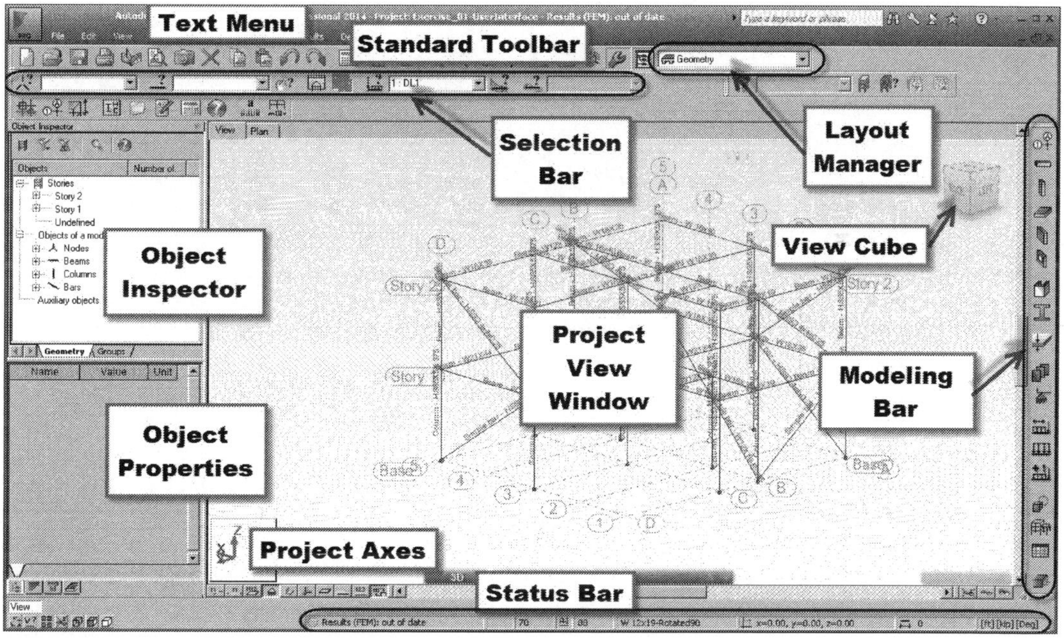

Project View Window: This is the 3D view of your model.

NOTE THAT THERE ARE **TWO TABS** AT THE TOP OF THE PROJECT VIEW WINDOW: "VIEW" AND "PLAN". THIS IS BECAUSE THIS EXAMPLE FILE WAS LAST SAVED WITH THE STRUCTURE TYPE SET TO BUILDING DESIGN. IT IS THE ONLY STRUCTURE TYPE SETTING WHICH PROVIDES BOTH OF THESE TABS. ALL OTHER STRUCTURE TYPES ONLY PROVIDE ONE DEFAULT VIEW FOR THE PROJECT ENVIRONMENT.

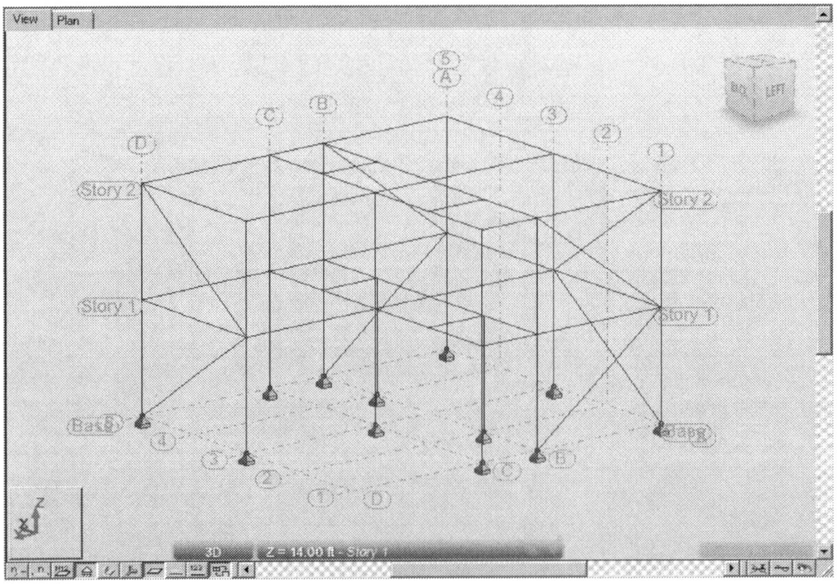

The Text Menu

The text menu provides access to all Robot commands through a traditional text based menu. We will refer to text menu items by first referencing the menu title followed by a ">" and then the sub-menu item. For example: **File>New Project...** will refer to the file menu and the first menu item "New Project".

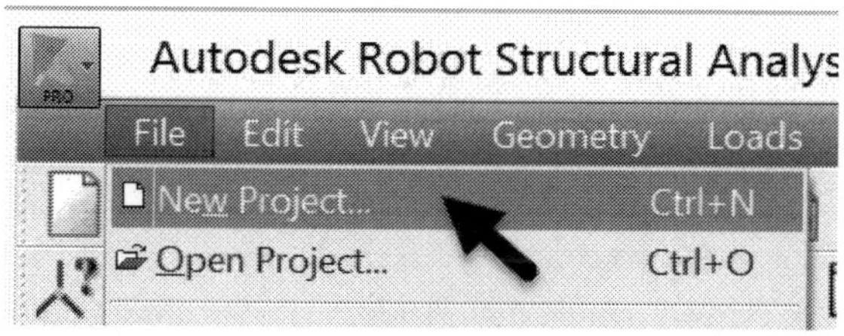

The Standard Toolbar

The toolbar consists of the following main parts:

Standard tools:

From Left to Right:

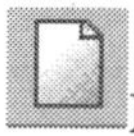New file: Close current project and create a new Robot Structural Analysis project

Open File: Close current project and open a different project

Save file: Save the current project. Save-As is only available from the File menu.

Print: sends the current view to the printer. Inactive for table views

Printout Composition: Launch printout composition dialog where you may select which data should be contained in your analysis and design calculation set.

Print Preview: Presents user with a preview of the current project view which will be sent to the printer.

Screen Capture: This is a **very powerful tool**. It can be used to not only capture views with members, results, etc., but also those views can be updated automatically as new results and model configurations are created and changed.

Delete: Delete selected element(s).

 Copy and Paste: Copy will put the selection on the clipboard. Paste will launch a special dialog to assist in locating the new copy in the project environment. In addition to copying selected items to Robot's clipboard, a snapshot of the project is added to the windows clipboard which may be inserted as an image in another windows program.

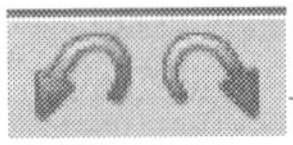Undo Redo: Standard undo/redo behavior

CALCULATION MANAGEMENT TOOLS:

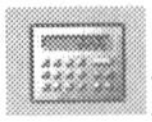Run Structural Analysis Calculations

Open Analysis Types Dialog to configure parameters of each load case or combination. See "Running Calculations" section for further information about the Analysis types dialog.

Lock Calculations: Prevents accidental modification of structural results.

The View Control Tools:

Zoom Window: Specify a window area on screen to zoom to.

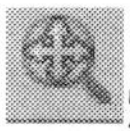Zoom to Fit: Zoom view such that all elements fit in the view window.

Pan/Zoom/Orbit tool: Sets view area to 4 "hot zones": Pan, Orbit, Zoom and Tilt. Use the escape key to exit this view navigation mode.

Redraw: Redraws the view. Usually not required but can sometimes help update or refresh a view, particularly a results view or a detailed results view.

Additional Toolbars and Object Inspector Toggle:

The first 4 buttons launch toolbars which contain the most used features of their commensurate text menus. From the text menus, take a look at the Edit Menu, the View menu, the **Analysis>Meshing...** menu and the Tools menus. Then launch each of these toolbars and hover over the icons to see which commands they represent.

Launch Edit Toolbar

Launch Meshing Toolbar

Launch View Toolbar

Launch Tools Toolbar

Toggle Object Inspector On/Off. The Object inspector is found on the left-hand side of the Robot window and contains information about elements. This button will turn that information window on and/or off.

Layouts:

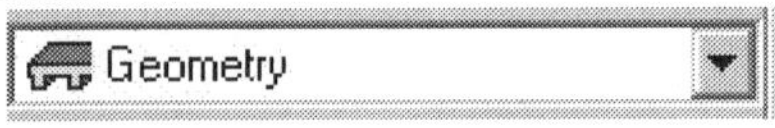

The Layout Control Manager is found at the far right of the standard toolbar (just below the text menus): Allows user to select between different automated layouts of the screen. The layout manager is a tool to help organize the project area for different tasks. Some of the configurations which can be selected from the layout manager can be created manually by opening different menus and dialogs. However, the layout manager creates a nice flow for taking a model from start to finish and manages a lot of the window/dialog opening, closing, and positioning for you. As we work through exercises, we will use both layouts and manually opened dialogs.

NOTE: IF YOU HAVE A LAYOUT "APPLIED" OR "SELECTED" VIA THE LAYOUT MANAGER, ALL DIALOGS AND WINDOWS MANAGED BY THE LAYOUT MANAGER CANNOT BE CLOSED WITH THE "X" IN THE UPPER RIGHT-HAND CORNER! ROBOT LAYOUT MANAGER IS "INFLEXIBLE" IN THIS REGARD. FOR MORE FLEXIBILITY YOU CAN OPEN SOME OF THE DIALOGS AND WINDOWS YOURSELF USING THE MENUS.

The Contents of the layout selector will vary depending on which structure type you currently have set for the project. Here are some of the main layouts and their Structure Model layouts:

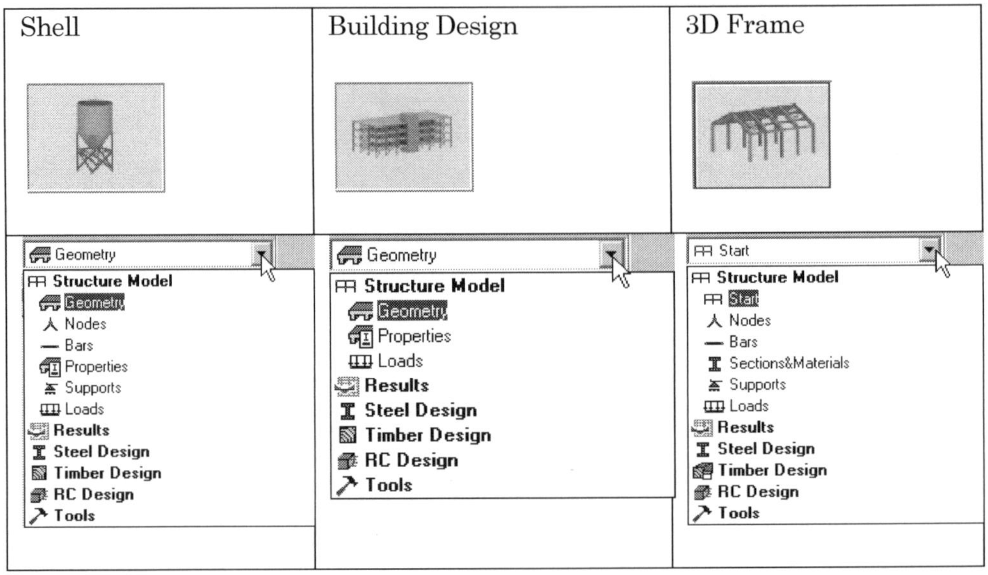

You can change the structure type at any time by selecting **Geometry>Structure Type...**

Selection Toolbar:

The selection Toolbar is just below the standard toolbar and the text menu:

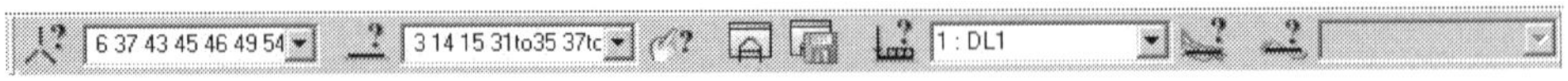

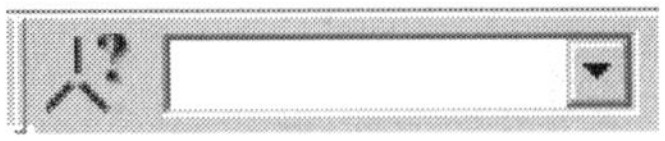
Node Selector: See "Making Selections" below.

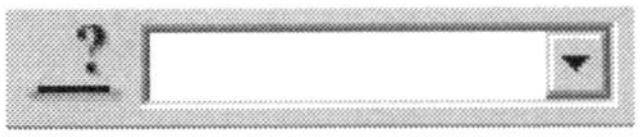
Bar Selector: For selecting bars, panels or objects. See "Making Selections" below.

 Special Selections Toolbar: Matches with menu **Edit>Special Select>** menu items.

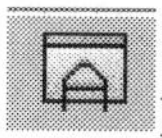 Edit Selection in New Window: **Very useful** for isolating a portion of the structure for editing or results viewing/screen-capturing. Once you have selected some elements or nodes, press this button and a new window will be opened with only these selected elements visible. This can be useful for viewing results for specific members in particular. (e.g., walls)

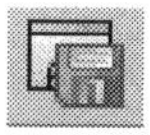 Save Substructure: Save the current selection as a new Robot file (.RTD file)

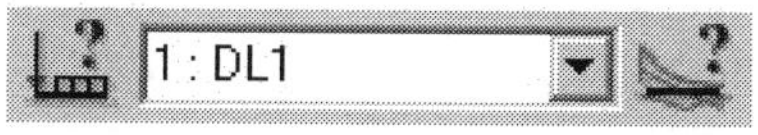

Load Cases and case component selector: Use first button to select multiple cases or the drop down list to select only one. If the currently selected case is an automated load combination, the component selector will allow you to choose from the components. Robot offers an automatic combination management which collapses all of the automatically generated load combinations into a case simply called "ULS" or "SLS" and individual load combinations are accessed via the "component" selector. More on this in the load combinations topic.

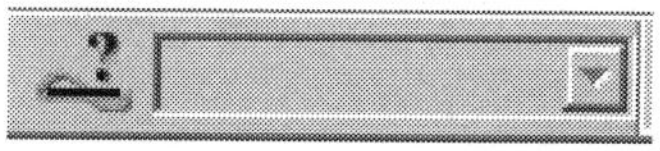 Mode Selector: When a modal load case is selected or a case based on a modal analysis, this selector will allow you to switch between modes as well as the summations of the modal contributions.

Other parts of the user interface will be covered in subsequent sections or exercised in tutorial sections so that you will gain familiarity with their operation.

Model View Navigation

There are a ton of ways to navigate and manipulate the project view in Robot.

Middle Mouse Button Model Navigation

Basic view manipulation is accomplished in large part by use of the middle mouse button and the shift key:

Pan = Middle Mouse Button Down + Mouse Move

Zoom in/out = Scroll Middle Mouse Wheel

Orbit = Shift + Middle Mouse Button Down + Mouse Move

Note: Orbit can only be performed in the "View" tab of the project view window. If the "Plan" view tab is selected, you will not be able to orbit the view. Take a few minutes with our Exercise_01-UserInterface.rtd project to try out these middle mouse button actions. You will find them identical to some other Autodesk products.

View Cube

Another method of orbiting the view is to either click on a hot zone of the view cube or to just grab the view cube (left-click on the view cube itself) and drag it to orbit the model.

Left click on areas of the view cube to see what it does. Explore the context menu by right-clicking on the view cube. Grab the view cube with a left-click and hold, then move the mouse and see what happens.

Project Global Axes View Tool

This is a very cool palette of tools which combines several menu options from the view menu. Click on the project axes icon to access this tool palette:

Clicking this Icon will launch the View Palette shown below:

This is more than just view manipulation, it also helps control which plane is the current work-plane for creating elements. If you are working in a 2D elevation view, you can use this dialog to control the structure plane in which you intend to create elements. Selecting 3D will give you the typical 3D view. Selecting 2D will orient the view to the plane selected in the next set of buttons (XY, YZ or XZ) and activate the depth selector (drop down just below the 2D button and the up/down arrows). The depth selector will set the view at various pre-set depths or allow you to type in a custom view depth. The 2D/3D button will put the view in 3D mode with a 2D plane specified by XY, YZ or XZ and a depth set by the depth selector. The drop down under the plane selector buttons (XY, YZ and XZ) offers the choice of standard 3D view orientations. The standard view orientations in this dropdown seem to be more-or-less incompatible with any view setting other than 3D. I find that I rarely use them. Try the exercise below to get a feel for how this dialog works.

The Dynamic View (**View>Dynamic View>Dynamic View**) adds transparent controls at the bottom of the view which give you access to exactly the same controls as in the View Palette with exception of the "standard views" dropdown. Here is an image of the dynamic view bar at the bottom of the project window:

Exercise 2: Understanding project axes and dynamic view

In the Exercise_01-UserInterface.rtd file, click on the project axes icon to bring up the view dialog shown below. Click the 3D button: Note that all structure axes and levels are shown. Now click the 2D button. Notice that one of the planes also becomes selected (XY, YZ or XZ) and the view is oriented to this plane.

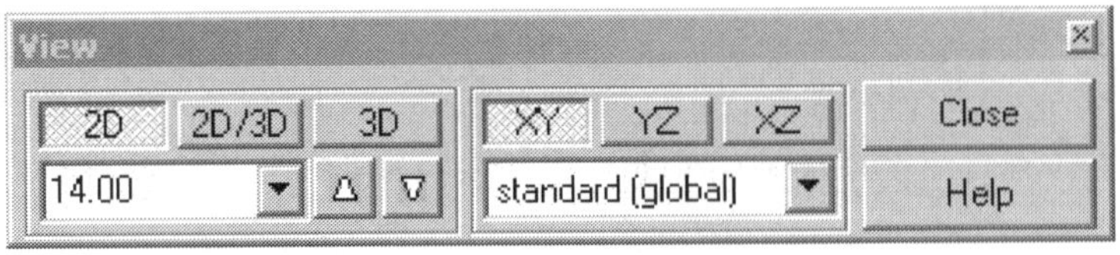

Try selecting XY, YZ and XZ respectively and notice that the view changes orientation to the indicated plane.

Now switch back to XY and notice that under the 2D button there is a dropdown accompanied by up/down arrows. Expand this drop down and you will notice all of the levels that we have input for the project (for Exercise1 file there are 3 levels) choose between these levels and observe what happens to the view: it will change what you see based on what level within the project is selected. Now switch to the YZ button and check the dropdown under the 2D button again. Now try using the up/down arrows and watch what happens to the view: it cycles through the available structure axes. Once it has gone beyond the structure axes it will continue in snap increments. You can also type in a specific value in this dialog. This will set the current work-plane for you to view or create new elements. Notice that the project axes icon now has up/down arrows on it too!! These work exactly the same way that the up/down arrows on the view tools palette work.

Lastly select the 2D/3D button. The view will change to a 3D view but the current work-plane will be aligned with the selected XY, YZ or XZ. Select through levels or axes in the drop down below the 2D button and watch what happens to the work-grid. Also notice the changes in the project axes icon, it will have up/down arrows and a square indicating which plane is parallel to the current work plane.

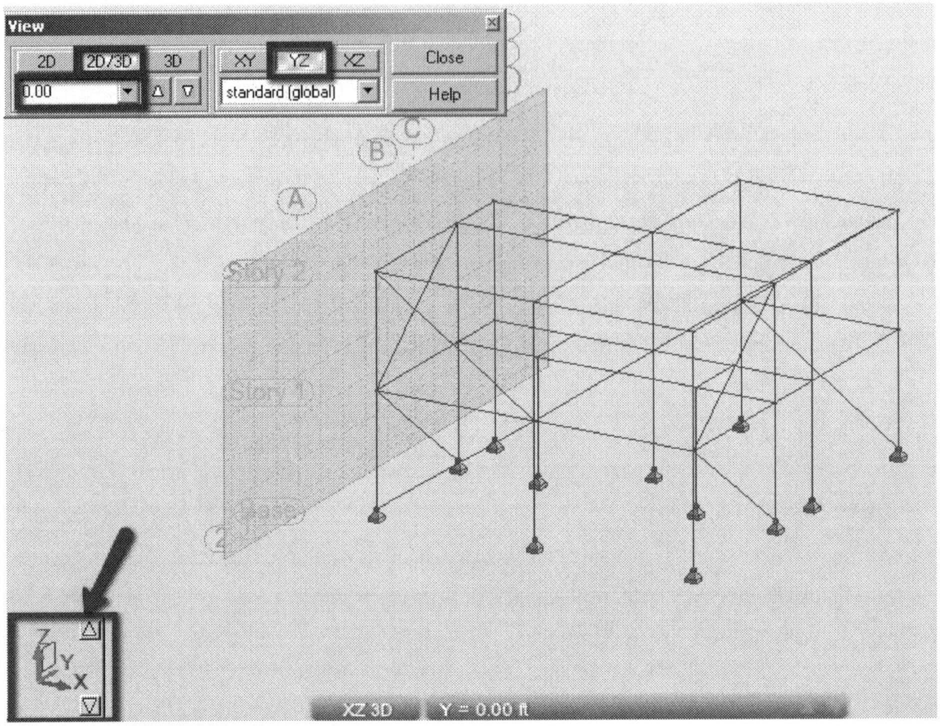

--End of Exercise--

Display Options

There are also a ton of view display options to control exactly what is visible in the view at any one time. You can toggle on and off text based information, graphical indications, and view elements as well.

Quick Access Toggles

Located along the bottom of each project view is the display information toggle:

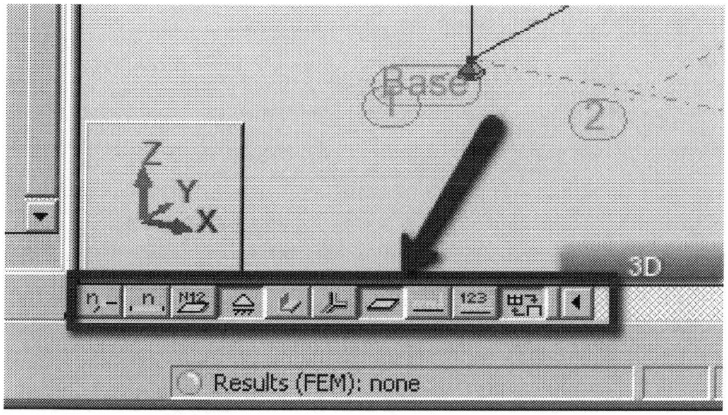

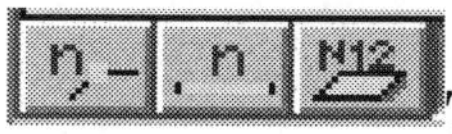These first three buttons toggle display of element numbers: Nodes, Member numbers and panel numbers.

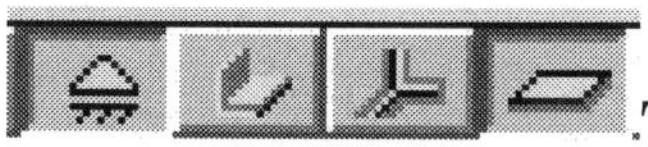.The next 4 buttons toggle display of graphical information: Support Graphics, Section Shape 3D visualization, Member Local Axes, and Panel Interiors.

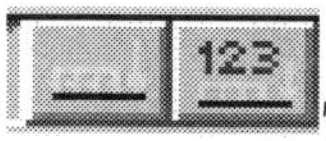The next 2 buttons control visibility of loads: Load Graphics and Load Magnitude Annotations.

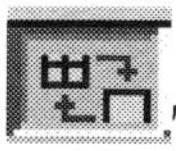.This button toggles display of the calculation model. Once a model has been meshed into many individual elements, changing this toggle will show and hide the meshed elements.

Way off to the right on the bottom of the view are some view state buttons. The first will restore view defaults. If you use the display options (see below) to change the visibility of the view and cannot figure out how to get back to the default, you can use the first button to restore the view to default. The next button will toggle on and off the display of the structure deformation for a structure

which has results available (calculations have been run). The third button will toggle display of the "presentation view" which is a rendered view with ground and sky which can be used to create more attractive presentation graphics to wow your architect friends. The cover image was created with this presentation view style.

NOTE: WHEN TEXT OR GRAPHICAL INFORMATION IS DISPLAYED ON THE MODEL THE PGUP/PGDN BUTTONS (PAGE UP AND PAGE DOWN) WILL INCREASE AND DECREASE THE SCREEN SIZE OF TEXT OR SOME OTHER GRAPHICAL ELEMENTS IN THE VIEW. THE TEXT WILL ALWAYS REMAIN THE SAME SIZE REGARDLESS OF VIEW ZOOM. YOU CAN GET FINER CONTROL OVER THE SIZE IN THE DISPLAY OPTIONS DIALOG (SEE BELOW).

Display options dialog

The Display options dialog may be accessed in two ways:

1. From the Right-Click context menu in the project environment

2. From the Menu **View>Display...**

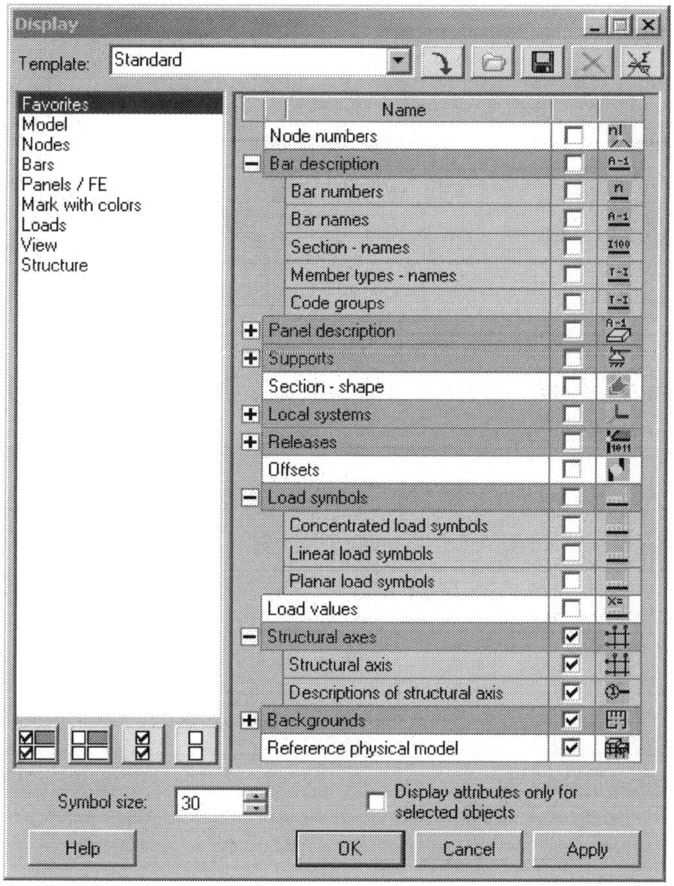

The settings in this dialog apply to the current view (later we will see how multiple views may be open at one time). The sheer number of available settings is staggering and will require the user to explore on their own in order to understand what each setting does. When you get to a view state that seems impossible to return from, you can always use the default settings button at the far bottom-right hand side of the screen to get back to the default. Important parts of this dialog are

Template Management Area:

Consists of Template selector, Template "open", Template Edit, Template Save, Template Delete and Save Default settings. You can view the settings associated with each template by selecting it with the dropdown list. You can type a new name in the Template selector dropdown, adjust the visibility checkbox settings and save it as your own custom new template. If you have adjusted settings for a template, you can re-apply the template by using the "open" template button. Edit template button will take you to a dialog where you can configure the category list displayed for the template. All edits to the toggle settings for the template are done in the display dialog, not in the edit template dialog.

Category List and checkbox toggle buttons:

Visibility settings are grouped into logical categories in the category list. Selecting a category on the left will show the associated visibility toggles on the right. This list of categories can be modified or customized! The template is more than just the toggle settings it is also the list of categories displayed on the left-hand side.

Checkbox Toggles, from left to right: Toggle ALL on, Toggle ALL off, Toggle on all in currently selected category, Toggle off all in currently selected category.

Additional Options:

While PgUp/PgDn will control the size of graphic information displayed on the screen, the Symbol Size setting gives much finer control over the size of symbols on the screen. The last checkbox sets the view to not display any of the information configured in the Display Options dialog until the user selects an element in the view. Selecting this checkbox will cause the entire view to be half-toned and no additional graphical information will be displayed until the user selects something in the view.

Exercise 3: Display Options and Templates

With the Exercise1 file, open the display options (right-click in project area and select Display… or choose **View>Display…** from the menu). Select the "Standard" template from the dropdown selector. Select the Bars category from the left hand side and expand the Bar Description item in the list to the right:

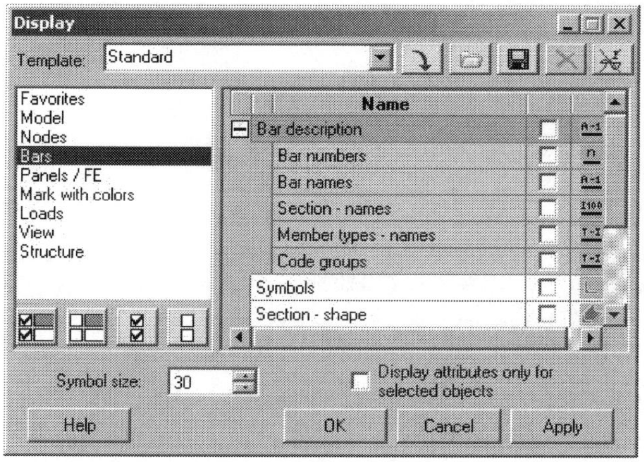

Check the box next to "Bar numbers" and next to "Section – names", then hit Apply and take a look at the project view. All of the quick toggles available along the bottom of the view window are also available in here. There are many, many options; take some time to experiment with the different view options and see what is available. We will highlight various options as they become important for various workflows further on in the text.

Next Type a new template name where it currently says "Standard" and then press the Save Template button  to commit the new template to the list. Next choose the edit template button ., then press the  New Category button.

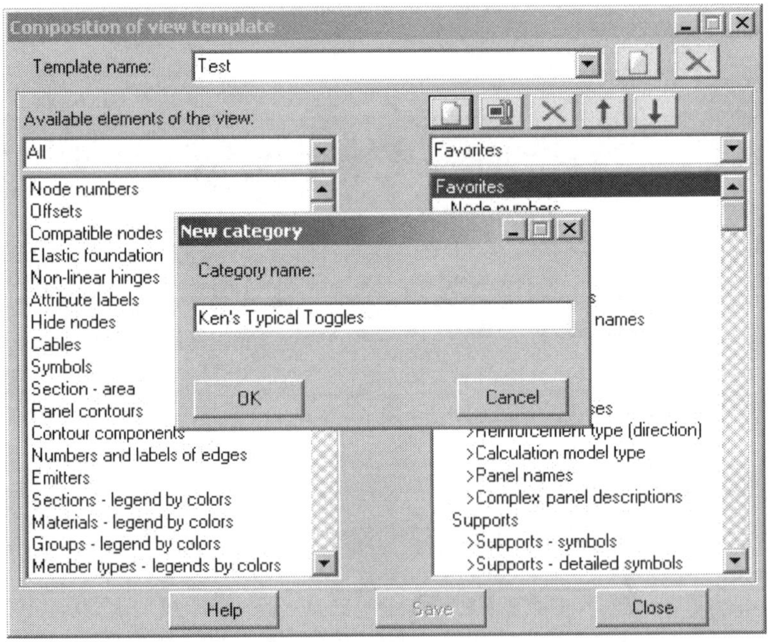

After typing in your own new category name, and pressing OK you can move the category around the list with the up/down buttons. Add new items to your new category with the add button. 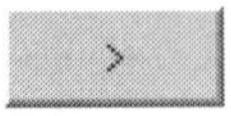 When you are finished adding new items to your category, click Save and Close. Choose some new toggle settings for your template in the display options dialog saving the template settings after you are done, then switch between the standard template and your template to see how easy it is to set up and use pre-defined view templates.

--End of Exercise--

Viewing Tabular Information

Tables in Robot are very powerful yet somewhat elusive. Before we get into the details of tables, I think it's important to mention that when we cover placing elements and viewing results, we will go into more depth on tables. For now, I think it is important to understand the basic functionality of tables so that navigating them later will be more enjoyable.

Types of tables in Robot

There are two general types of tables in Robot: Data Tables and Results Tables. Data Tables may be used to view and edit model data. Results tables are used for digesting calculation results including viewing enveloped results as well as global extremes for calculated values. Results tables will be discussed in more depth in the results section.

Opening tables

Tables are represented as windows in addition to any open project windows. They may be opened in one of two main ways. (In reality tables will be accessed from several locations, especially in the tabular results from the results menu). Either right-click in the project window and select Tables... or from the View menu **View>Tables...**

This opens the Tables: Data and Results dialog. From here, you may select (or unselect) one or multiple tables. Tables of results will typically be greyed out until calculations have been successfully run and results are available.

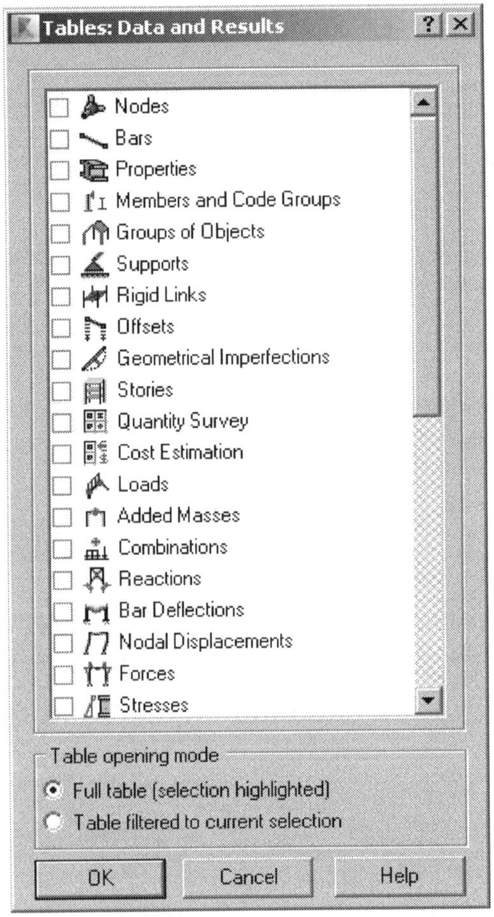

The "Full Table" option means that there will be no filtering applied to the table when it opens. Any elements currently selected will be highlighted in the table.

"Table filtered to current selection" option will apply an element filter to the table which limits the table view to only the elements selected in the project.

The image below shows the UI after selecting the bars table to open unfiltered or "Full table" option.

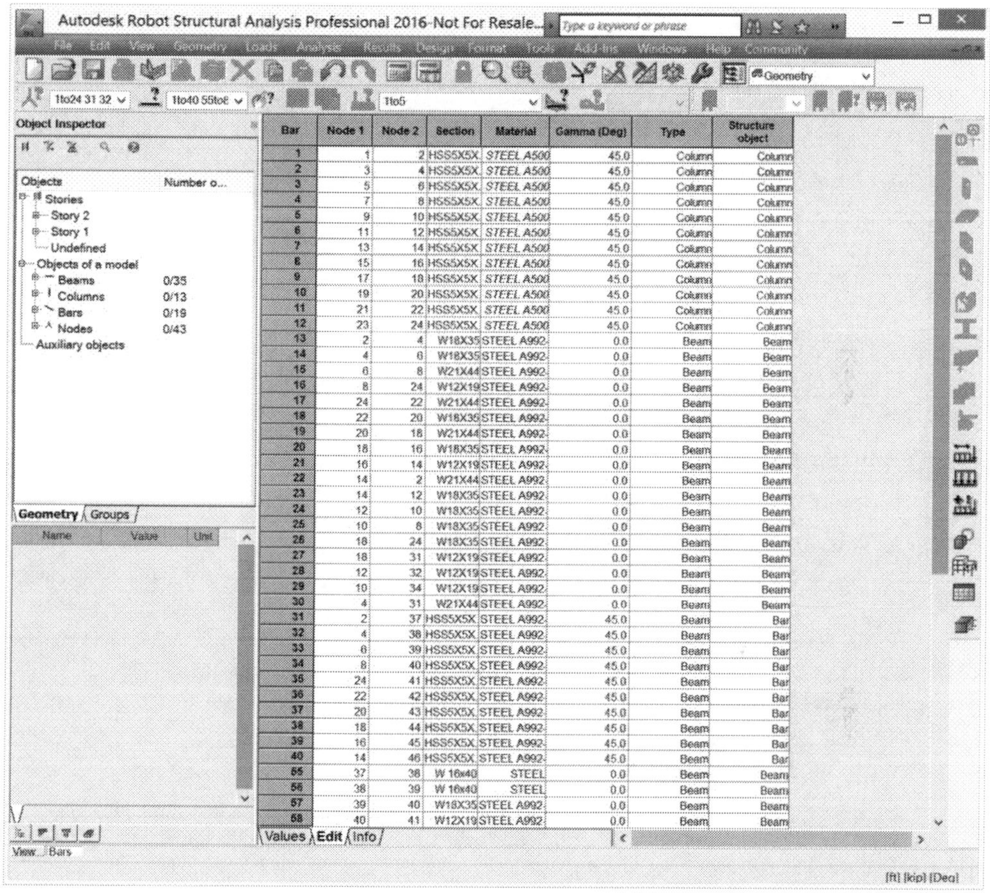

Unless you have already "Restored Down" the main project window, the table view is now maximized in the project area covering up the model view as shown here.

Robot has a unique way of managing view windows including table views. Take a look at the lower left hand corner of the user interface:

All open views are listed here in the open windows section of the UI. To switch between windows just click on the title of the window you would like to see. Click "View" to return to the model view.

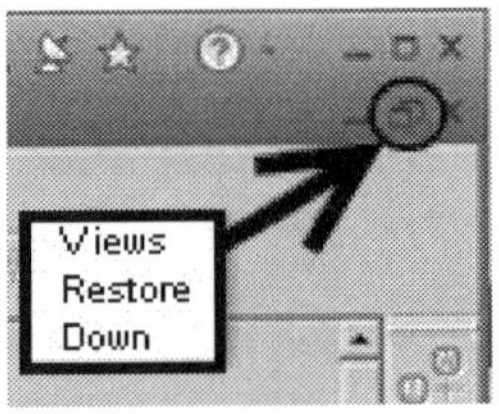

Also notice in the upper right-hand corner of the Robot window that there is a "Restore Down" option for windows management. Selecting "Restore Down" will allow you to manage the view windows individually. Also from the Windows menu you may select **Windows>Tile Vertically** or **Windows>Tile Horizontally**, to automatically arrange all the currently open views in the project view area. As simple as it seems now, I found this very confusing when I first started using Robot.

Understanding Tables

Notice that at the bottom of the bars table there are three tabs:

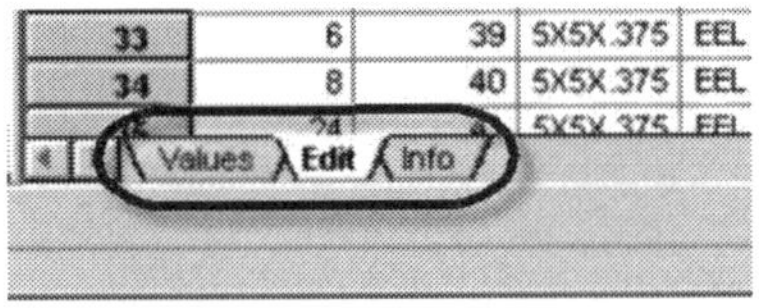

Values: Presentation information which may be screen-captured with the screen-capture tool:  If you are not in the Values tab, the screen-capture tool will not be available.

Edit: The edit tab will provide an editable spreadsheet like interface to the table. Any values that may be edited will be editable in this tab.

Info: The info tab catalogs information about the current table. How many elements are displayed, how many are currently selected.

NOTE: IN RESULTS TABLES YOU WILL ALSO FIND ENVELOPE AND GLOBAL EXTREMES TABS. IN THE LOAD TABLE YOU WILL FIND YET OTHER TABS. THE TABLES ARE STREAMLINED FOR THE TYPE OF WORK YOU ARE DOING IN EACH. INSTEAD OF FORCING YOU TO CONFIGURE EACH DIFFERENTLY, THE UI INTENDS TO PRESENT THE MOST USEFUL INFORMATION AT THE RIGHT TIME FOR YOU TO USE IT.

Editing Values in a Table:

First make sure you have the Edit tab selected at the bottom of a table. Individual cells in the edit tab may be edited by either selecting from a drop down list for properties which have discrete settings or by typing into the cell.

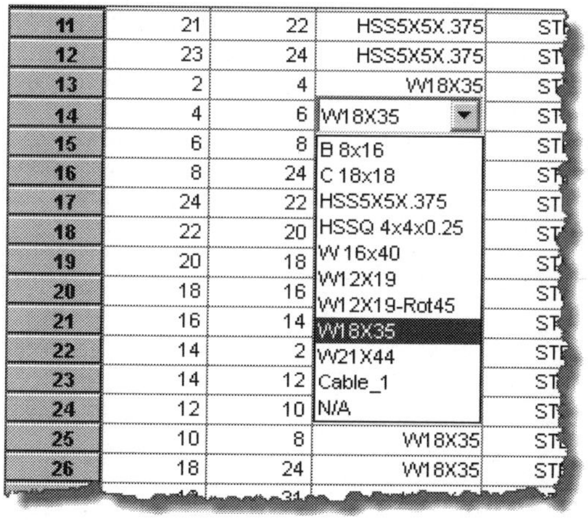

Multiple cells may be filled using the Fill Special… tool from the Right-Click menu in the table. Note that Fill Special will not be available in the "Values" tab of a table.

The key to using Fill Special is to select a crossing of rows and columns you wish to fill.

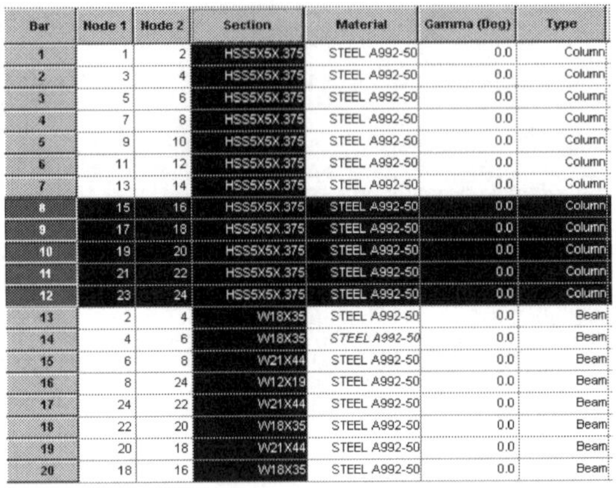

Bar	Node 1	Node 2	Section	Material	Gamma (Deg)	Type
1	1	2	HSS5X5X.375	STEEL A992-50	0.0	Column
2	3	4	HSS5X5X.375	STEEL A992-50	0.0	Column
3	5	6	HSS5X5X.375	STEEL A992-50	0.0	Column
4	7	8	HSS5X5X.375	STEEL A992-50	0.0	Column
5	9	10	HSS5X5X.375	STEEL A992-50	0.0	Column
6	11	12	HSS5X5X.375	STEEL A992-50	0.0	Column
7	13	14	HSS5X5X.375	STEEL A992-50	0.0	Column
8	15	16	HSS5X5X.375	STEEL A992-50	0.0	Column
9	17	18	HSS5X5X.375	STEEL A992-50	0.0	Column
10	19	20	HSS5X5X.375	STEEL A992-50	0.0	Column
11	21	22	HSS5X5X.375	STEEL A992-50	0.0	Column
12	23	24	HSS5X5X.375	STEEL A992-50	0.0	Column
13	2	4	W18X35	STEEL A992-50	0.0	Beam
14	4	6	W18X35	*STEEL A992-50*	0.0	Beam
15	6	8	W21X44	STEEL A992-50	0.0	Beam
16	8	24	W12X19	STEEL A992-50	0.0	Beam
17	24	22	W21X44	STEEL A992-50	0.0	Beam
18	22	20	W18X35	STEEL A992-50	0.0	Beam
19	20	18	W21X44	STEEL A992-50	0.0	Beam
20	18	16	W18X35	STEEL A992-50	0.0	Beam

1. Select the rows for which you wish to fill the value of some column of data. If you wish to fill the entire column, then skip this step and just select the column by left clicking once on the column header.

2. Holding the Control Key, select the column header for the column of data you wish to fill. (You can only select one column for this tool to work) This will create a crossing which indicates which values you want Robot to fill for you.

    a. You may select multiple rows in sequence or randomly by using the control key or shift+control just like standard windows selection. When selecting the column, be careful to use the Control key or the behavior may be somewhat unexpected though not unpredictable. The column must be selected after rows are selected.

3. Right-click and select Fill Special…

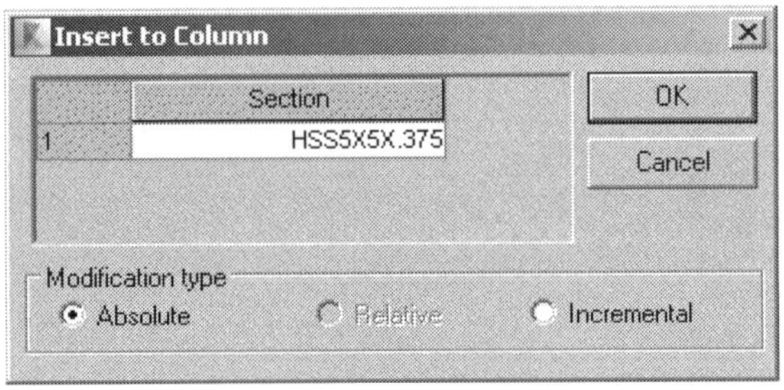

In the Insert to Column dialog, configure the value you wish to be inserted in the cells.

<u>Values with discrete settings</u> (e.g., Section, Material, Type) the Absolute option is the only valid option. (I believe that Incremental being available is a bug but may have some functionality yet unbeknownst to me).

For freeform numerical values:

Absolute: put exactly this number in the indicated cells

Relative: Take whatever number is currently in the cell and add this number to it. Example: **Relative 2.0**:

Before	After
4.0	6.0
12.0	14.0
7.0	9.0
2.0	4.0
20.0	22.0
11.0	13.0

Incremental: Set the first value in the range of cells to its original value plus this incremental value, second cell in the range will have 2 times the number added to the current value of the second cell, the value of the nth cell will have n times the number indicated added to the nth cell's original value. So

Example: **Incremental 3.0**:

Before	After	
6.0	9.0	Adds 3 to original cell contents (6)
12.0	18.0	Adds 6 to original cell contents (12)
5.0	14.0	Adds 9 to original cell contents (5)
30.0	42.0	Adds 12 to original cell contents (30)
3.0	18.0	Adds 15 to original cell contents (3)
130.0	148.0	Adds 18 to original cell contents (130)

Table Columns

There is a wealth of additional information available in tables. When a table is opened, it is set to a default configuration. In any table view with either the values tab or the edit tab selected (or in Results tables you will also find Envelope and Global Extremes tabs), Right-click and select Table Columns…

This will bring up the columns manager associated with the particular table which is currently active. It is important to note that different tables will have different column settings and the options will even vary depending on which tab of the table is currently active. This screen-capture of the "Table Columns" dialog is from the bars table on the edit tab. Notice that "Length" is greyed out!

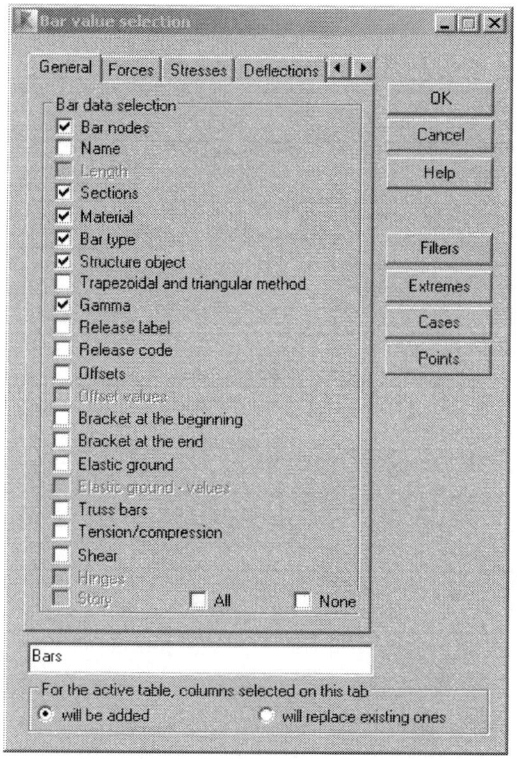

However, if you close this dialog and switch to the Values tab and again right-click and select Table Columns… these are your options:

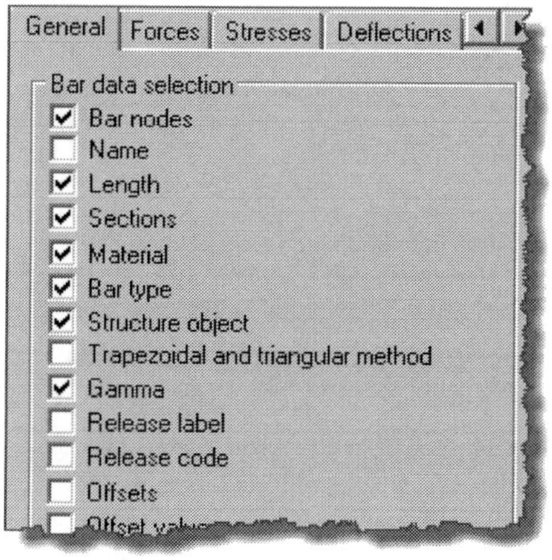

Notice that now Length is available and even checked. So the tabs can also display different information between Values and Edit tabs. At first this seems strange, but in reality there is nothing editable about the member length so there is really no reason to have it available in the edit tab. If you want to use a piece of information to identify an element, selecting it on the values tab will also scroll the edit tab to the same row so that when you switch over to the edit tab, your selected element is waiting for you to make some edits. The general rule of thumb is that the Values tab will display all selected information, the edit tab will only display editable information.

Saving your column settings: Once you have modified your table columns to display additional information you find valuable for your design or verification process, you can save your settings as the default for the table. With a table view active and having customized the table columns, from the View menu choose **View>Save Table Template**. This is a singular operation with no options. The next time you open the table the column configuration will match your previous settings. If you decide you no longer like your new default you can always restore Robot's default by selecting **View>Default Table Template** which will not change your current view but will reset to default template the next time you open the table.

NOTE: TABLE COLUMNS IS VERY POWERFUL AND WILL HELP ESPECIALLY IN CONFIGURING RESULTS TABLES TO DISPLAY SUMMATION RESULTS ESPECIALLY FOR REACTIONS ETC. WE WILL COVER IMPORTANT FEATURES OF TABLE COLUMNS IN SUBSEQUENT CHAPTERS.

Making Selections

I think that understanding basic selection functionality is very important to grasping the general operation of Robot. The selection dialog shows up in so many places in Robot and is an incredibly powerful tool for assembling a selection of elements. The selection dialog can operate on nodes, bars, plates, load cases, load combination and other information to build highly specific and refined selection sets. It allows you to quickly build a selection where simply picking graphically on the screen would either be too tedious or easy to miss an important selection. In this topic, we will cover general window and click graphical selection, selection via the selection dialog as well as special selections and have an exercise to help drive home the usage of the dialog and selection tools.

Pick and Window Selections

In the project environment, elements including nodes, bars, and panels, may be selected by single click with the left mouse button or by dragging a box or crossing window selection. Selection sets may be built by either adding or subtracting elements from the set using control or shift keys while selecting elements.

NOTE: WHEN ROBOT INITIALLY OPENS A PROJECT IT IS, BY DEFAULT, IN ROTATE/PAN/ZOOM MODE. YOU WILL NOTICE THAT THE CURSOR CHANGES SHAPE IN EACH QUADRANT OF THE SCREEN. THIS IS SIMILAR TO THE "STEERING WHEEL" AVAILABLE IN OTHER AUTODESK PRODUCTS LIKE AUTOCAD AND AUTODESK REVIT. PRESS "ESC" TO EXIT THIS MODE.

Single click selections: The cursor will appear as a hand with a pointer finger and hovering over elements will show some information about them. Highlighted information will typically be the element number as shown here.

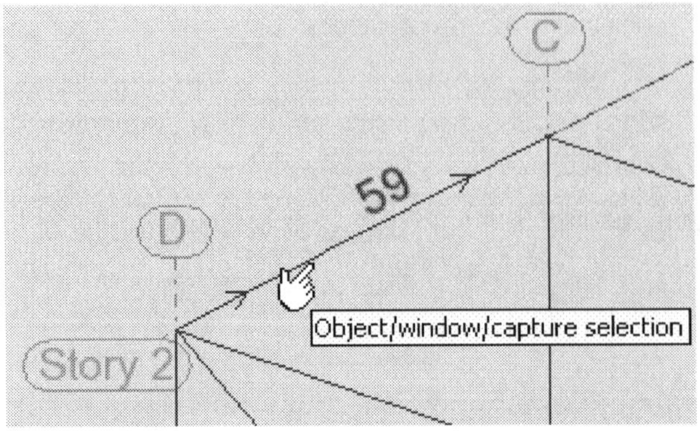

Selections in Robot are series selections meaning that clicking another item will clear the previous selection and select the new element instead. This applies to both single click as well as crossing and box selections.

NOTE: THE BIGGEST DIFFERENCE BETWEEN ROBOT AND OTHER AUTODESK PROGRAMS IS THAT SELECTIONS ARE NOT CLEARED BY PRESSING ESC. SELECTIONS ARE ONLY CLEARED BY CLICKING IN AN OPEN AREA OF THE PROJECT WINDOW THEREBY SELECTING "NOTHING". THERE IS ANOTHER WAY TO CLEAR THE SELECTION WHICH WE WILL COVER LATER WHEN WE DISCUSS THE SELECTION BOXES.

Crossing and Box selections: Elements may be selected using standard crossing and box selections as shown below. Box selections are made from left-to-right and select only elements which are completely contained in the boxed area. Crossing selections are made from right-to-left and select any element which is either inside the boxed area and any element which crosses the boundary of the selected area.

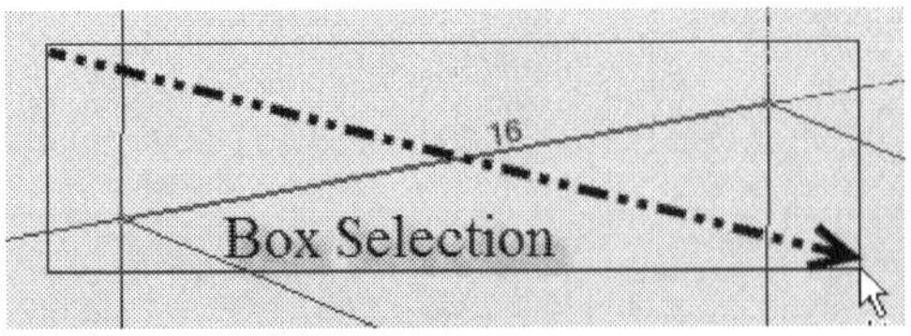

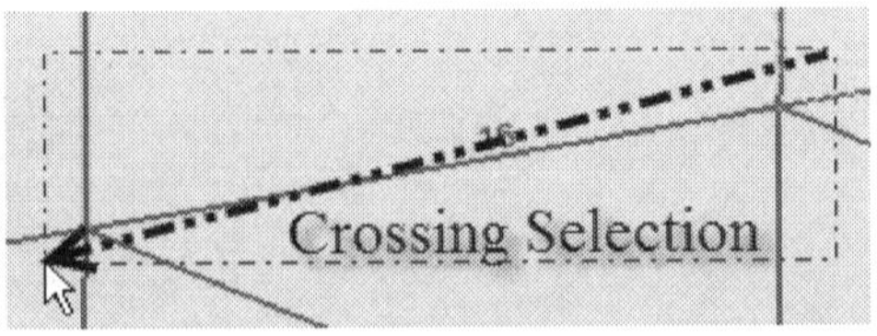

Building a Selection Set: Adding and Subtracting from click selections is accomplished by holding the control key while making selections whether single-click or box/crossing window selections will add or subtract items from the selection set. Holding the control key and selecting an unselected element will add it to the selection set whereas selecting an already selected element will remove it from the selection set. Similarly, holding the Shift key while making selections will add newly selected elements to the current selection (indicated by either single-click or box/crossing window selection).

Using Datums: Single-click selections can also include selecting datum labels: Single-clicking on a datum label will select all elements on the datum. Datums include structural axes as well as structural levels. If a datum label (e.g., "A", "1", "Level 1") is box selected by itself (*only* the label name), then all elements which lie on that datum will be selected (similar to a single-click on the datum label). If a datum label is crossing selected, then any element which touches that datum plane will be selected. When box and crossing selections are used with a datum, only the datum label can be in the selection window. If there are elements also included, Robot ignores the datum label and only selects the elements.

Node and Element Selection Windows

Node Selection Window

Element Selection Window

The node and element selection windows can be used to view the current selection, select elements manually by typing in element or node numbers, remove categories from selection sets, and to create selections of entire categories.

Viewing the Selection: As elements are added to a selection set, the element or node numbers will begin to populate in the selection windows.

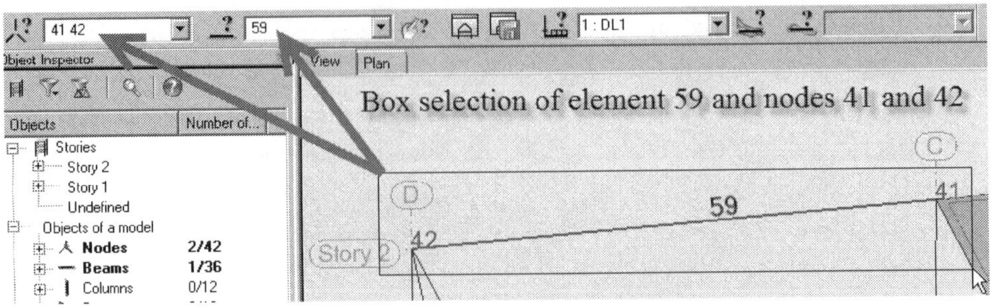

The drop-down in the selection window can also be used to perform selections of "all" of an element type or "none" of an element type. Choosing a "None" selection, if multiple elements are selected, will clear that particular category from the selection leaving the other categories but "all" will clear the entire selection and add only "all" of the newly selected category. (e.g., if you had typed "all" into the bars selection then subsequently selected "Panels – none" only the panels in the project would be removed from the selection, leaving all "Bars" and "Objects")

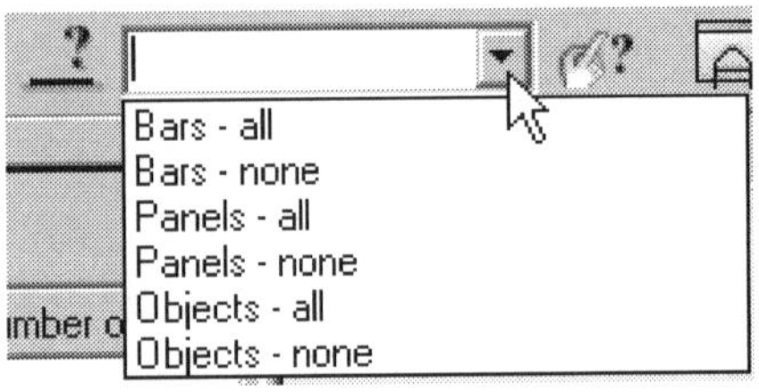

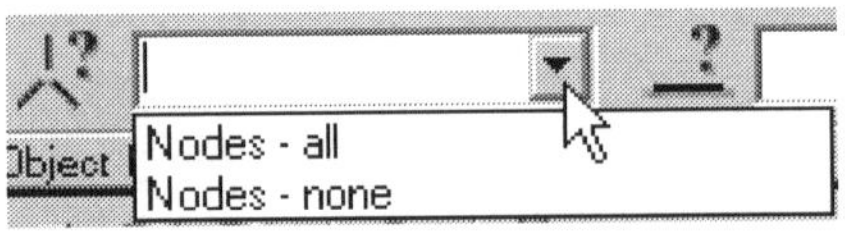

Manual Selections: You may also type element numbers into the node and/or element selection windows. Press <enter> after entering selection to select elements in the project. Selection syntax allows creative selection creation:

Keywords: [all, to, by]

Separators: <space> or <,> separate successive element numbers.

Examples:

1. 1 2 3 4 5 selects: 1,2,3,4,5
2. 1to5by2,8,10to13 selects: 1,3,5,8,10,11,12,13
3. 1to13by3 selects: 1,4,7,10,13
4. All selects: all elements or all nodes.

Element Selection Dialog

The element selection dialog is a very powerful tool, but can be a bit confusing at first. This tool is used to build selections of nodes, bars, panels, finite elements, load cases/combinations, code groups, objects as well as current mode when modal analysis results are available.

Launch the selection dialog by using either the node selection, element selection or case selection buttons respectively or by using the menus **Edit>Select...**:

There is effectively no difference between these three buttons other than the Element category selector will initially be set to either nodes, bars, or cases.

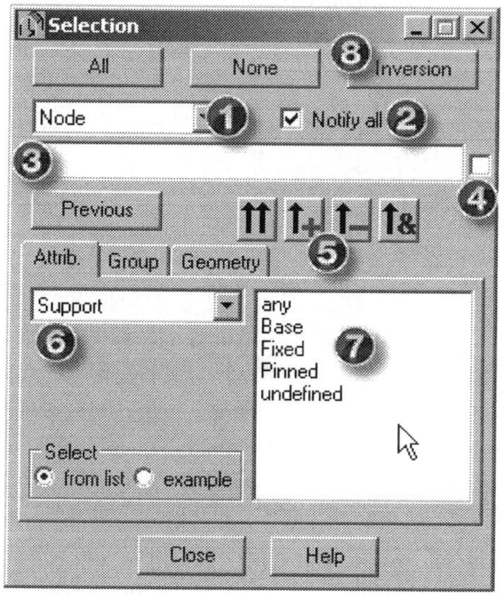

1. Element Category Selector: Controls which type of elements will be added to the selection.
2. Notify All: Typically set to checked. Checking will automatically populate the selection windows and other dialogs with selections built in this dialog. Unchecking will allow you to configure a selection in this dialog and manually cut and paste to other dialogs.
3. Selected Element Numbers: Numbers of elements selected will show here.
4. Show Selected/Show Criteria: If checked, the criteria used to build the selection will be shown. If unchecked, the actual selected element numbers will be shown.
5. Selection manipulation buttons: (more detail later in this section)
    a.  Replace current selection with new items
    b.  Add new items to current selection
    c.  Remove new items from current selection
    d.  Find the union of the current selection and the new items.
6. Criteria Selector
    a. Attrib: Select by an attribute of the current category
    b. Group: Select by or configure a new selection group
    c. Geometry: Use structural axes as selection boundaries
7. Attribute value for selected attribute.
8. Inversion: Invert current selection. I.e., select all elements except the currently selected elements.

Exercise 4: Using the Selection Dialog

1. Open the Exercise_01-UserInterface.rtd file
2. Press <Esc> to exit Pan/Zoom/Orbit mode
3. Use the display settings to show member sections and member types:

   a. View>Display…
   b. Select bars in the left hand categories
   c. Expand "Bar Description" on the right
   d. Check "Section – names" and "Member types – names"
   e. Click "OK"
4. Open the Selection dialog by clicking on the bar selection icon or from **Edit>Select...**
5. Select "Bar" in the element type dropdown list
6. In the Attrib. Tab select "section" from the list.
7. On the right hand side choose W16x40 and press
8. The selection area should say: 55 56 60 62 65to68 and you will notice that all W16x40s in the project have been selected.
9. Add W12x19s to this selection by selecting W12x19 and pressing
10. The selection area should now show: 16 21 27to29 55 56to62By2 63 65to71
11. Change the selection instead to only W21x44s by selecting W21x44 and pressing
12. This will clear the previous selection replacing it with only W21x44s: 15 17 19 22 30 verify the member numbers by using the member numbers toggle at the bottom of the view window.
13. Build the same selection in the following manner:
   a. Add any section by clicking on "any" and
   b. Remove HSS5x5s, HSSQ4x4s, W16x40s, W12x19s and W18x35s by selecting each and pressing the button after each.
14. We have some mismatched member types that we need to clean up. We will use the union button to find all columns which have a beam member type set.
   a. Clear the current selection and select HSS 5x5x.375 using the button. (1to12 31to40)
   b. Change "Section" to "Type" on the Attrib. Tab and choose "beam". Create a union selection by pressing the button.
   c. Notice that only HSS5x5 columns with the "beam" member type are selected. (If we were working on this model for real, we would want to

change their member type to "column" for design. More on that Member Types later in the design chapter.)

--End of Exercise--

Chapter 2 - Preferences and Project Setup

Preferences and project preferences can be accessed from the Tools menu or from the tools toolbar. From the standard toolbar select 

Essential Robot Preferences

Open the preferences dialog from the menu **Tools>Preferences** or tools toolbar button 

To save a new template use the template tools:

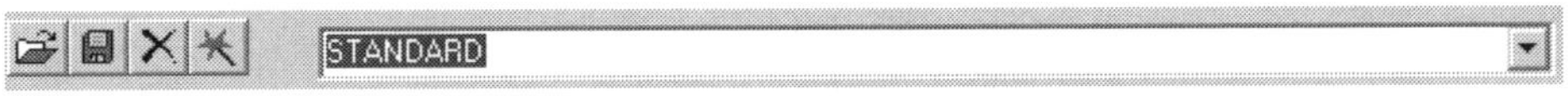

To create a new template type a new name where "STANDARD" is currently shown, and then use the save button to save this new settings template.

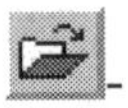- Opens a previously saved preference file

- Saves the current preferences to a file allowing you to specify a name

- Deletes the current preference file

- If you have made changes and want to return to the defaults of the current file.

Languages

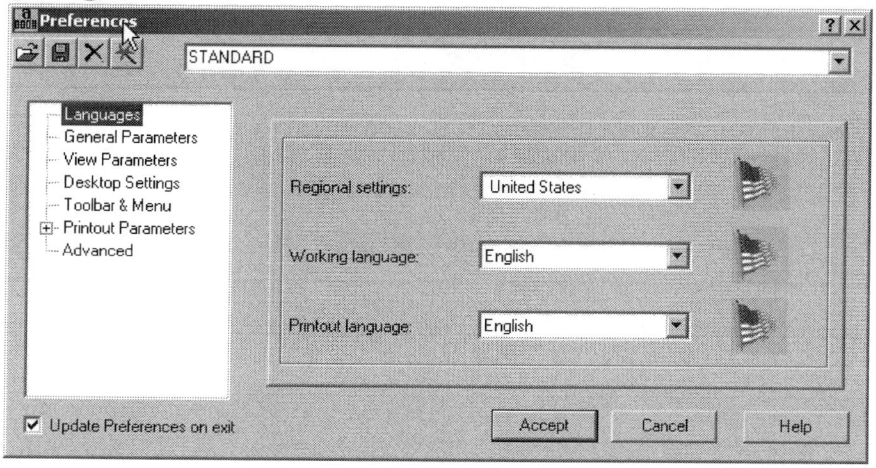

Selecting Languages in the Robot Preferences dialog shows the settings for region, working language and printout language. Selecting a "Regional Setting" will also preconfigure defaults for working language, printout language as well as certain project settings. The recommended practice is to set your regional settings, then adjust design codes on a project by project basis. Due to the automatic changes it is not recommended to change your Regional setting once you have started a project.

NOTE: CHANGING THE REGIONAL SETTING WILL ALSO AFFECT THE INITIAL PROJECT UNITS, MATERIALS, SECTION/MEMBER DATABASES, DESIGN CODES AND LOADING CODES WHICH ARE ALL CONFIGURED ON THE PROJECT SETTINGS DIALOG. CHANGING THE REGIONAL SETTINGS THEREFORE CHANGE A LOT OF INFORMATION CONTAINED IN A DIALOG WHICH IS NOT VISIBLE AT THE TIME OF THE CHANGE.

You can work in a different language with Robot. Choose your working language from the drop down list. If you are working with a windows machine which isn't yet configured to display foreign characters, you will probably see something like the question marks shown here:

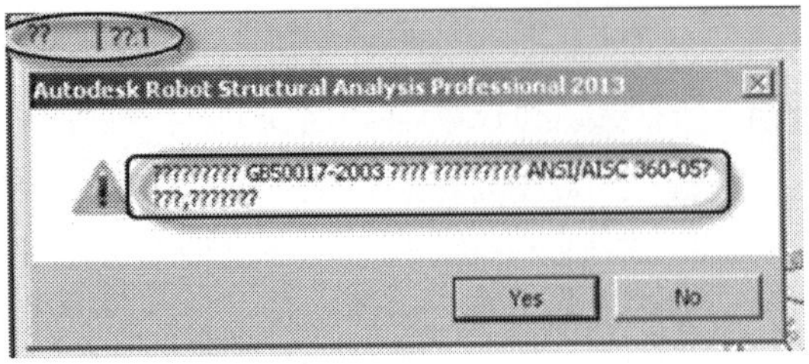

In this case you can go to **Control Panel>Clock, Language, and Region>Region>Administrative Tab** and find the "Current Language for Non-unicode Programs"

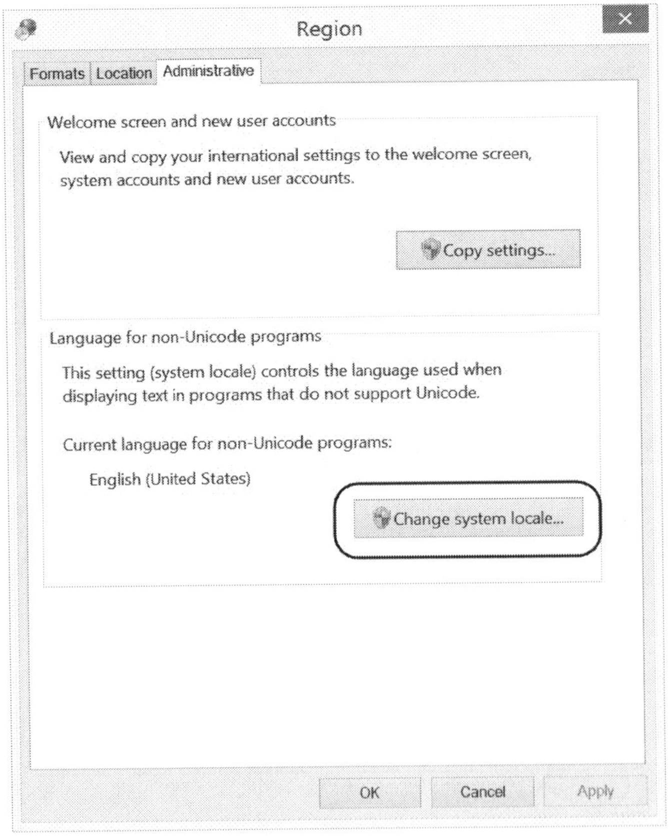

Then select "Change system locale…" and select the corresponding language.

NOTE: IT IS POSSIBLE TO WORK IN A DIFFERENT LANGUAGE THAN YOU HAVE CONFIGURED FOR PRINTOUTS. HOWEVER, IT IS NOT A WELL-TESTED CONFIGURATION AND ISSUES MAY ARISE PARTICULARLY ON DRAWING PRINTOUTS OR CALCULATION NOTES. IT IS RECOMMENDED THAT BOTH THE WORKING LANGUAGE AND THE PRINTOUT LANGUAGE BE SET TO THE SAME SETTING. IF YOU CHOOSE TO WORK IN A DIFFERENT LANGUAGE THAN THE LANGUAGE YOU INTEND TO PRINT, JUST SPEND A BIT MORE TIME REVIEWING YOUR PRINTOUTS TO ENSURE THAT THE PROPER LANGUAGE IS USED.

General Parameters

General parameters options include configuration options around file open and save, automatic backups and automatic saving and reminders.

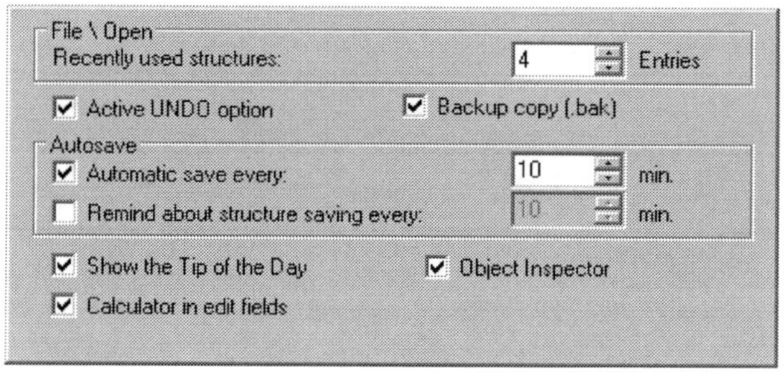

The settings here are fairly self-explanatory with possible exception of "Calculator in edit fields". In any edit control where you can reasonably type a number (e.g., coordinates, load values) double-clicking with the left mouse button in that edit control will launch a calculator shown here:

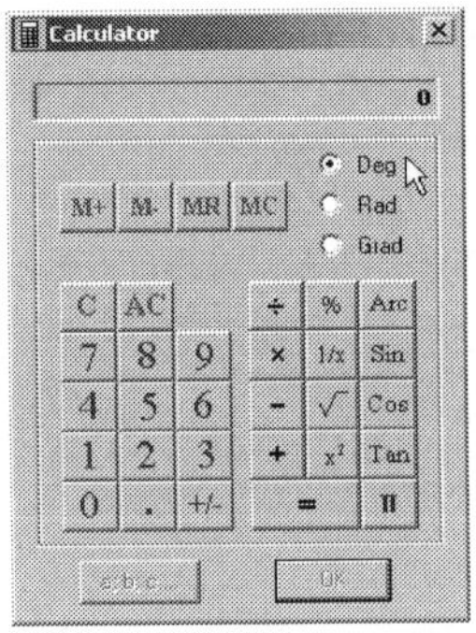

The calculator can be used to do some quick math to figure out what value you wish to populate in the edit control. The calculator is also available at any time from the tools menu **Tools>Calculator...** If launched from an edit control (text box), the OK button will take the current result and paste it into the edit control. If you select the a;b;c;... button, the edit control will be populated with its current value plus a comma and the current value in your calculator. It's a pretty handy little tool.

Desktop Settings

Many of Robot's default settings for view item colors can be changed in this tab. There are several preconfigured settings available from the drop-down list. You can also create your own custom scheme. There are so many things which may be configured, it is best to experiment on your own. You can always restore the default or use one of the preconfigured templates. However, sometimes it is nice to add your own style to the display window.

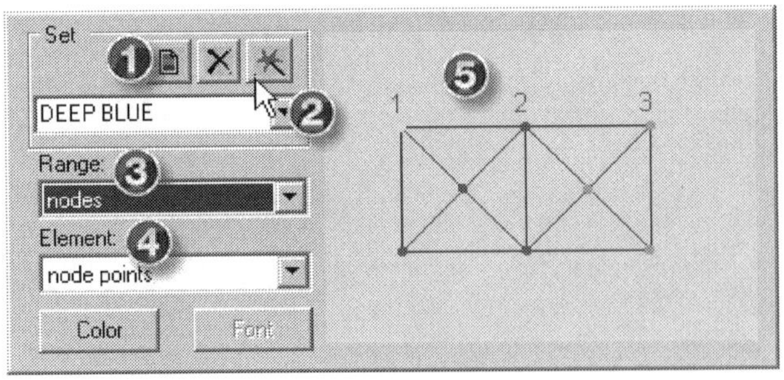

1- Template management area, save, delete and restore defaults

2- Template selector (try a few to see what they look like on the selected range)

3- Range selector: Changes elements available in the "Element" dropdown and updates the preview area to show the effects of the current settings on the range. The template controls settings for all elements in all ranges. The range selector is simply a tool to break down the elements into logical groupings and to configure the preview display.

4- Element selector: Allows you to choose the specific element of the view or display for which you would like to set the color and/or the font.

5- Preview window: Gives you a preview of the settings for any selected range.

Printout and Parameters

This is similar to the desktop settings, but allows you to configure styles, colors and fonts for use in drawings and calculation notes.

Project specific settings

Tools>Project Preferences or tools toolbar button 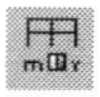

Units and Formats

In Robot, for each project, you have total control over the units used for each different type of dimension used.

On the first pane of the Units and Formats, the zero format controls how Robot will display a value if is equal to 0.0. For instance, you can enter "zero" here and in Robot tables anytime a value is zero you will see the word "zero". The "Imperial" and "Metric" buttons are big hammers which drive default Imperial or default Metric units through the subsequent panes: Dimensions, Forces, and Other.

DIMENSIONS

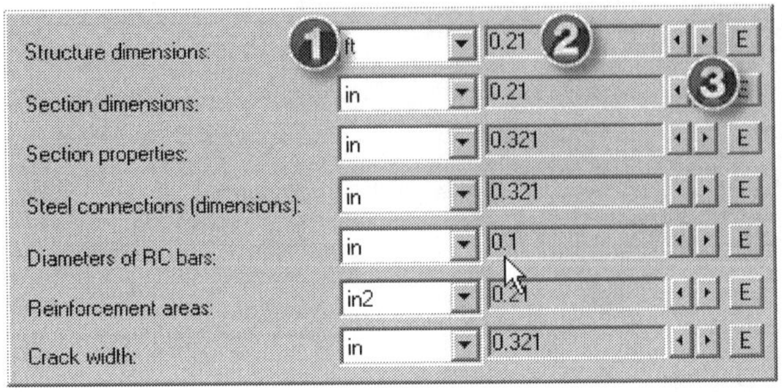

1- Select unit for display

2- Value precision preview

3- Precision control. Left and Right arrows decrease and increase displayed precision. "E" changes format for this unit to scientific notation (e.g., 1.123x10^3)

FORCES

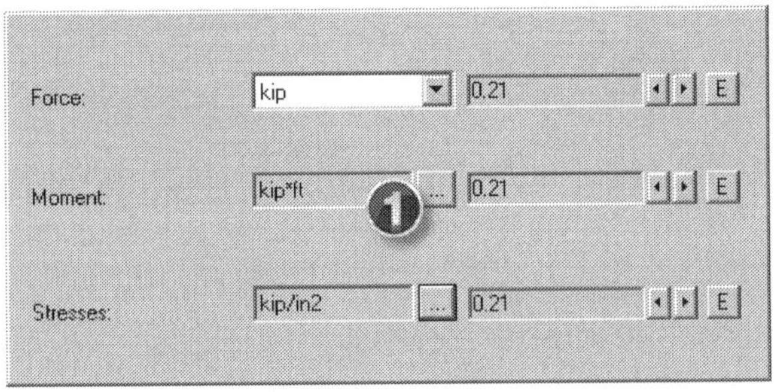

1 - In the Forces section, everything works the same except that for compound units you need to use the ellipsis button to launch individual configuration dialogs to set units for each quantity. E.G.:

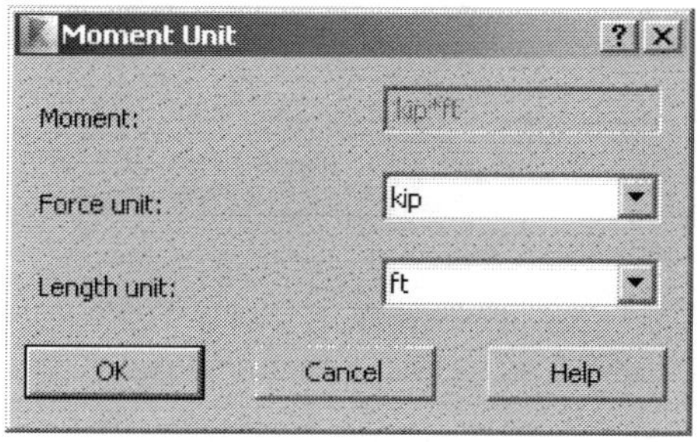

OTHER

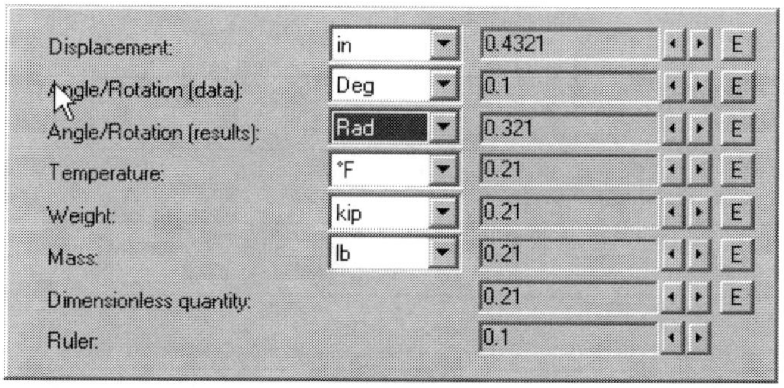

Nothing new in the "other" tab.

UNIT EDITION

Unit edition is a bit of a misnomer. Unit edition means "Create your own units". These units are based on previously configured units and allows you to create virtually any unit naming convention you wish. If there is not an existing unit you would like to use, you can create your own Length, Force or Mass units using this tab. All units are based on Meters, Newtons and Kilograms. Notice that the inch unit is 0.0254 meters in this screenshot. Each segment of the tab works independently: Length unit, Force unit and Mass unit.

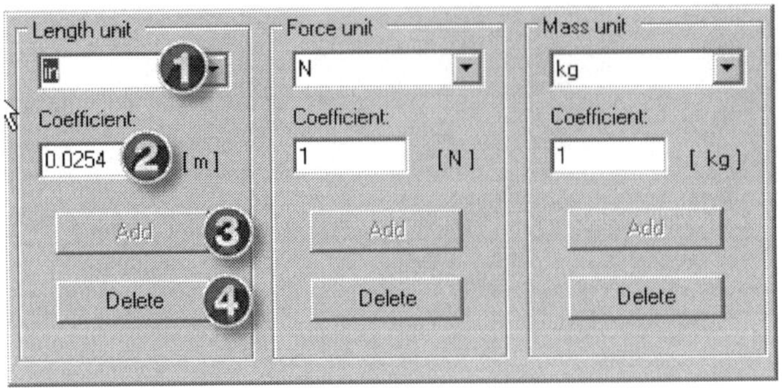

1- Type in a new Unit name here, or select an existing one to modify

2- Provide the coefficient of meters to convert from meters to your new unit

3- Add the unit to the available units list or modify an existing one you are editing

4- Delete the unit from the list of available units.

Here is an example of some very unique unit settings. Just some fun with creating custom units:

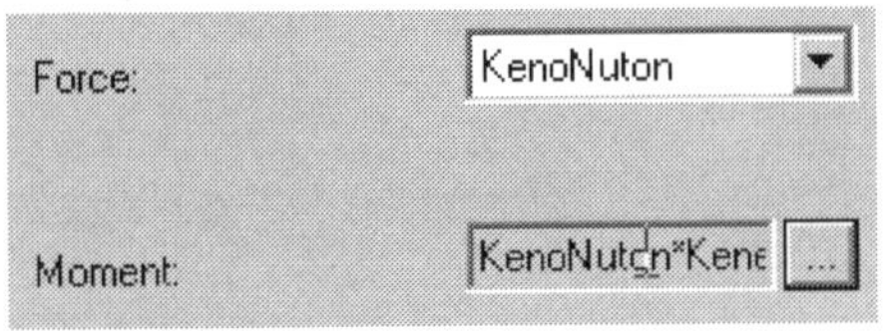

Note, Robot does not like unit names longer than 9 characters :)

Materials

The regional settings define the defaults for materials in this dialog, but you can use any materials in your project.

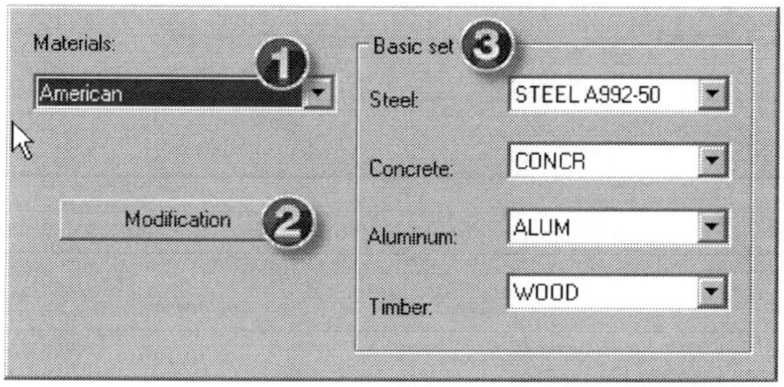

1- The first dropdown "Materials" offers a list of materials which have been created for many different geos.

2- Modification launches a dialog shown below where you can view the settings for materials in the database selected above (1) and modify or create new materials in that database.

3- Configures default settings for newly created elements in the project. As you create bars of different materials, these are the materials which will be assigned by default. They can always be modified later. It is best to set the most commonly used materials for structural elements here, then modify specific ones later. (e.g., A992-50 for structural steel)

Once you enter the material modification dialog, you can review current settings, modify the settings for existing materials and add new materials to the library.

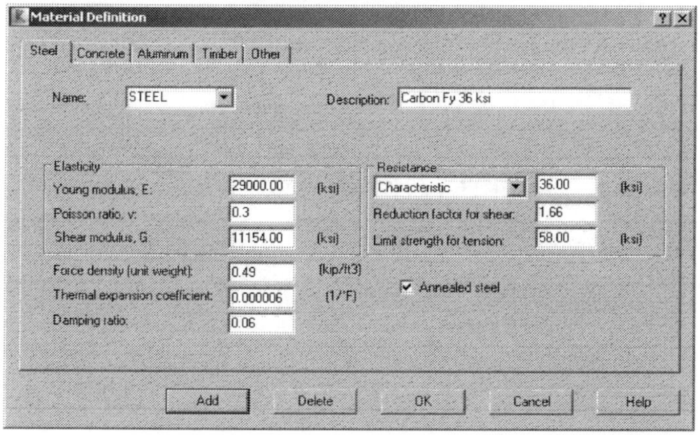

To view or modify an existing material select the material with the name dropdown. Review the settings for the parameters of the material and to save any changes you make, use the "OK" button.

To delete a material from the library, select it in the name drop down and press "delete"

To create a new material, type a new name into the Name dropdown, configure material parameters, then use the "Add" button to add this new material to the library.

The settings are slightly different between the different material tabs as you might expect.

Databases

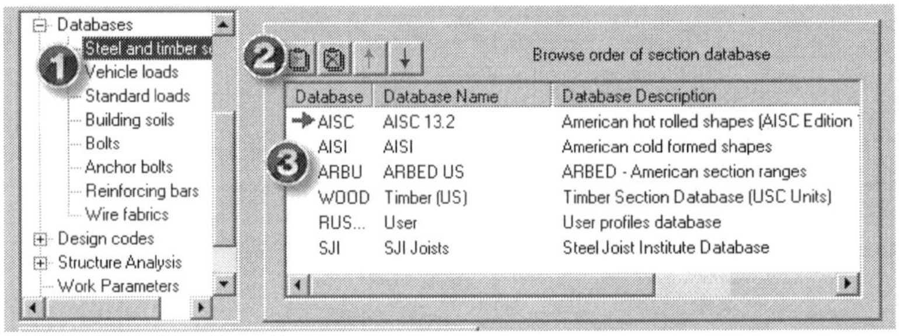

1- Database group: Select different ones to load or manage active databases.

2- Add, Remove, and re-order buttons for managing active database list below.

3- Active Database list: Red arrow indicates currently selected table row.

If there is a section (or other type of) database you would like to have available for use in your project, you can select the "load database" button. In the example above, I have the Steel Joist Institute catalog of standard joists as well as some heavy sections and Russian (RUS) sections available. This list is completely configurable. When you launch the Add database dialog:

This is useful if you typically work with AISC shapes, but perhaps need to have other shapes available in your work.

The position buttons allow you to configure the ordering of these databases in any list of databases when they are presented in the user interface.

Other database lists work similarly.

Design Codes

The design codes tab allows you to configure the design code to be used in member code checks and design modules. Not all design modules support all geo design codes. The example below indicates "Undefined Code" for both Steel Connections and Timber because there are only a limited number of design codes available for steel connections and timber design. Those codes are mostly European, French and Polish codes.

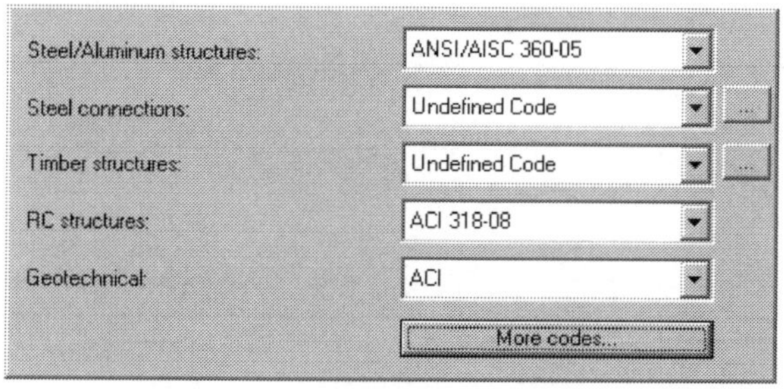

To find out what codes are available for the materials listed and to manage the list of codes for each, click "More Codes" which will launch the Configuration of Code List dialog:

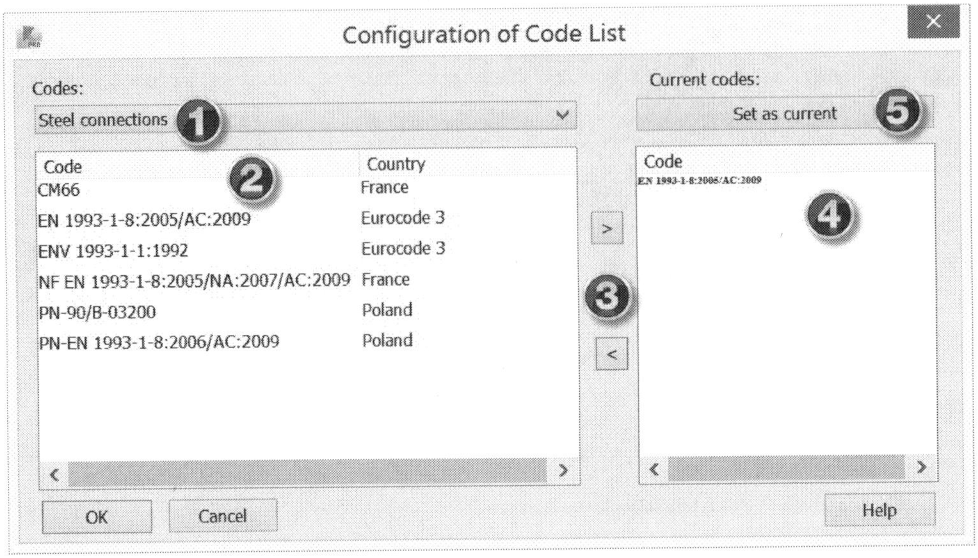

1- Code Group Selector. Choose from steel, steel connections, timber, concrete, geotechnical, and also the various loads: Load Combinations, Wind/Snow, and Seismic.

2- List of available codes for the selected group. List can be sorted by clicking on the column header.

3- Add/Remove code from Active code list.

4- Active Code list: Current default is set in bold face type

5- Change current default to selected code.

The idea behind this dialog is that you don't want to have to scroll through the entire list of all possible design codes each time you want to select a design code. This dialog allows you to manage that list complexity by only showing you the codes in the Active list when you configure design codes for your project. Back in the **Job Preferences>Design Codes** tab you will only see codes from the active list in each dropdown for each material type or loading code.

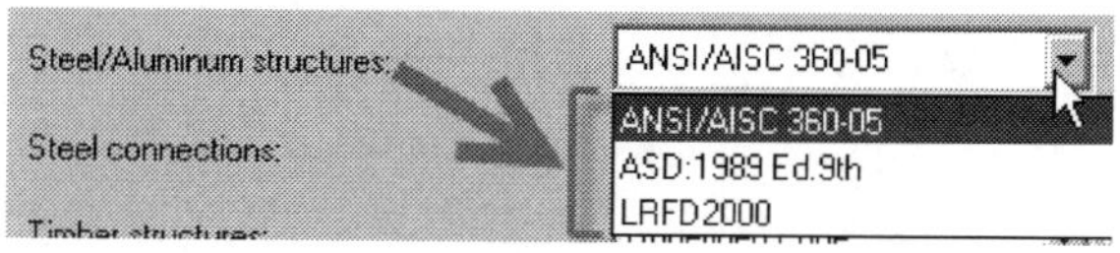

Structure Analysis

This tab contains setting for method of solution of the equation matrix, settings for modal analysis and settings for non-linear analysis.

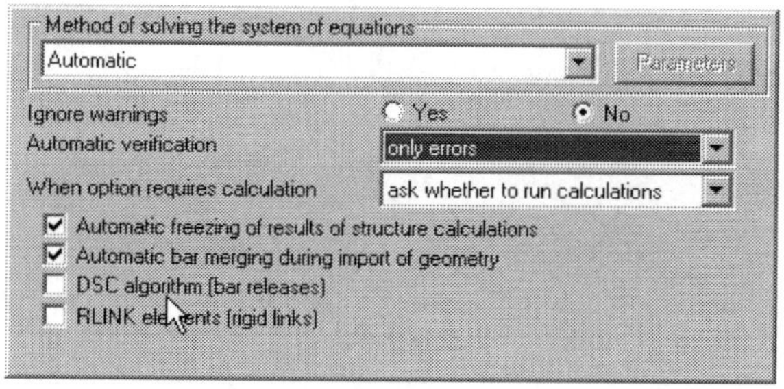

Choice of solver (Method of solving the system of equations)

For most work and initial analyses, it is recommended to leave the choice of solver set to Automatic. Different solvers have been added over time and most are different methods of exploiting standard structure stiffness matrix properties to accelerate the solution as well as to take advantage of computer memory savings by reducing storage requirements. Selecting different solvers can have advantages or disadvantages depending on the number of equations, the stability of the matrix and other factors. We will treat the particular advantages or disadvantages in a more advanced text. Additionally, information on solution algorithms and their methods of solution can be found in texts on the subject of finite element analysis. Please refer to the Further Reading Section for information on Daryl Logan's "A First Course in the Finite Element Method". Specific information on advantages and disadvantages and solution methods can also be found in the Robot help files.

In general, it is best to ask Robot to Not Ignore Warnings so that as you learn the product you can understand and deal with any issues in your model and also understand how they may affect your results. Along these lines, it is recommended to set Robot's automatic verification to Errors and Warnings so that you can make conscious decisions about how your model is analyzed and how it is behaving.

DSC and RLink elements are options for advanced analysis facilitating elastic releases and rigid links which can support modal analysis. You can read more about them in the help files but we will leave discussion of these options to an advanced text.

Other sub-sections of this dialog contain settings for advanced analysis options. In this text we will focus on methods of working with Robot for the beginner. Advanced topics such as modal analysis, and analyses based on a modal analysis e.g., spectral, harmonic, dynamic will be left to an advanced text where we can treat those topics in a more robust manner.

Project Properties

Project properties provides a location to identify the current project and project team members. Access the Project Properties dialog from the file menu: **File>Project Properties**:

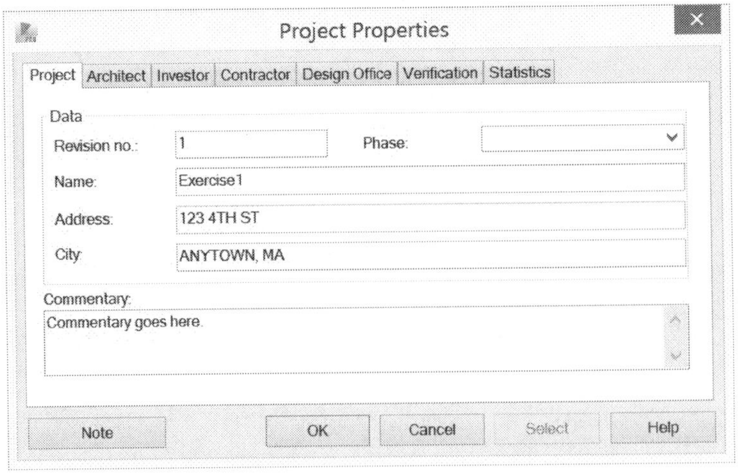

This dialog provides a location for recording information about the project, the parties involved as well as a tab for viewing information about the project (Statistics tab). Unless the "add to note" box is checked in other tabs, only the first tab (Project) information will be added to the report. Take a look at the "Architect" tab:

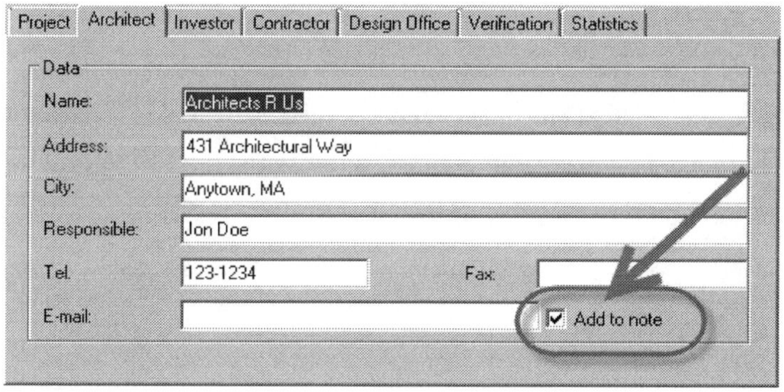

Add to note is not referring to the e-mail, but rather to the information on this tab in general. Checking this box will add the architect's information to the project properties note:

The statistics tab is for informational purposes and cannot be added to the note. Some of the information on this tab will be available from the calculation notes. (**Analysis>Calculation Report>Full Note**)

Chapter 3 - Basic Structural Modeling

This Chapter will cover the basic modeling techniques from creating and managing project datums to modeling bar elements, creating load cases and adding loads to your model. Once you have completed this chapter, you will be able to run your first analysis and will be ready to learn about Robot's results exploration capabilities.

Project Grids

Project grids are not required for modelling. They can be used to assist project layout and geometry configuration or they can be ignored completely. They do have limited ability to manipulate geometry coincident with the datums, but it is not a necessarily robust relationship. We will discuss grid modification later.

Making sure that Robot is ready for model creation/manipulation: Check the Robot layout selector and make sure that it is set to the "Geometry" option under the "Structural Model" group:

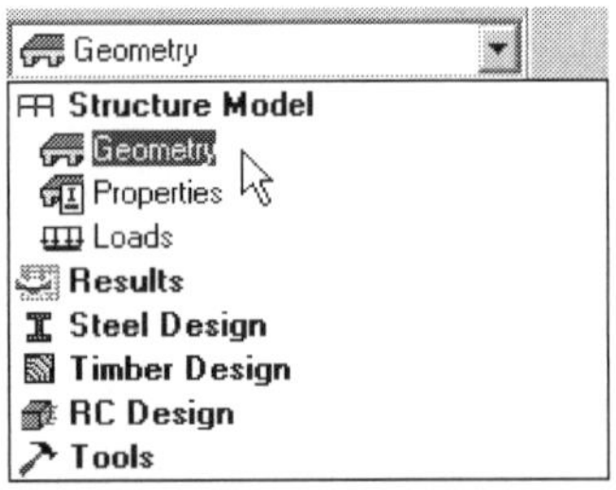

As we have mentioned before, the Robot layout manager can really facilitate working in Robot with preconfigured layouts for performing various tasks. Since our task at hand is to create the structure model geometry, we select Geometry from the layout selector.

Project grids or "Structural Axes" dialog can be launched from the Geometry menu: **Geometry>Axis Definition...** or from the tools toolbar with the Axis Definition button 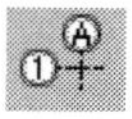

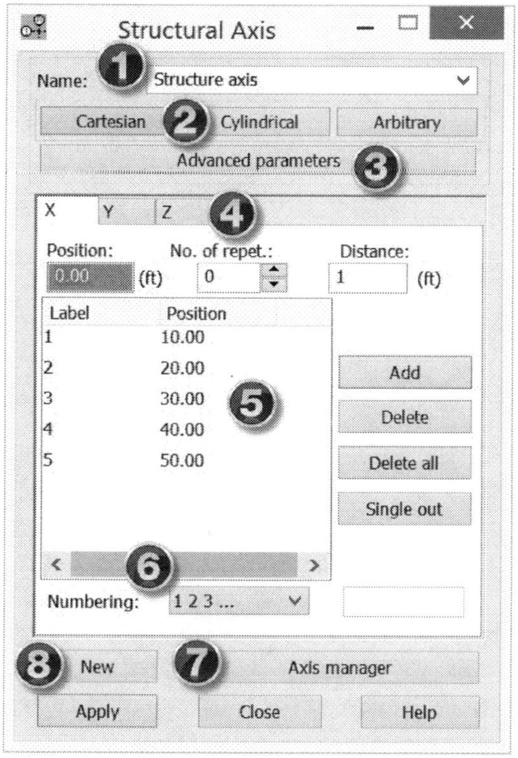

1- Structure Axis Name: You can have many structural axes sets in Robot and have any number of them visible at any one time. Either select an existing axes set to modify in the drop down or type in a name for a new set. (or use the "New" button 8)

2- Each grid set can be of one of these coordinate types. Cartesian grids may be orthogonal to the project axes or be at an arbitrary angle. They may also be relative to a point in the model. Cylindrical grids may also be relative to a point in the model: initially 0,0,0

a. Cartesian 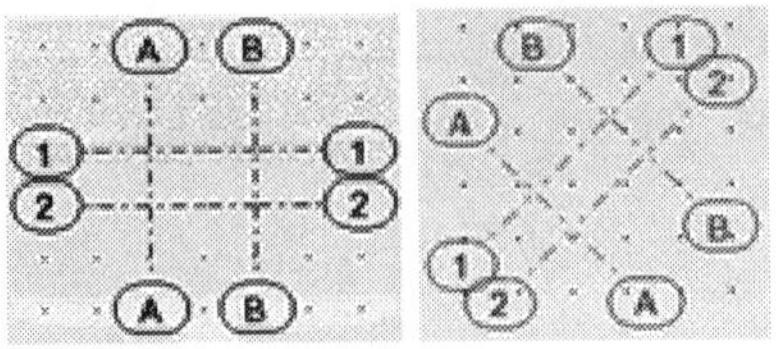

b. Cylindrical 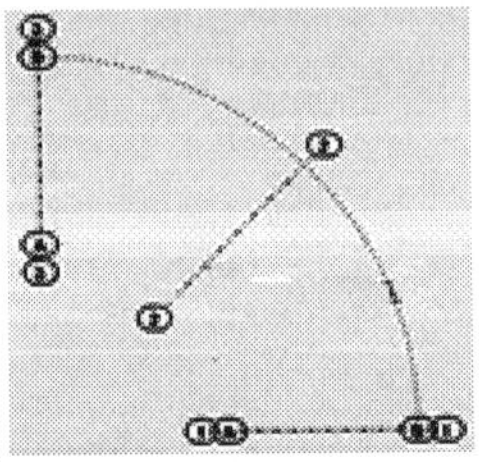

c. Arbitrary. 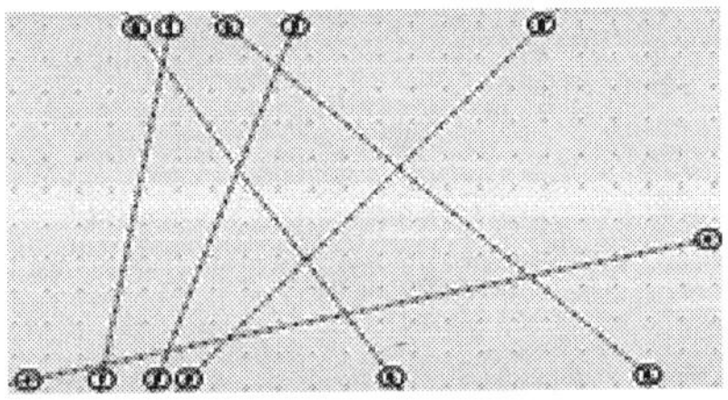

3- Use the "Advanced Parameters" button to configure skew angle and/or relative positioning of the grid set. Any coordinate point in Robot may be specified by clicking on the screen. Simply place the focus on that edit box by clicking once in the edit box, then select a location on screen. The coordinates of the selected location will automatically populate in the edit control. Use Axes relative to a point to place the origin of the axis set at the location of the selected point. It is the origin (0,0,0) by default. An axis set may be rotated about any of the three project axes (X, Y, or Z) allowing for many possibilities for structural axes set configurations.

a. Cartesian Parameters: 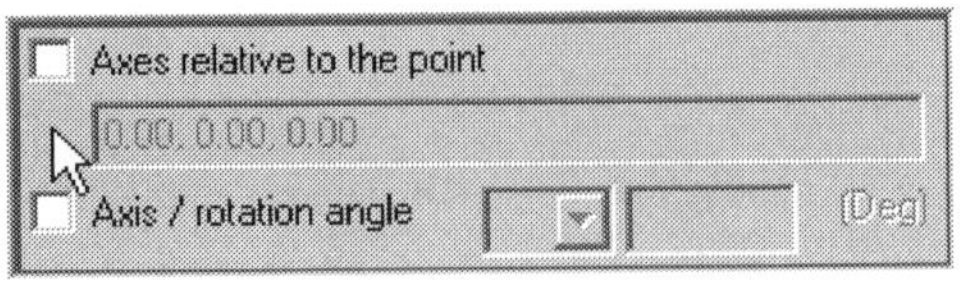

b. Cylindrical Parameters: 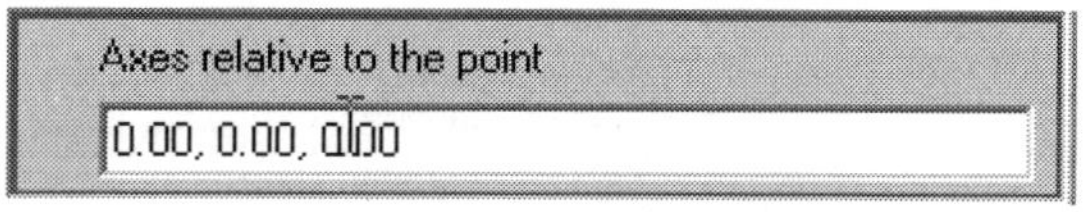

4- Select to define grids along corresponding axis. When "X" tab is selected, grids will be created perpendicular to the X axis with grid increments measured along the X axis. Same for Y and Z axes. Each tab is the same. The Z tab should only be used if you do not wish to use Stories for your structure. If you intend to do building analysis and seismic analysis in particular, you should opt for using Stories and leave the Z axis settings blank for structure grids.

5- Axis set labels: Lists the currently configured axes along the current axis (specified by the tab, e.g., X, Y, or Z). Individual axes labels can be modified by clicking them, but to change the location of the axis, either use the object properties in the project environment or delete the grid and re-create it here.

6- Numbering can be numerical or alphabetical (1,2,3 or A,B,C) which is automatically incremented as you add new grids to the dialog. You can also have the label be the geometric position of the grid i.e., "Value" or you can "Define" custom labels by manually typing the desired name for the next label to be created into the edit control just to the right of the "numbering" selector.

7- Axis Manager is where you manage axis sets and their visibility in the project. You can have any number of axis sets that you require to help layout your project. Unchecking them in this dialog will turn off their visibility in the project environment, deleting them from here will remove them completely from the project.

a. 

8- Create a fresh new axis set in addition to any previously created sets. Previously created sets may be viewed or modified with the dropdown selector (Item 1 above).

Exercise 5: Creating Structure Axes:

1. Start a new project in Robot and select Shell structures:
2. Select 
3. 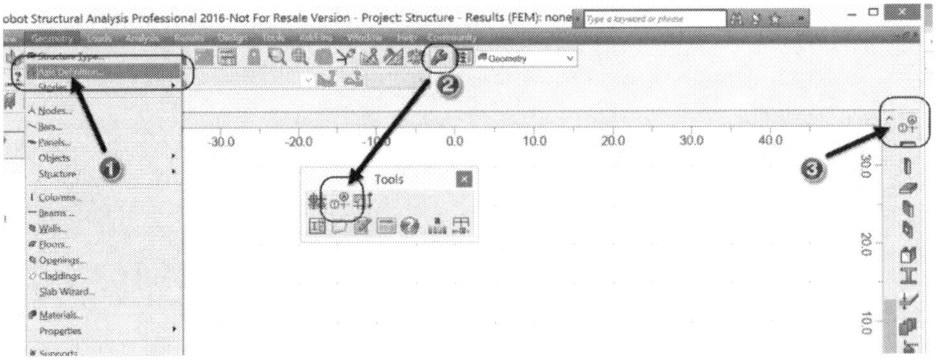

a. Launch the Axis Definition dialog from the Geometry menu(1), the tools toolbar (2) or the structure geometry toolbar on the right hand side of your screen (3)

4. 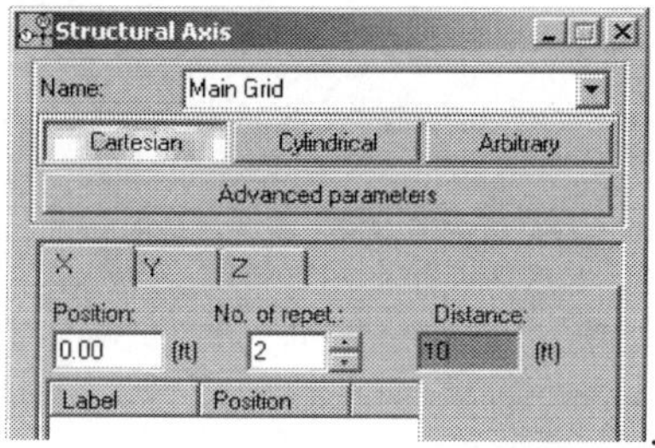.

a. Type "Main Grid" (or something similar) in the Name edit control.

b. Select "Cartesian"

c. Select the "X" tab and enter 0.00 for position, 2 for number of repetititons and 10.0 for Distance.

5. Press the "Add" button. Your X grid list should look like this:

6. 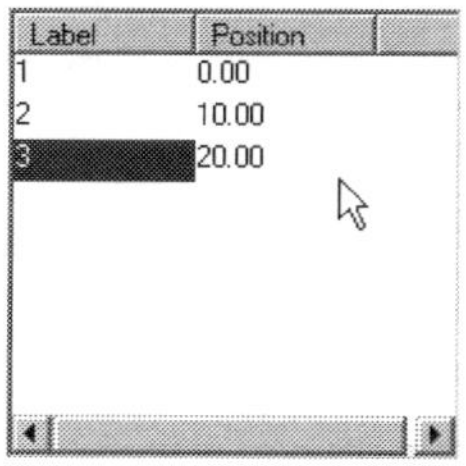

7. Switch to the Y tab and enter the same values but change the "numbering" to A,B,C…

8. 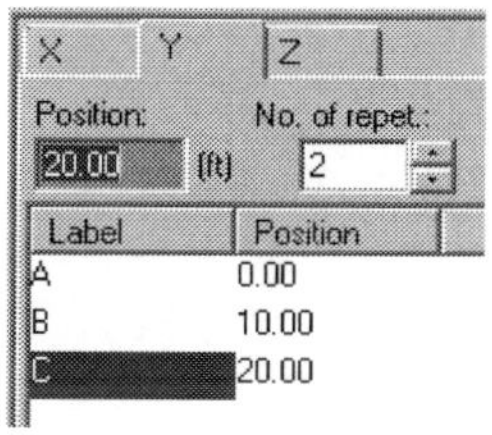

9. Press "Add", then "Apply" at the bottom of the dialog:

10. If you started the Shell type structure, the default orientation for the project view is the ZX plane so all of our created grids will not be visible yet. Click on the project axes icon in the lower left hand corner, select 2D and XY as shown here:

    a. 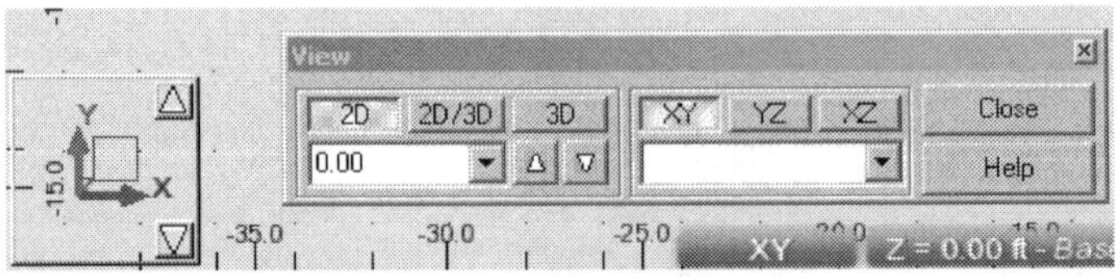

11. 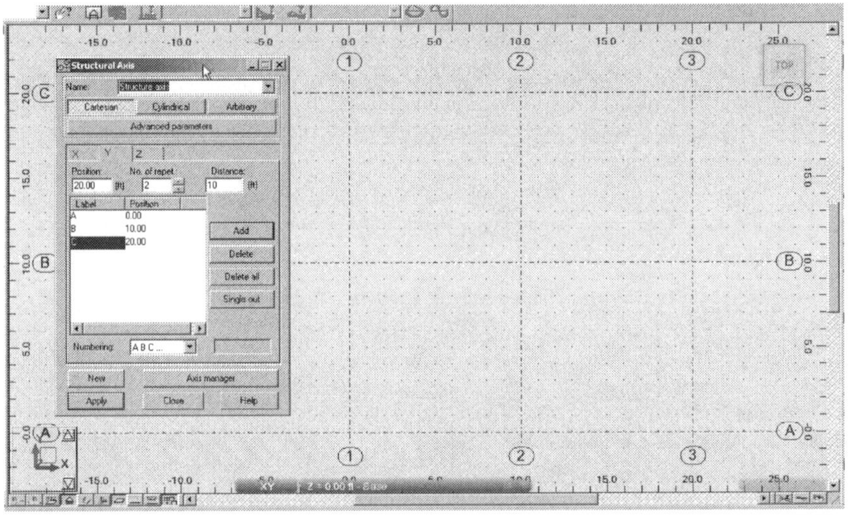

12. Now Press the "New" button to start a new set.

13. Choose "Cylindrical".

14. Type "Arc Grids" or something similar in the name field

15. Press the "P" button below to reveal the grid set origin

    a. 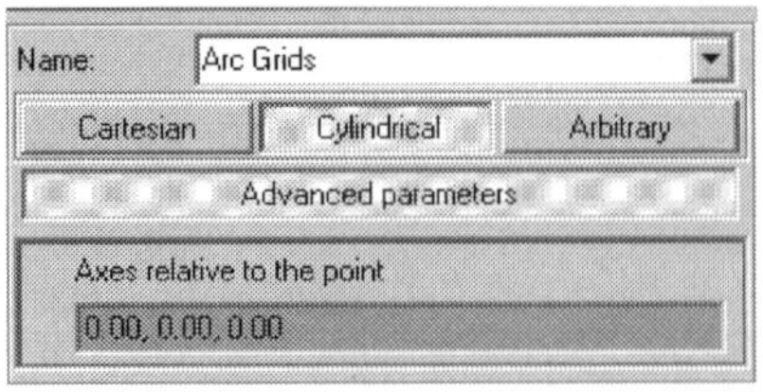

16. Left-click the mouse in the coordinate entry edit control and it will turn green as shown above. We could type in coordinates here, but this time we will pick a coordinate from the project environment.

17. Hover near the intersection of grids 2 and C and click to select the coordinate location

    a. 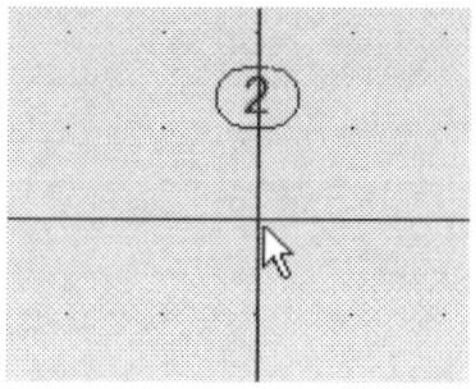

18. The edit control will now have the coordinates of the selected point in the project view.

    a. 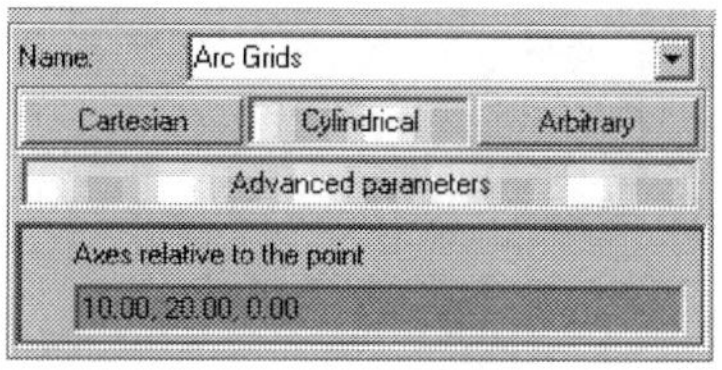

19. Select the Radial tab and set "position" to 5.00ft, set Number of repetitions at 1 and Distance to 5.00 then press "Add".

    a. 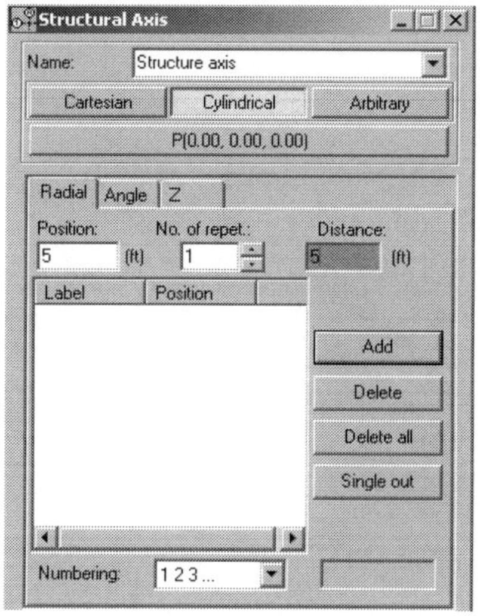

20. Switch to the "Angle" tab and set position to 20 (deg) and repetitions to 7 and Distance to 20(deg) Then Press "Add" and then "Apply"

a. 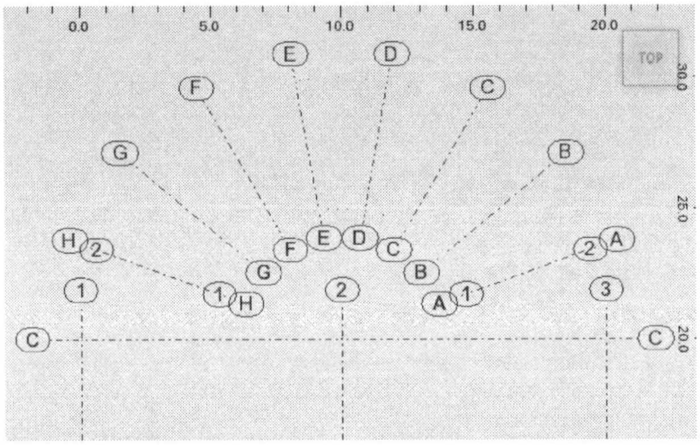

21. From the lower right hand corner of this dialog, launch the Axis Manager and notice that both of our newly created grid sets are available in the list. The checkbox controls visibility in the project. Try turning off the Main Grid by unchecking the checkbox and pressing "OK". Then reverse that process to turn it back on.

a. 

--End of Exercise--

Stories

Instead of using the "Z" axis settings, it will help any analyses based on a sense of the story structure of a model (e.g., seismic) to have stories defined. Stories can also be used to filter the view to only elements associated with a particular story. This filtering only works in the building design configuration from the **Geometry>Structure Type...** menu.

Stories can also be used to select elements which either lie on the story or are part of the story. To configure stories for the model, select **Geometry>Stories>Stories...**

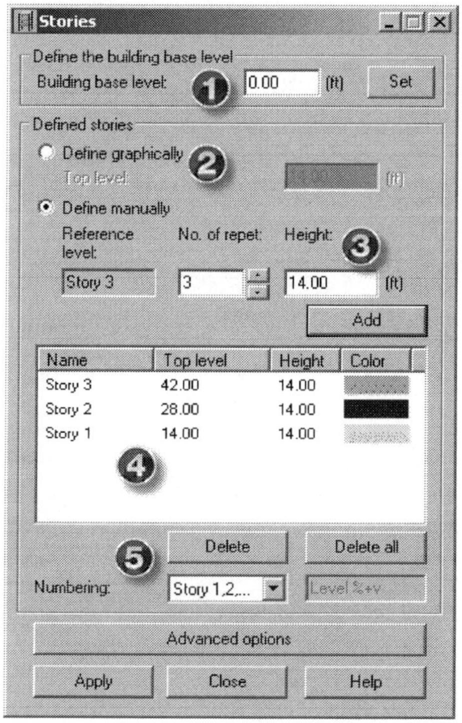

1- The elevation in the Z-coordinate of the base level of the building. This will be used in seismic analysis.

2- Story levels may be created by clicking on a location or element graphically. While you may use this graphical definition feature in any view orientation, this is most easily accomplished in a view oriented to the XZ or YZ planes.

3- Specify the height of the next story(s) to be created and a number of repetitions. Pressing "Add" will create the new stories.

4- Story list: Story name, the top level elevation for the story, the height of the story and the color which will be used to mark elements assigned to that story if "Mark by colors">Stories – Legend by colors is selected in the Display properties (**View>Display...**)

5- Numbering: Default is "Story 1,2,3...". The Define option allows you to configure a naming style. If the last character in the template is a letter then levels will be alphabetical, if it is a number then they will be numerical. All other characters before the last one will simply prepend the incremented value (the last character). If you use %+v or %v the story name will have the elevation value as the suffix of the name. All characters to the left of %v will make up the first part of the name.

Assigning Elements to Stories

Once you have your stories defined, elements (floors, walls, beams, columns, braces, etc.) will need to be assigned to their respective stories in order for any seismic analysis to be performed. If you model your structure one story at a time, setting your workplane as you go, you will find that elements are automatically assigned to stories as you create them. However, if you model one story and copy it up through the structure, you will find that the copied elements all think they are part of "Story 1" the default story. Robot offers a quick tool to help out in a situation where you may have copied elements or other situations where you may not have elements properly assigned to stories.

First we'll take a look at the stories defined and use the "Filter by story" tool for that. First, make sure you are in "Building Design" type by selecting Geometry>Structure Type... and verifying that you have "Building Design" selected. You should notice in the toolbars area, a toolbar for stories commands:

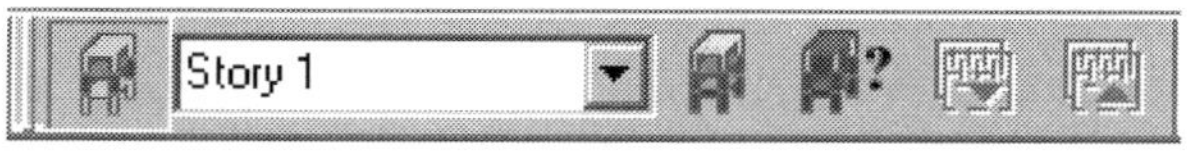

The first button is the most important one, which turns on filtering the view by story. Here we can see that all of my element in this building model are currently assigned to story "Level 2":

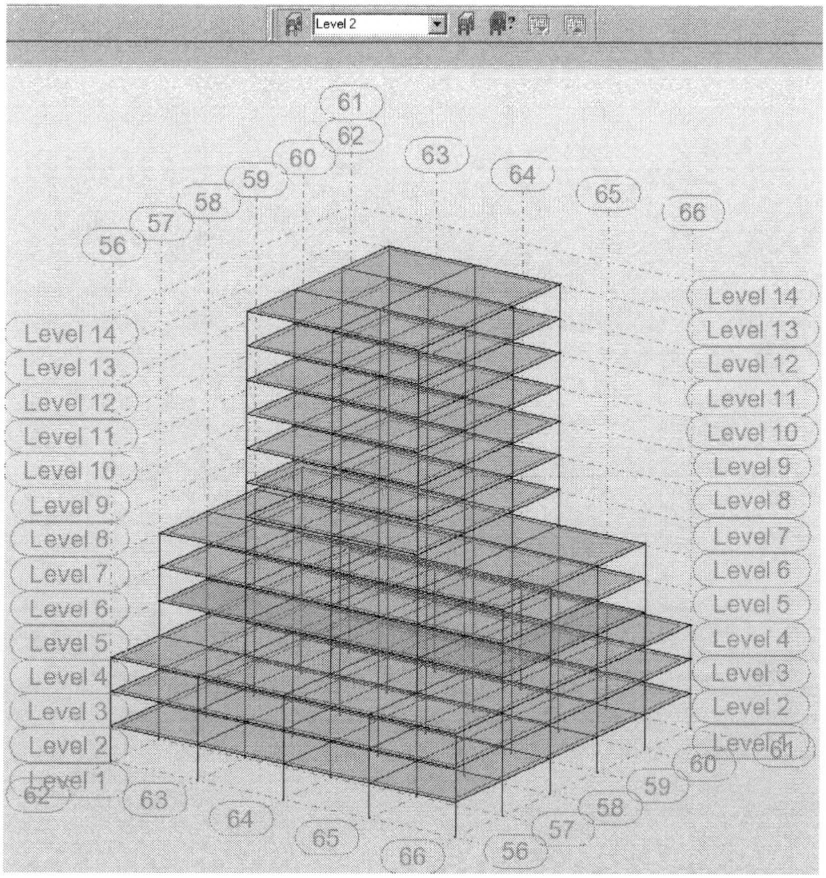

The next task is to create the stories that we need and then assign the elements in the stories to those stories for proper accounting in seismic analysis. Start the Stories tool from Geometry>Stories>Stories…and then (or before) click on the project axes to set the view to 2D, XZ projection like so:

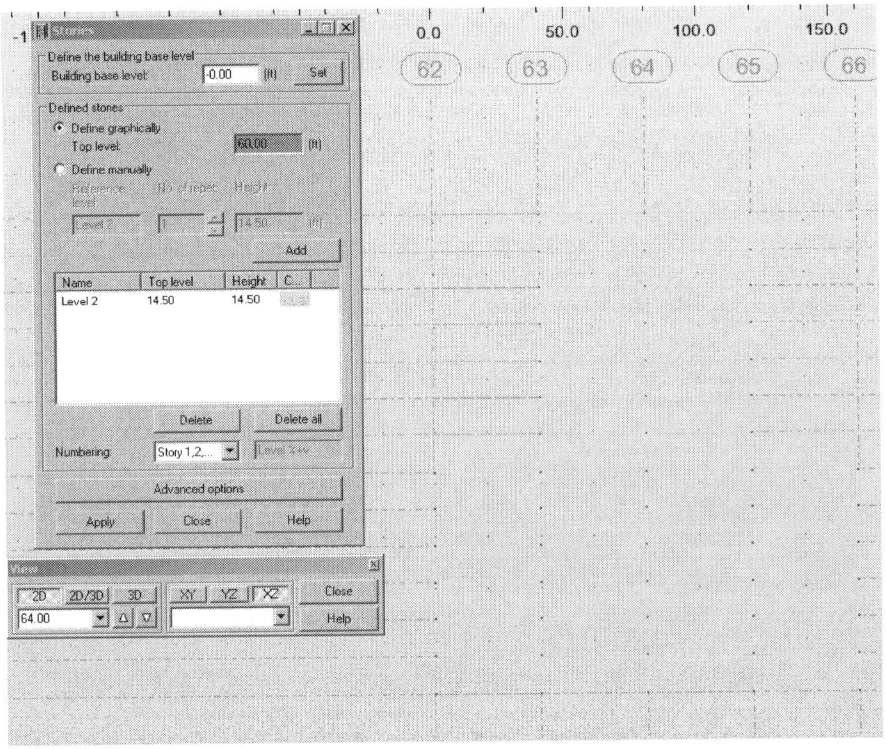

Select "Define graphically" and then click at each elevation of the structure in your project:

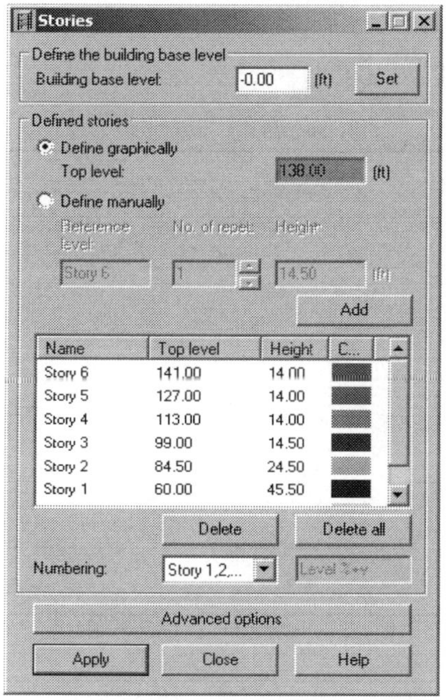

All we have done here is to define the stories, now let's assign elements to them: Open the geometrical assignment tool from Geometry>Stories>Assign Geometrically…which will open the Assign Geometrically dialog:

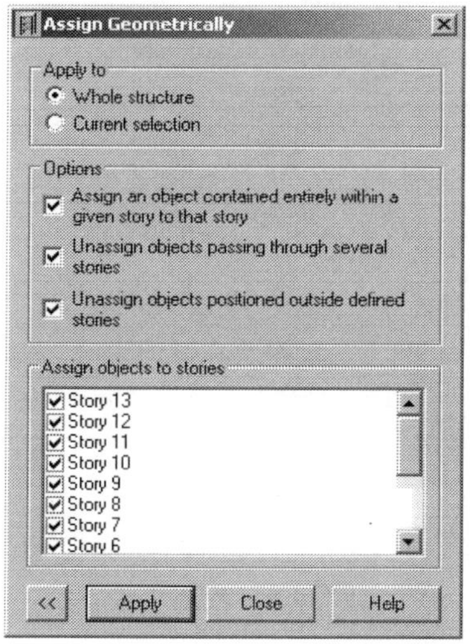

Select "Whole Structure" and leave all options checked for this operation. You can go back and fine-tune later. Check all stories that should be assigned to and then press "Apply".

Now our view should be filtered to any story we choose and we can go through each story to visually validate which elements are considered part of that story:

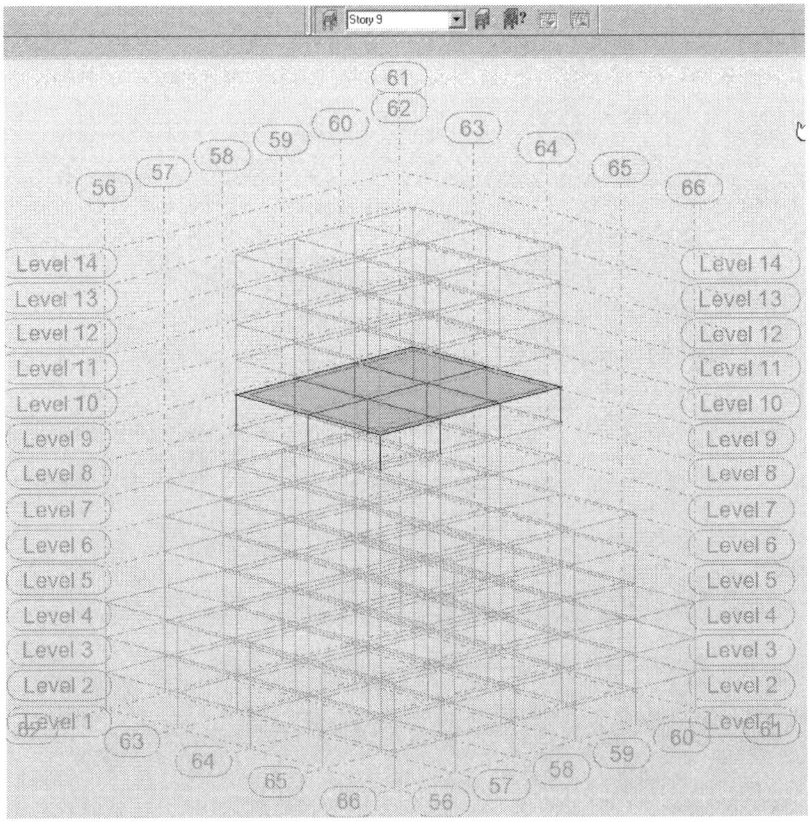

Defining Nodes

NOTE: ROBOT CREATES NODES AUTOMATICALLY AS BARS, COLUMNS AND BEAMS ARE CREATED SO THERE IS NO NEED TO BEGIN BY CREATING NODES EXPLICITLY. THIS TOOL IS SOMEWHAT OF A HOLDOVER FROM A TIME WHEN STRUCTURAL MODELING BEGAN WITH SPECIFYING NODES BY COORDINATE VALUES AND THEN MEMBERS WHICH REFERENCE THOSE NODES. IF YOU NEED TO ADD NODES ALONG A MEMBER YOU CAN USE THE EDIT> DIVIDE... FUNCTIONALITY.

Placement dialogs in Robot can be thought of as placement editors and the Nodes Dialog is no exception. While the dialog is open (or active), you are in the "Placement Mode". The cursor behavior and selection behavior is all geared towards placing whatever you happen to be in the middle of placing. You cannot exit a placement mode with the <Esc> key. The *only* way to exit a placement editor is to either close the current placement dialog or, if you are using layouts, to switch the layout to a different layout.

After starting the nodes dialog from either **Geomtery>Nodes...** or using the nodes button 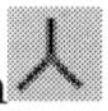 on the structure toolbar you are presented with the Nodes Dialog:

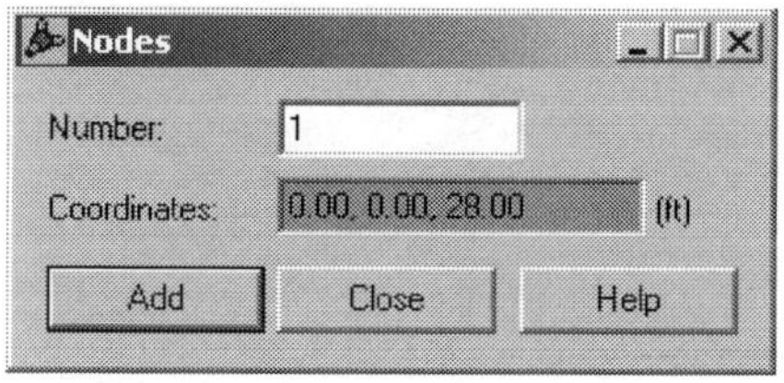

You may begin manually entering node coordinates and "Add"ing new nodes or place the cursor in the Coordinates edit control (edit control turns green) and begin clicking locations in the project environment. Robot will attempt to snap to grid/story intersections to facilitate this process.

Using the Nodes Layout

If your structure type is set to Shell, 3D Frame, 2D frame you can also use the nodes layout. From the layout selector, choose Structure Model and then Nodes:

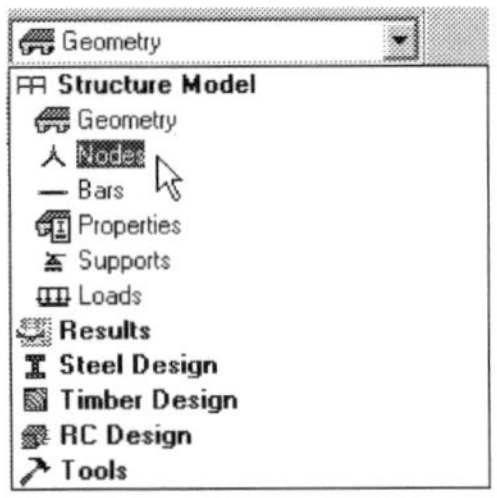

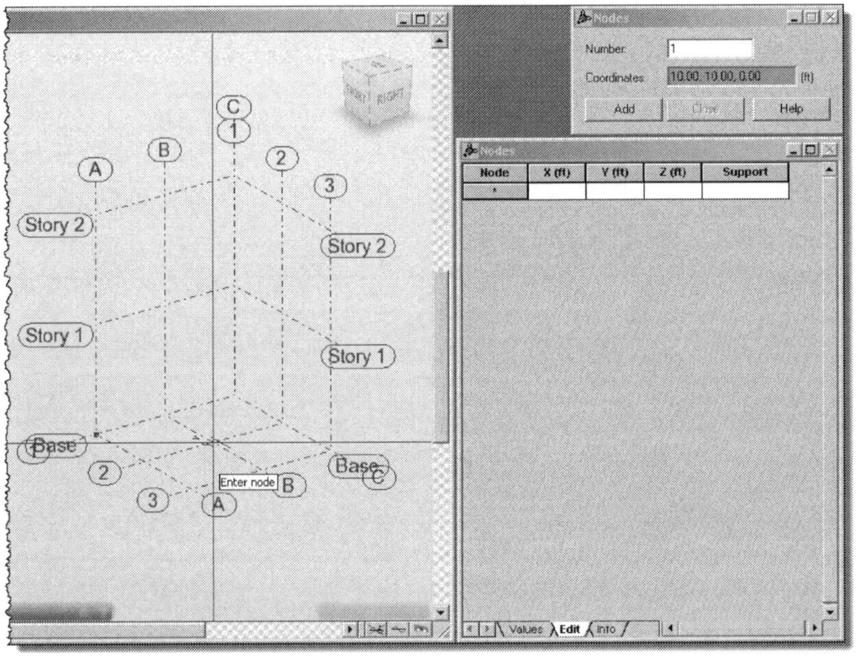

Notice that Robot has automatically arranged your windows and also opened the nodes table for you. As you add nodes, you can see them populate in the nodes table which gives great feedback on what you're doing within the nodes dialog. Alternatively, you can manually add rows to the table in the "edit" tab (at the bottom) and/or cut and paste in information to this table or use Fill Special to populate the table with nodes.

Probably the most jarring aspect of using the layout is that the "x" buttons in the dialog and the table are greyed out because a layout is applied. You could have opened the nodes dialog yourself from **Geometry>Nodes...** as well as the nodes table by **View>Tables...** and selected "nodes" in the tables list. The layout provides a simple management of the interface to facilitate working with Robot. You will remain in the nodes placement editor until you change layouts. To get back to normal, just select Structure>Geometry from the layout selector.

Modeling Bars

Bar modeling in Robot means any linear element. Be it a beam, column, or brace it is considered a bar. Modeling bars in Robot serves as not only geometry definition, but is also the confluence of several different settings. Each bar in Robot will have 2 main properties, known in Robot as "Labels", assigned to it: Bar type label and Section Type label.

Launch the Bars dialog either from the menu **Geometry>Bars...** or by selecting the bars layout if you have the structure type set to Shell or 3D/2D Frame.

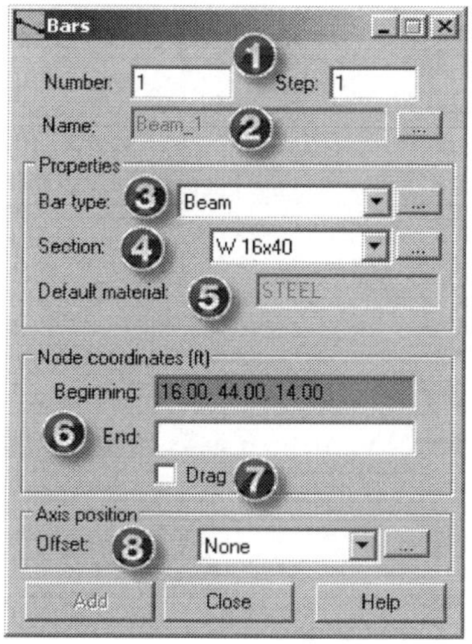

1- Number and Step: As bars are created the next bar will have member number set to "number" and subsequently created bars will have their number automatically set to a number of the previously created bar +Step. Generally speaking, letting Robot manage this is the best option.

2- Member Name: This is an automatic Naming for members which will be used in member code-check verification and design to help you identify the member. Using the ellipsis button to the right of name will allow you to adjust the naming scheme for the bars you are creating. See the Bar Name section just below for more information

3- Bar Type label: This label stores all properties regarding member code checking: E.g., unbraced lengths for bucking checks, service criteria for deflection, allowable percentage of slab to use as a part of a T-beam in a concrete slab. See the Bar Type section below.

4- Section Label: This label stores information about the structural section type. (E.g., W16x40) See the section label section below. The values available in this dropdown are directly dependent on the bar type selected above in the Bar type label dropdown. Only elements that could be assigned the Bar Type selected will be displayed in the section dropdown. This helps avoid, for instance, applying Timber calculation properties to a steel member and vice versa.

5- Default material shows you the current material configured for the section in the Materials dialog (**Geometry>Materials...**). New sections added to the project in the section labels dialog will have their default material set to the project settings default (**Tools>Job Preferences...>Materials** under the "Basic Set")

6- Enter coordinates defining the member beginning and end or place the cursor in the edit control (it will turn green), then select a location in the project window directly. The edit control focus will switch to the "End" once you have selected a start point in the view.

7- Drag will automatically set the next start point to the current end point when selecting points in the project view. This allows you to quickly place beams without needing to select the start point again if your start point for the next beam was the end point of your last one.

8- Offset allows you to specify an offset which you have configured previously in the Offsets Dialog. Discussion of using offsets will be dealt with in a future text.

Bar Name

If you are interested in controlling the naming of your bar elements (visible by selecting the option in **View>Display...**>Bar descriptions>Beam-Name) use the ellipsis button to the right of the Name field, launch the Names of Bars/Objects dialog. These names may be modified at any later point in the process via the Geometry>Names of Bars/Objects... Menu.

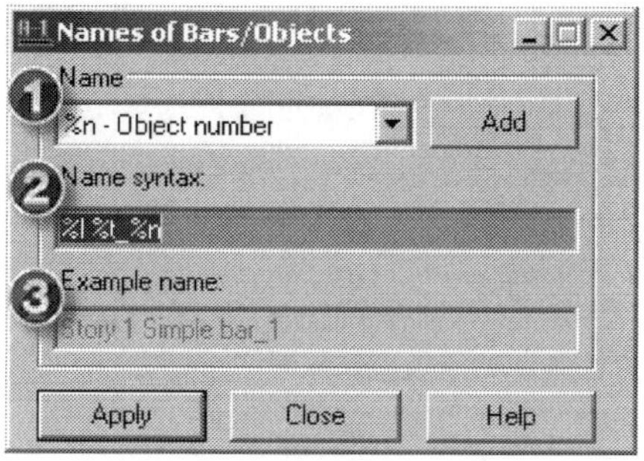

1- Variable Selector. The "Name" caption is misleading. This dropdown allows you to select from Robot's preconfigured variables. In this image %n is the Object number or "member number". Select a variable you would like to use and press "Add" to add it to the Name Syntax field. You can also type these variable names directly into the Name Syntax box but this is a convenient way to make sure that you have the variable syntax correct.

2- Name Syntax: You can add any characters you desire in and around the variable names to create a unique naming scheme for your elements. As you configure your name syntax, see an example of how it will look in the "Example Name" below. Try out a few different styles and see how this works.

3- Example Name: As you configure your name syntax this control shows you a preview of the name.

Bar Type

Labels in Robot are a means of creating sets of parameter settings which may be referenced by elements in the model. This is very powerful in that members only reference a parameter set (label) so changing the parameters of the set (the label) all members which reference that label will have the new properties. Labels can be applied later in the process and are visible in the bars table

The bar type label selector in the Bars dialog not only sets the member type for code checking and design, but also controls what values are available in the Section dropdown in the Bars dialog. We will discuss the particular settings for steel and concrete in their respective design sections. Every member created must have a Bar Type I.e., member code checking parameters. These code checking parameters are specific to the code you have selected in **Tools>Job Preferences** for the type of material. The examples in this book will use the US standards of AISC 360 for steel and ACI318-08 for Concrete. Access the Bar Type label manager from the associated design menu: **Design>Steel Members Design – Options>Code Parameters...** or for concrete **Design>Required Reinforcement of Beams/Columns – Options>Code Parameters...**

To configure Bar Type labels while using the bars placement dialog, simply click the ellipsis button next to Bar Type dropdown:

If the Bar Type selected at the time of pressing the ellipsis button was a steel bar type, then the steel Member Definition – Parameters for your selected steel design code will be shown. In my case, I have selected AISC 360-10 as my steel design code so my dialog indicates that in the title. Notice that this is where all the settings used for determining member capacities are found.

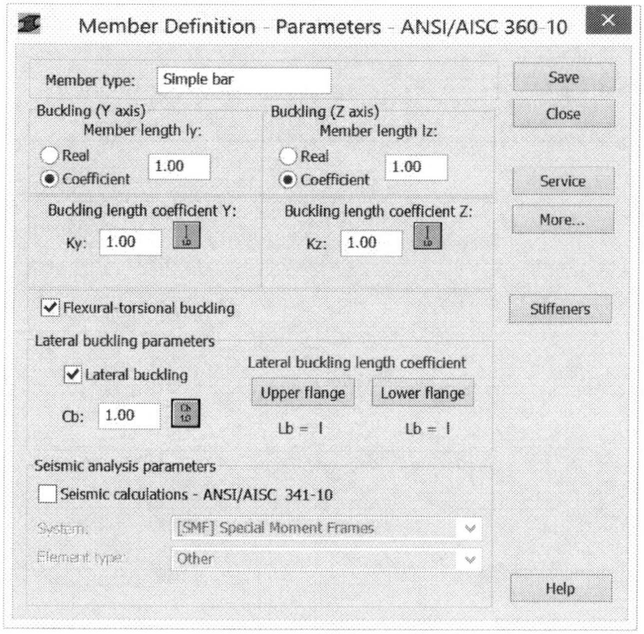

NOTE: ROBOT PROVIDES SOME DEFAULT MEMBER TYPES FOR STEEL, CONCRETE, AND TIMBER. YOU ARE NOT ALLOWED TO MODIFY THE PARAMETER VALUES OF THE DEFAULT MEMBER TYPES. FOR STEEL, THESE ARE BEAM, COLUMN AND SIMPLE BAR. IF YOU MODIFY SOME SETTINGS AND TRY TO SAVE THEM YOU WILL RECEIVE A WARNING THAT THE ONE YOU ARE ATTEMPTING TO MODIFY IS A DEFAULT MEMBER AND CANNOT BE MODIFIED. TO ENABLE MODIFICATION, TYPE A NEW NAME TO CREATE A NEW MEMBER TYPE DEFINITION AND UPDATE WITH THE DESIRED CHANGES.

To create a new Member Type Definition, change the Member Type name at the top of the dialog, adjust parameters (Note that additional parameters live under “Service” and “More” here), then use “Save” to save this new type definition. If you are modifying a member type definition, you will be asked if you intend to overwrite the existing definition. Note that many members may already have this member type applied and by changing the settings, they will be changed for all members which have this member type definition label.

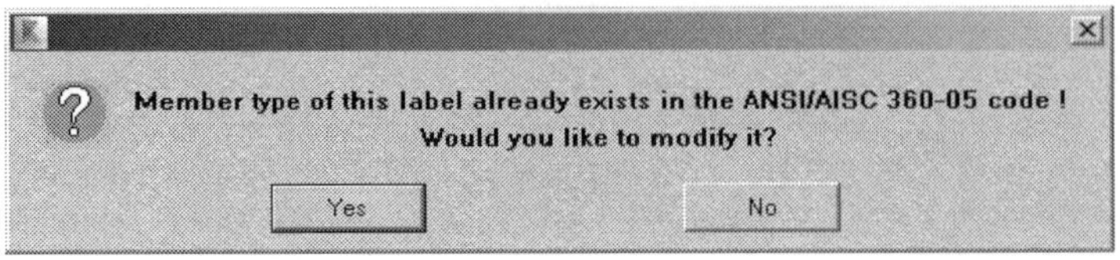

Your new label will now be available in the list of Bar Types in the Bars placement dialog. Refer to Chapter 8 – Basic Steel Design for more information.

Sections

Selecting the ellipsis button next to Section will launch the New Section dialog. In this dialog, you can configure new sections of all types and add them to the list of sections available in the Section dropdown in the Bars dialog.

If the section in the dropdown was a steel section, then the New Section dialog will open to the Steel Member configuration:

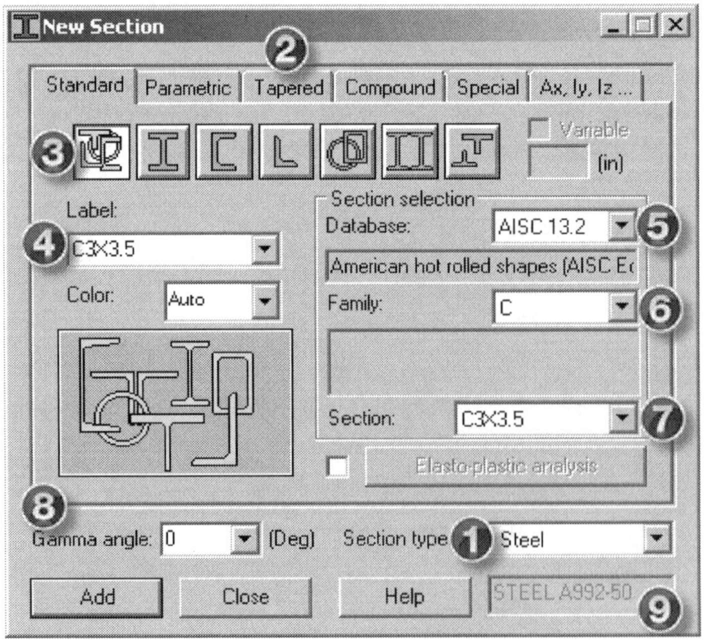

I mention this specifically because the most important part of this dialog is down at the lower right, or #1 in this image.

1- Section type selector. Options here are

   a. 

   b. This is where you tell Robot which type of section you intend to configure. The entire dialog will change based on this setting. If you configure a new section of the type you had selected in the Bars dialog, then once you close this dialog that new section will be available. If you change Section type, then the new section type will be added to the list of section type labels but will not automatically populate in the Section dropdown in the Bars dialog. You first need to change the Bar Type dropdown to a bar type which applies to the new section you have configured.

2- Type to create tabs: Robot can create, manage, and analyze many different section types. We will deal with Standard Section types in this book, but it is worth noting that parametric, tapered, compound (built-up) sections can all be specified and analyzed. The active tab at the time of pressing "Add" is the section properties which will be added to the section labels list.

3- Family Filter: Select one of these options to reduce the number of families shown in the Family dropdown (#6). Select the left-most for "All"

4- Label and Color: Offers options for how to display the label (e.g., W12X26 or W12x26) and if you would like to specify a color which can be used in the Display options or allow Robot to automatically assign a color.

5- Database: Allows you to select elements from various databases which you have configured in the **Tools>Job Preferences...**>Databases>Steel and timber sections

6- Family: Select the shape family from which you would like to select a section. E.g., W, S, HP, M, C, L. This will change the options available under the Section dropdown below.

7- Section: Choose the member section you intend to define for the project.

8- Gamma angle: The angle of rotation of a member is a property of the section label. If you want to use a C section as both a girt and a beam (e.g., stair stringer) you will need to define two sections of the C profile with different gamma angles.

9- Default material. This shows you the default material set to the project settings default (**Tools>Job Preferences...**>Materials under the "Basic Set")

It is probably worth noting that there is also a checkbox next to "Elasto-plastic analysis" which allows you to specify that this member will be considered to have elastic-plastic properties and will allow you to configure those parameters by pressing "Elasto-plastic analysis". This advanced analysis and associated settings will be left to an advanced text.

After configuring parameters, press "Add". Closing the dialog will not save any changes and will not create the new section label for you to use in your project. If you wish to create several labels at once, simply pressing the Add button after configuring each will add them to the list of available sections.

NOTE: SECTIONS ARE USED FOR MODELLING, YOU DO NOT NEED TO HAVE ALL POSSIBLE SECTIONS ADDED FOR USE WITH DESIGN. ROBOT WILL USE A LIST OF SECTIONS WHICH YOU CONFIGURE AS POTENTIAL DESIGN CANDIDATES. IF YOU SELECT ONE OF THOSE CANDIDATES AFTER A DESIGN THE NEW SECTION WILL BE ADDED TO THE LIST OF SECTIONS FOR YOU.

Section Labels, similar to Bar Type labels are settings applied to bars. Sections are added to the project for use in modeling via the sections dialog. Using the ellipsis button next to the Section dropdown in the Bars dialog is a shortcut to adding a new section to the list of current sections in the project. To access the section labels dialog, use **Geometry>Properties>Sections...** or use the sections button from the modeling toolbar: 

Exercise 6: Modeling Bar Elements

1. Start a new project with the Building Design UI Configuration
    a. 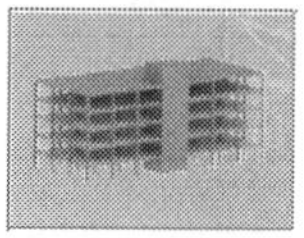
2. You will be in the "Plan" tab of the view and you will notice that at the bottom of the view it indicates that you are at Z=14.00 ft – Story 1. (Your default elevation may vary) The template is preconfigured with this story. If you are not in the Plan tab, select that tab now. If you don't see a plan tab, check that you have the Building Design structure type active (**Geometry>Structure Type...**)
3. As a note: the Building Design UI configuration has both the standard bar placement tool (**Geometry>Bars...**) and some tools specifically configured for placing beams and columns.
4. Make sure that the grid and ruler are on and snapping to grids is turned on:
    a. **View>Grid>Turn-On/Turn-Off** is pressed
    b. **View>Grid>Ruler** is pressed
    c. **Tools>Snap Settings...** has all items checked.
5. Start the column placement tool: **Geometry>Columns...** or press the columns button:  from the modeling toolbar on the right hand side of the view
6. In the "Column" dialog (the Bars dialog configured for placing columns) Select the ellipsis button next to Name and create a new naming convention "%o_%n-

%l" by first clearing the name syntax edit control, then one at a time select the following variables: Structural Object(%o), Object Number(%n) and Story (%l), from the dropdown list and add them to the Name Syntax. Finally add an underscore between %o and %n and a dash between %n and %l (or try out your own unique combination)

a. 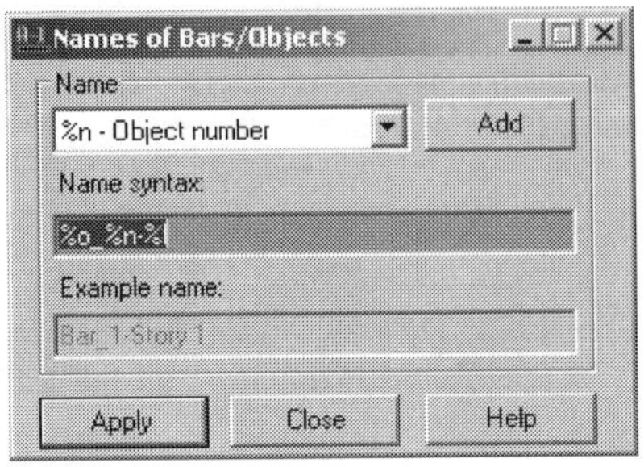

7. Click the "Apply" button in the Names of Bars/Objects dialog.
8. Select "Steel Column" from the Section Type dropdown (this is the Bar type dropdown in the generic bars dialog). There is no opportunity to configure settings for this member type from the Columns dialog as there was from the bars dialog.
9. For Section, select the ellipsis button next to the section dropdown which will open the New Section dialog
10. Select the Standard tab, select AISC 13.2 from the database selector. (if you do not see this database or are using a different standard library configured from **Tools>Job Preferences...**>Databases>Steel and timber sections. Then select that database here)
11. Select "W" for the family (or similar family if you are using a different database)
12. For Section, choose W12x65 (or again, if using a different database, choose a section that is suitable for use as a column)
13. Press "Add" and then "Close"

a. 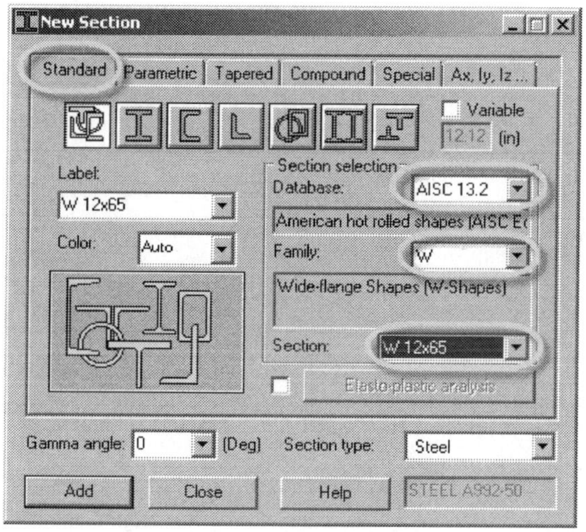

14. Now click in the "Beginning" edit control and your column dialog should now look like this:

a. 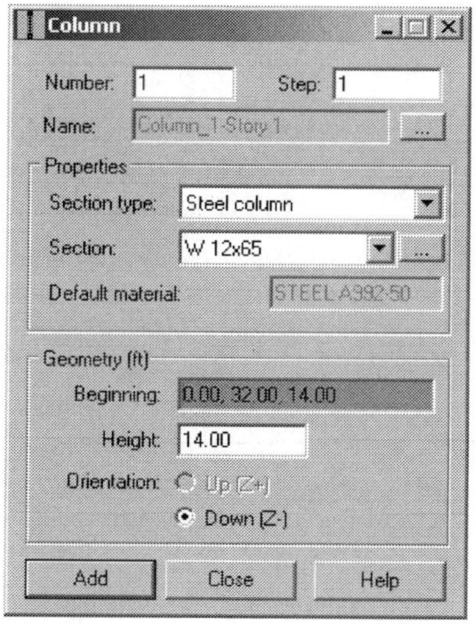

b. Note that the default material comes from settings in **Tools>Job Preferences**>Materials

15. Move your mouse around the view window and notice that the coordinates of "beginning" update based on your mouse location. The mouse is also snapping to grid locations. You can see current cursor coordinates in the status bar at the bottom as well:

a. 

16. Use the visual ruler and grid snapping to place columns (elevation is managed by the work-plane) at 0,0,14 – 0,10,14 – 10,10,14 and 10,0,14 by clicking those locations in plan view. Columns will be created vertically down 14' if your height is also set to 14' as mine is.
17. Your plan view should now look like this:

a. 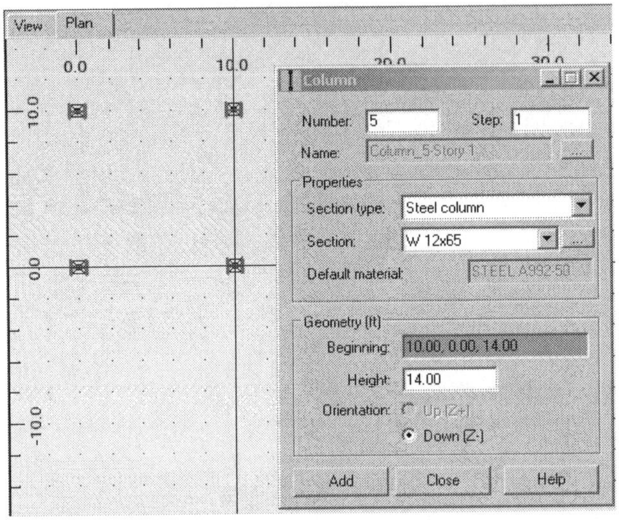

18. Notice that Robot has gone ahead and placed some boundary conditions (nodal supports) at the bottom for us. We may want to change those later but for now, they are fine.
19. Close the Column placement dialog and open the Beam placement dialog **Geometry>Beams...** or use the Beam button on the right hand side: 
20. This dialog is almost identical to both the Columns dialog and the Bars dialog.
21. Our naming scheme will be the same as it was previously. Select "Steel beam" as the Section type and use the ellipsis button next to the Section drop-down to add a new W18x35 section to the active sections list.
22. Place the cursor in the "Beginning" edit control and then click in plan to place beams across the top of our columns: (You can use the "drag" option to make this more efficient)

a. 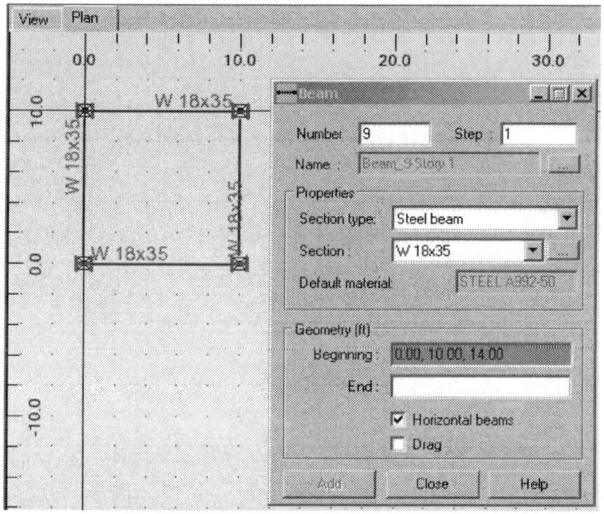

23. Close the Beam placement dialog and select the "View" tab at the top of the view and then select the shape toggle at the bottom of the 3d view  to admire our handiwork.

a. 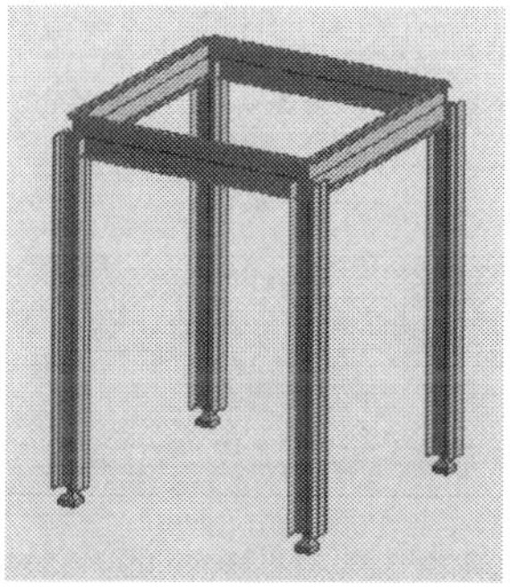

--End of Exercise--

Modeling Nodal Supports

Nodal supports in Robot are effected as Support Labels assigned to nodes. The method of creating labels and applying them to elements in the project is a very standard interaction pattern in Robot that you will see repeated again and again. The idea is that we create sets of parameters called labels and then let elements in the model refer to those sets of parameters (labels) similar to the way a referential database would work. Think of labels as keys into tables.

The Support Label dialog

Launch the Support Labels definition dialog from either **Geometry>Supports...** or by using the supports button on the structure toolbar on the right side of the screen: 

The labels dialog is very similar to other labels dialogs E.g., Sections, Offsets, Member Types.

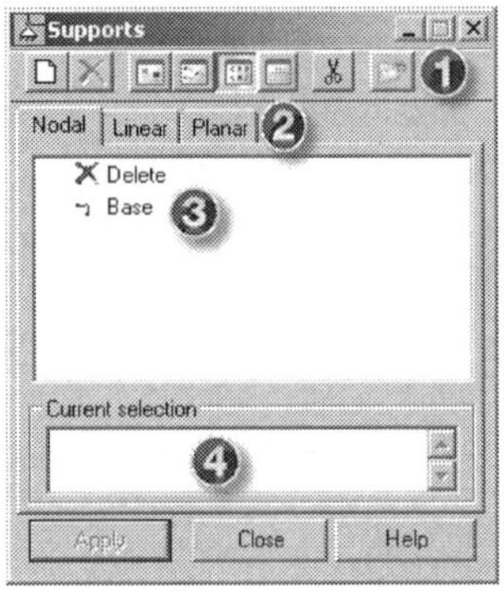

1- Labels and list view management tools

a.  New/edit label

b.  Delete selected label

c.  List Display options (large icons, small icons, list, detail)

d. 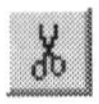 Delete all labels not currently in use (Purge Unused)

e.  Launch the Label Manager (a centralized location for configuring all labels in Robot. This is also available from **Tools>Label Manager...**)

2- Type of support to apply: to apply linear or planar (area) select the corresponding tab before selecting elements. NOTE: linear supports cannot be applied to bar elements. Use **Edit>Divide...** to add nodes along the member and then add nodal supports.

3- List area: displays all currently configured supports in the project. Selecting "Delete" before applying will remove a defined support from a node or edge. If you want to delete a support label, use the delete button mentioned in number 1.b above.

Creating a new support label

You can either press the new label button  to create a completely new label or you can double-click on any existing label to launch the configuration dialog and give it a new name. Either route will present the Support Definition Dialog:

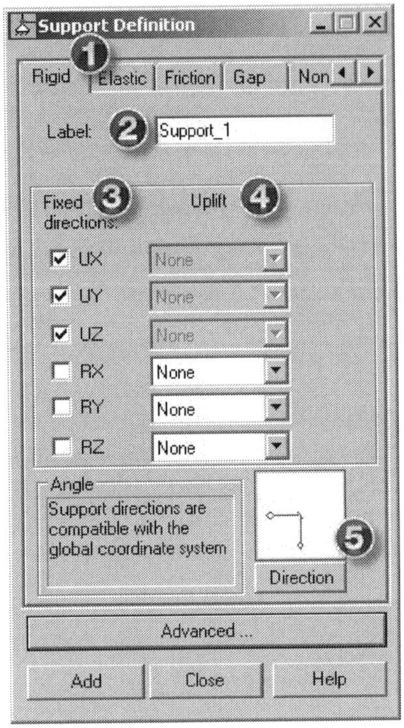

1- Support parameter selector. We will cover Rigid and Elastic as other support types will require and understanding of non-linear analysis techniques. Advanced support types will be covered in an advanced text.

2- Support Name: If you use the same name as a previously defined support you will be prompted to overwrite that support definition. Doing so will affect any nodes, edges or planes which currently have this support type defined. Accept the default supplied name or provide your own custom name. It is recommended to provide a name which will help you identify the support settings later e.g., Pinned, XYRoller, ZSpring10kft

3- Fixity directions: These are all based on the project coordinate system for nodal supports however, Linear and Planar supports may be oriented relative to the member local coordinate system. "U" indicates translation along the specified axis and "R" indicates rotation about the specified axis.

a. 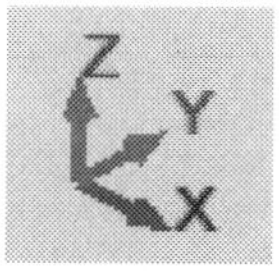

4- Uplift is a way to release one direction but not the other. For instance, if you want gravity support at the base of a structure but you do not want to consider any uplift resistance then choosing UZ+ will leave any displacement in the positive Z direction free from restraint. Correspondingly, the structure will not be allowed to displace in the negative Z direction. This will also require non-linear analysis so we will not cover the specifics here but since the option is available for Rigid supports and also works with elastic supports I wanted to include it here. These are the options

a. 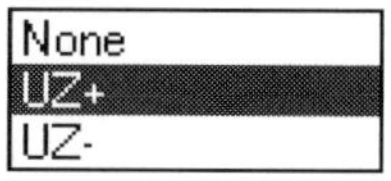

5- Direction: This will launch the Support direction dialog which allows you to configure the orientation of the support relative to the project axes, a specified vector or a specified node.

a. 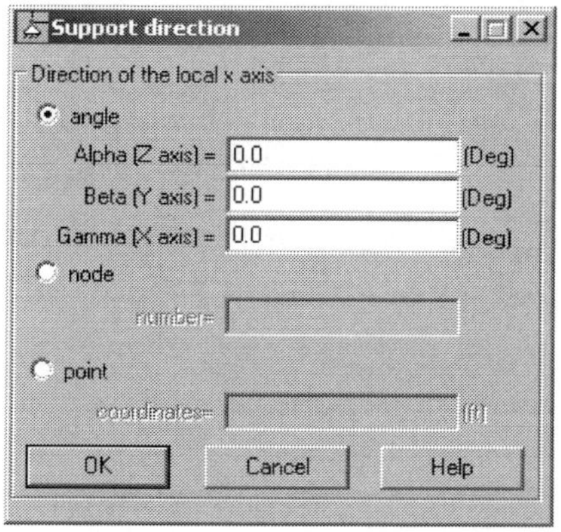

b. For Angle, rotations are measured about the specified axis so Alpha indicates rotation in plan (about the Z axis)

c. Node means that the X axis of the support will be aligned from its nodal position towards the node specified. No matter where you apply this support it will orient its X axis toward the node you specify no matter where it is in the project.

d. Point is the same as node but is not bound to a node. However, the X axis of the support will always align itself from the supported node toward the specified coordinate location. Obviously node and coordinate could provide some interesting support configurations but more frequently if you have an inclined support you will either use the angle options or create a special support label for each situation where only specifying a node or a coordinate will work well.

Editing an existing label

To edit the parameters of an existing label simply select the label in the list and double-click to launch the definition dialog. You can also select the label and press the new label button  which will simply open the definition dialog for the currently selected label. Re-configure parameters as necessary and leave the name the same. You will be prompted to overwrite the label when you press "Add".

Applying support labels to nodes

The lower half of the support labels dialog is the selection edit box. You can simply type in your node or edge selection, you may make your selection graphically in the project or you can simply click nodes in the view to apply the support. We will take each method in turn. Note that if you had previously selected nodes or edges/planes prior to launching the support labels dialog, the selection will already be pre-populated in the "Current Selection" edit control.

Click to apply: In the Support labels dialog, select the appropriate tab at the top Nodal, Linear or Planar. You will only be able to apply support labels to the element type which corresponds to the tab you have selected. Next, select the label you would like to apply (or the delete label if you wish to remove supports). An arrow will indicate the currently selected label:

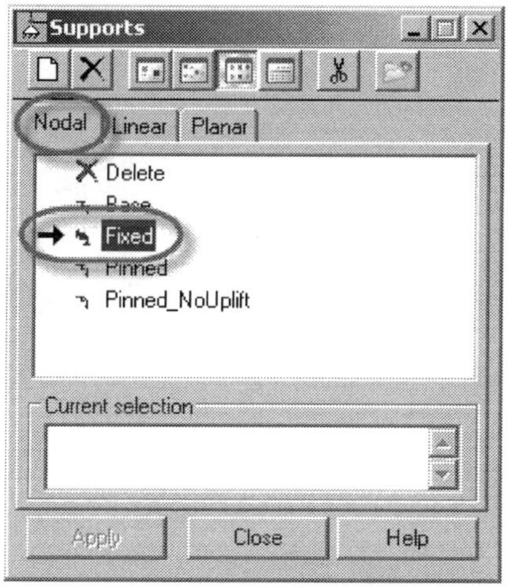

When you move your cursor out into the project area, the cursor graphic will change to represent the support label you are about to apply.

You may now select nodes, edges or planes by clicking them directly in the project. The label is applied as soon as you click.

Type in selection: In a similar way to the Click to Apply method above, select the appropriate tab (Nodal, Linear or Area) select the label you would like to apply. Next click in the "Current Selection" area and type in the node or edge numbers to which you would like to add the supports.

<u>Keywords:</u> [all, to, by, p_EDGE(n)]

<u>Separators:</u> <space> or <,> separate successive element numbers.

Examples:

1. 1 2 3 4 5 selects: 1,2,3,4,5
2. 1to5by2,8,10to13 applies label to nodes 1,3,5,8,10,11,12,13

3. 1to13by3 applies label to 1,4,7,10,13
4. 10_EDGE(1) applies label to only edge 1 of element (panel) 10
5. 10_EDGE(1) 10_EDGE(3) applies label to edges 1 and 3
6. 10 (if 10 is a panel) applies the support to all edges of 10

NOTE: EDGE SPECIFICATION CAN ONLY WORK WHEN THE "LINEAR" TAB IS SELECTED. IF YOU TYPE IN AN EDGE DESIGNATION AND ARE ON THE NODAL TAB, ROBOT WILL SIMPLY IGNORE THE SELECTION WHEN YOU PRESS "ADD". JUST SWITCH TO THE LINEAR TAB BEFORE PRESSING APPLY.

Create Selection Graphically: In a similar way to the "Click to Apply" method above, select the appropriate tab (Nodal, Linear or Area) select the label you would like to apply. Next Click in the Current Selection edit control to indicate that you wish to create a selection and move your cursor out into the view. You will notice that the cursor changes to the selection cursor instead of the apply cursor we saw above in "Click to Apply":

This does not always work… If you find that you keep being presented with the apply cursor then either reselect the label and then click in the "Current Selection" dialog again or, while the cursor is over the project environment, right-click and choose "Select…" which should show you the selection cursor. Now you may use single click, box and crossing selections as we have discussed in "Making Selections" above. The one caveat here is that if you select an edge of an element you will be asked if you wish to select the edge or the entire element:

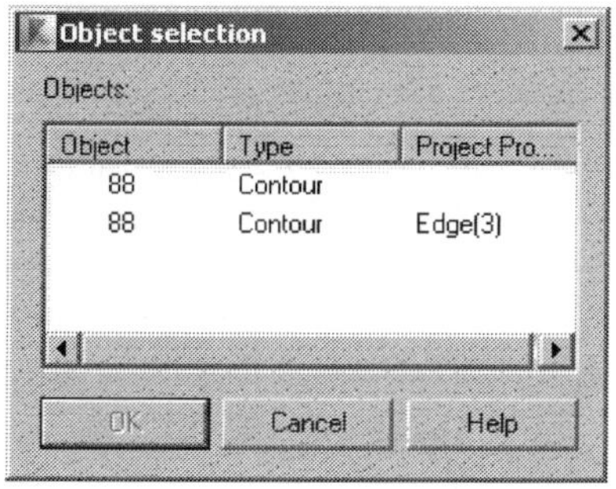

Simply select which object you intended to select and press OK.

Using the Selection Dialog: Use the same process as in "Click to Apply" method above: select the appropriate tab (Nodal, Linear or Area) select the label you would like to apply. Next, click on the node selection button on the selections bar: 

This will launch the Selection Dialog:

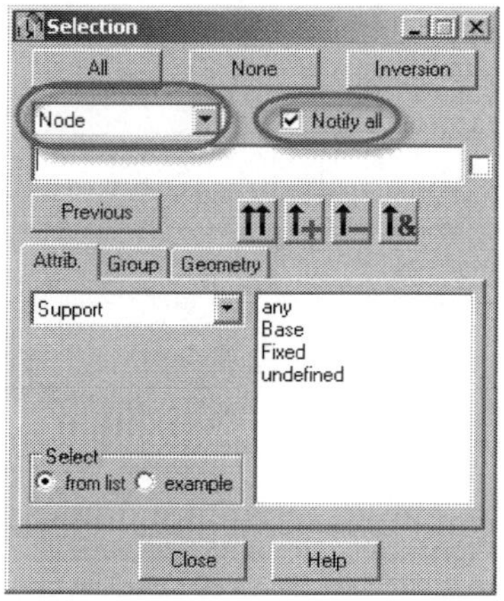

Node is the current selection so if you wish to select panels you can use this dropdown to select panels instead. Notify All has no significance in this instance. The workflow here is to use the attributes or groups or geometry to select nodes (or panels) or to simply type in node numbers and then close the selection dialog. When you click in the "Current Selection" box again, whatever you configured in the Selection Dialog will populate in the edit control (automatic copy and paste). For more information about the use of the Selection Dialog please see "Making Selections".

Exercise 7: Applying Support Labels

1. Open "Exercise_07-SupportsDefinition.rtd" and switch to the "View" tab at the top of the project window:
   a. 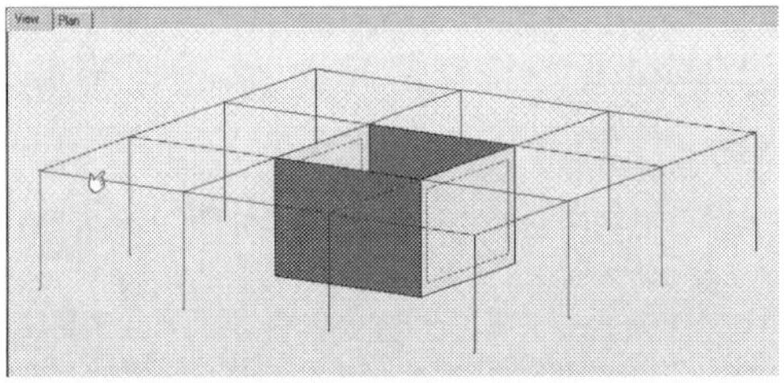
2. Turn on node numbers using the toggle at the bottom of the view window
3. Go to the Display properties (**View>Display...**) and select the Panels category then check "Numbers and labels of edges"
   a. 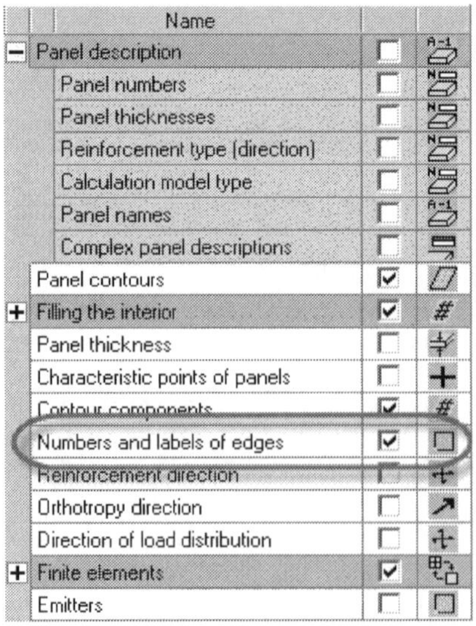

4. You should now have a view which looks something like this:

a. 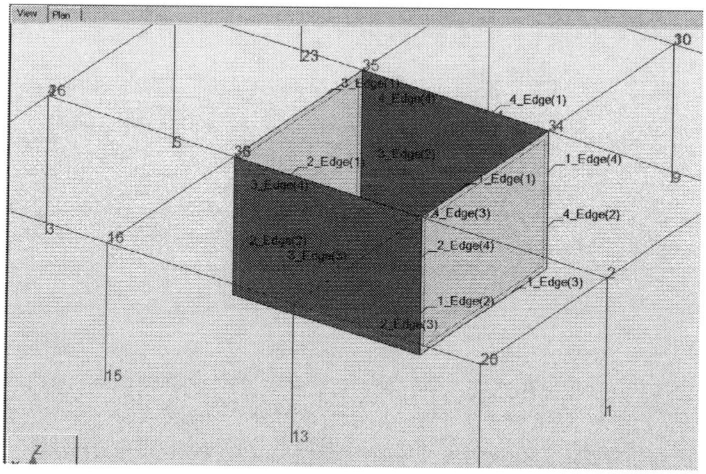

5. Launch the Supports dialog from **Geometry>Supports...** or from the modeling toolbar on the right hand side: 

a. 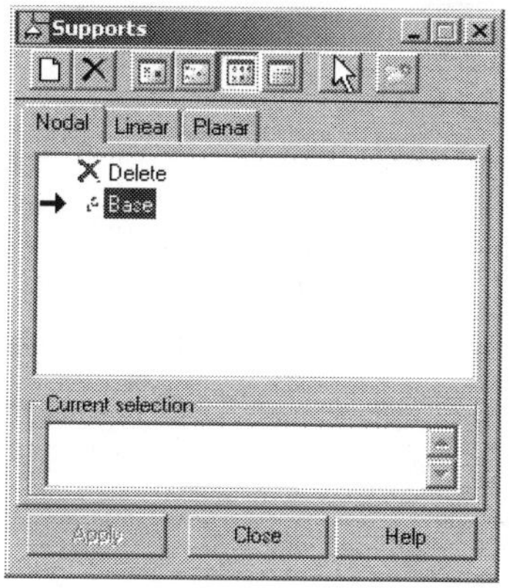

6. Create a new support by clicking the New Support button: 
7. This will launch the Support Definition Dialog

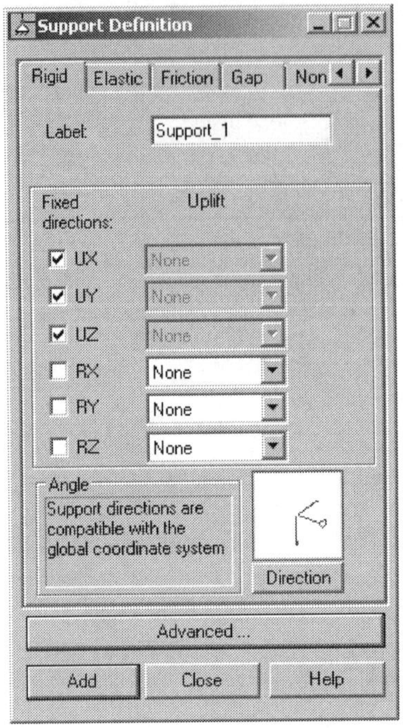

a.

8. Select the "Rigid" Tab and type "PinnedFreeRotation" in the Label Field.
9. Next, check the UX, UY and UZ checkboxes and uncheck RX, RY and RZ
10. Press "Add"
11. Create another by simply changing the name in the Label Field to "XYRoller"
12. Uncheck UX and UY (un-constrain movement in the x and y directions)
13. Press "Add" and "Close"
14. Notice that there are now two new labels in our Supports Dialog:

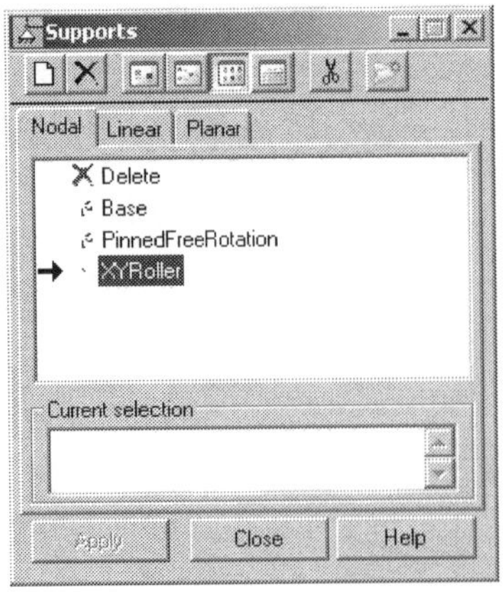

a.

15. Select PinnedFreeRotation and move your cursor out into the project
16. Your cursor should change to a black arrow with a graphic of the support you are about to apply. Go ahead and select several nodes by left-clicking on them.

    a. 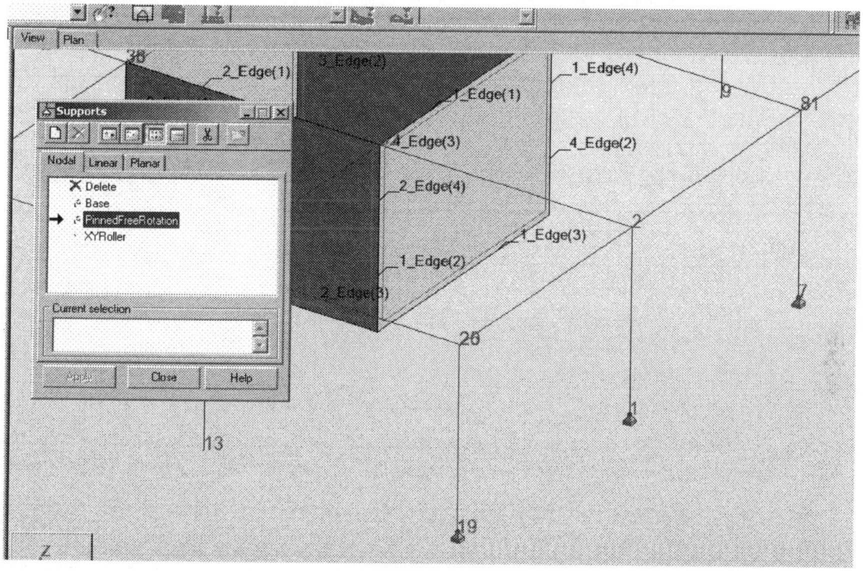

17. Now left-click in the "Current Selection" edit control and notice that the cursor changes to a hand when moved over the project view. Go ahead and select several other nodes by box selecting. Then Press "Apply". Select "XYRoller" and repeat the process to add some more supports.
18. Switch to the "Linear" tab at the top and then select the PinnedFreeRotation label then click in the "Current Selection" edit control. Type in 1_EDGE(3) 2_EDGE(3) and press "Apply"

a. 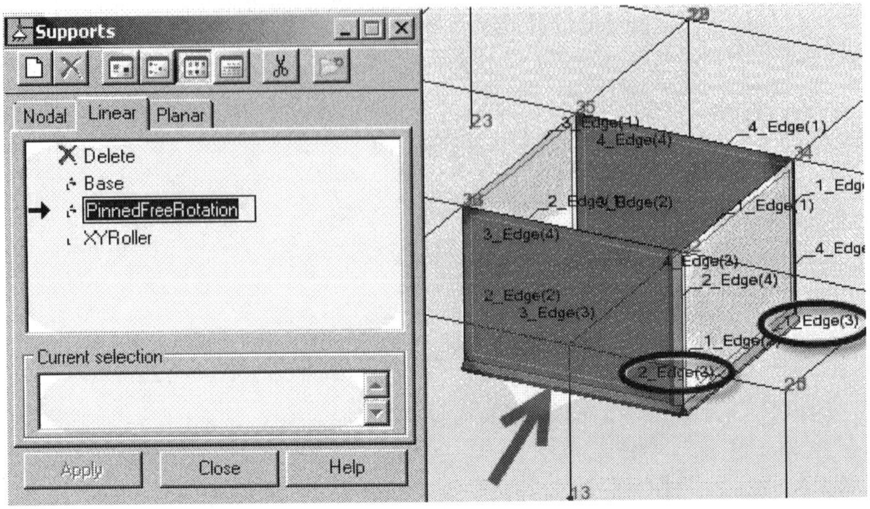

19. Orbit the model around (hold the <Shift> key and the middle mouse button or use the Pan/Zoom/Orbit tool:  So that we're looking at the other side of the model
20. Select PinnedFreeRotation again and move the cursor into the view and click the lower edges of the other panels in the core wall.
21. Close the Supports Dialog.

--End of Exercise—

Modeling Member End Releases

In order to model the effects of simple connections at the ends of members it will be necessary to apply a release label to your bar. The first thing we'll want to do is turn on visibility of release codes and/or symbols. When you are in the Release Label tool, the view will automatically switch to display codes but you can also do this manually. They are not part of bar properties in the default organization of the display settings so you'll find them under "Model". Start the display settings **View>Display...** and then select "Model":

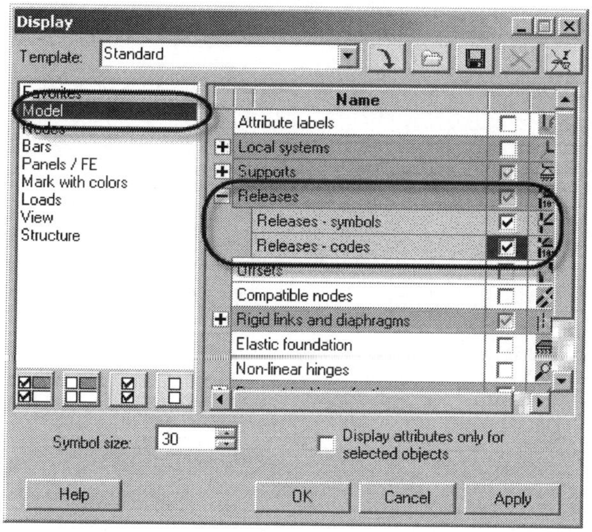

Then close that dialog and go ahead and open the release labels dialog from the geometry menu: **Geometry>Releases...**

This will open the Release Labels Dialog:

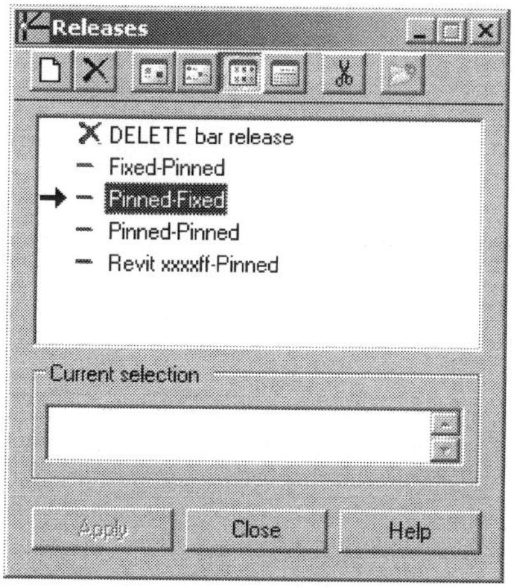

This is another Robot Labels dialog and can be used to assign release code labels to bar elements. The settings for each label are organized according to member start and end. This will be important to consider in applying this label later.

First let's take a look at the New Release Definition dialog. Press "New Release Definition" button from the top of the Releases dialog 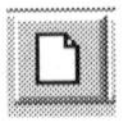 which will open the New Release Definition dialog:

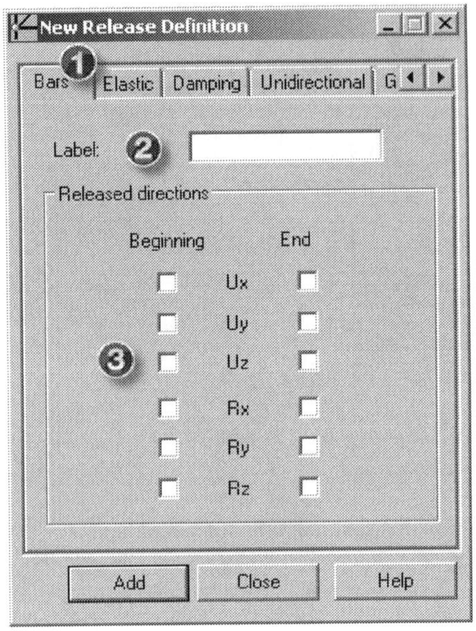

1- Across the top of this dialog are tabs to define different advanced properties of releases. In addition to a simple free release, the released direction could have elasticity, damping, behave only in one direction (unidirectional), behave as a gap, and/or have non-linear behavior. These are very powerful capabilities allowing advanced modelling of structural response. We will only cover basic releases here.

2- Label: Each release label must have a "label" or more plainly, a name. Many people name release labels with some indication of the released properties at start and end such as "Pinned-Fixed" for a simple support or maybe even "Pin_xff-Pin_xxf" for a more complicated release pairing. "Xxf" mimicking Robot's codes for release conditions (See below).

3- Releases: For each translation direction (Ux, Uy, Uz) or rotation (Rx, Ry, Rz) select whether this direction/rotation is released at the start and/or the end of the member. A check in a box indicates that this displacement/rotation is to be released. Unchecked means that it is to remain fixed.

After configuring a name for the release pair and selecting released directions/rotations, press "Add" to add it to the list of available release labels. Note that if the label already exists, Robot will ask you if you want to overwrite the label definition. Choose carefully as overwriting a release label will affect any member to which this label has already been assigned.

NOTE: LABELS MAY NOT BE DELETED FROM THE LABELS DIALOG IF THEY ARE CURRENTLY APPLIED TO ANY MEMBERS IN THE MODEL. IF YOU WANT TO GET RID OF ONE, USE THE "BARS" TABLE AND TABLE COLUMNS TO SHOW THE "RELEASE LABEL", SORT ON THAT COLUMN AND FIND THE MEMBERS WHICH HAVE THIS LABEL TO REMOVE IT OR SET IT TO ANOTHER LABEL.

Applying Release Labels

To apply a release to a bar, in the Releases dialog, select the label you wish to apply. You now have two options. You may move your mouse out into the project environment and "click to apply" the label or you can click in the "Current Selection" dialog and either enter or create a selection in the "Current Selection" window.

CLICK TO APPLY

After selecting the label you wish to apply, when you move your mouse out into the project environment and hover over an element you will notice arrows along the bar. Similar to load application, these arrows are indicating the direction of the bar start to end which will be assumed by Robot for application. Because the label is particular to start and end, this is an important distinction. Regardless of the actual start and end of the element in the model (which may be viewed by turning on member local axes: **View>Display...** Model>Local systems>Local system of bars)

When you move your mouse out into the project environment, it will be in the click to apply release label form:

As you hover over elements, you will see the arrows indicating the assumed direction of start/end for click-to apply mode. If you move the mouse to the other end of the element, the direction of the arrows will flip allowing you to apply the reverse of the label to the element.

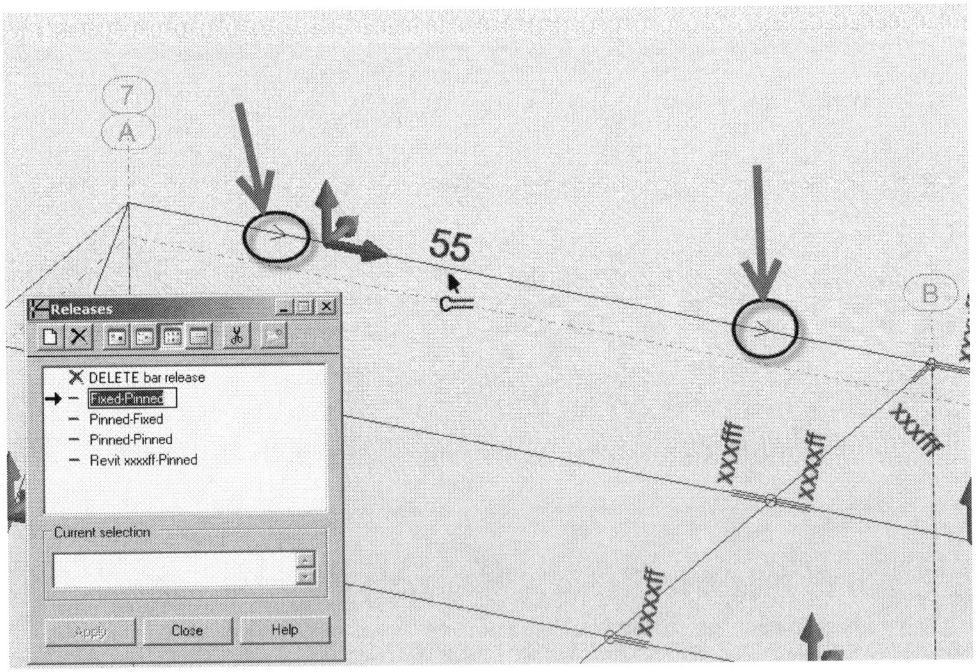

Notice that hovering near the other end results in the arrows flipping:

This is contrary to the local member axes direction so check to be sure you have the correct orientation (whichever you wish) when applying and take a look at the codes (see below) to be sure that you have the correct releases applied.

APPLY BY SELECTION

If, instead of moving the cursor out into the project environment, you click once in the "Current Selection" dialog, you will find that the cursor (in the project environment) now is the selection cursor: 

You may now create a selection in any one of the three main selection creation methods (See section "Making Selections" in "Getting Started"). Typing in a selection of bars, creating a selection by clicking elements in the project environment, and/or using the selection dialog.

Once your selection has been created, pressing the apply button will apply the release labels to the selected elements and take the sense of start/end from the local coordinate axes orientation of the members selected.

Release Label Codes and Symbols

Robot displays information about the releases applied to bars with two different methods. You can display "symbols" and/or "release codes". The Robot release symbol looks a lot like a "hairpin" and is shown here:

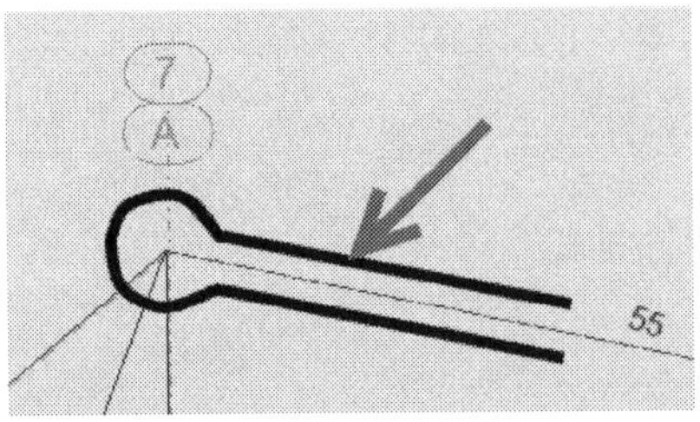

Member 55 has a release condition at this end and is indicated by this symbol. The symbol ONLY indicates a release, there is no way to tell which displacements or rotations have been released.

However, release codes will give you exact information on which displacements and rotation have been released. **The codes use an "x" for a fixed and an "f" for free** (or released). There are 6 letter codes for each end and are arranged in this order:

Ux, Uy, Uz, Rx, Ry, Rz

So a code xxxfff would indicate that the three displacements are fixed but the three rotational degrees of freedom are released. Here is what it looks like in the project environment:

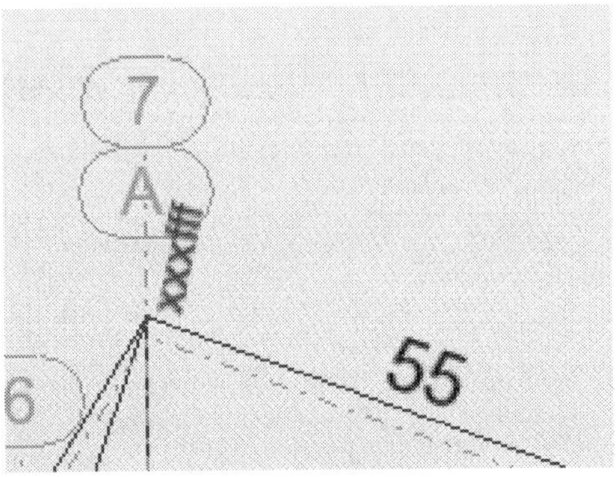

A note about releases:

Releasing degrees of freedom in the model can lead to model instabilities when it comes time to solve the system of equations. Here are some typical situations that you should be careful about.

Releasing Rx at both ends of the member: You can think of releasing Rx (axial rotation, or rotation about the member local x axis) at both ends of the member as though you put a wire through a straw: The member can "spin" freely about its own local x axis.

Releasing rotational degrees of freedom at a pinned support. This is not necessarily intuitive but instabilities will arise if you release rotational degrees of freedom at ends of members which frame into pinned supports. Here is a visualization pulled from the final section of this book on resolving instabilities.

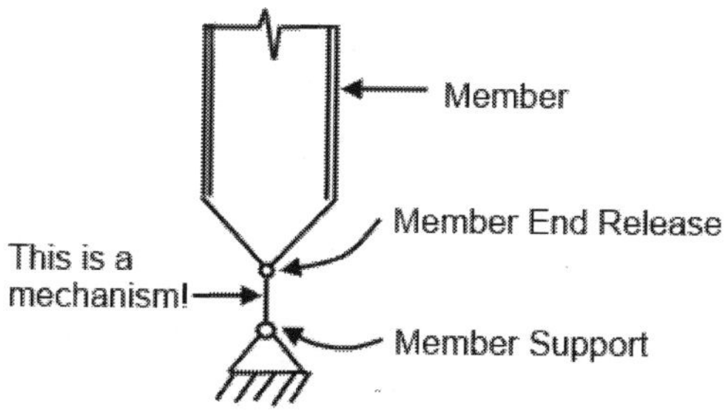

Now this isn't really the situation physically speaking but computationally speaking this will cause problems. BEWARE: Robot will present you a dialog which will allow you to continue the solution. That dialog looks like this:

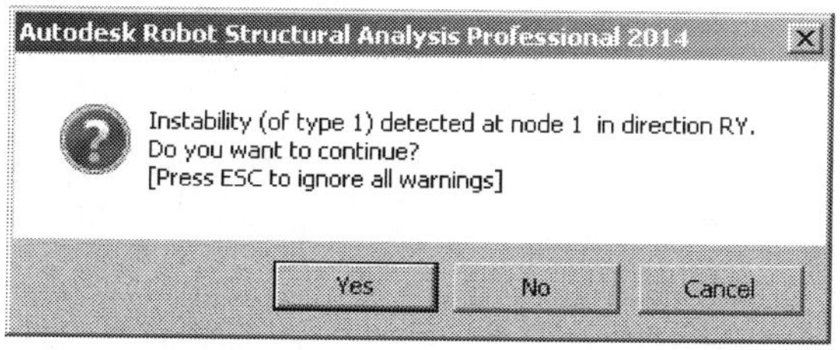

Pressing "Yes" or "Esc" will cause Robot to introduce small stiffnesses to allow the model to be solved. In other words, Robot found a zero on the diagonal and introduced a slight stiffness which negates the zero; however, the model will very likely be incorrect in this state. Which degrees of freedom Robot chooses and why it chooses them is not immediately clear. If you choose to use releases in your model, it is recommended to begin with a few at a time, check the model can be analyzed then continue adding and checking. **The first place to begin looking at your model is in the displacements to see if there are any drastically unexpected displacements.** Also be watchful for unexpected force diagrams, and look out for unexpected torsion showing up along your members. These are indications that there is something wrong in your model. Helpfully, when instabilities are detected and ignored during calculations, the Calculation Messages dialog will show, letting you know what issues occurred during calculations so that you can go back and track them down:

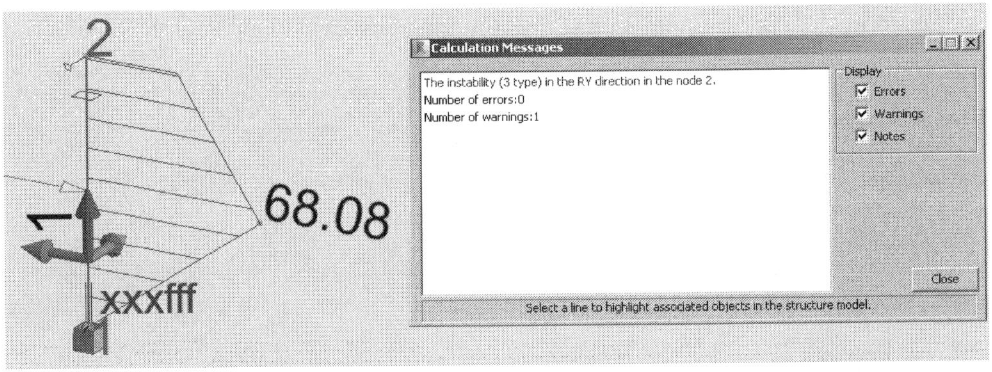

In this model above I've released too many degrees of freedom (Note that the column is rotationally free at the fixed support and the top has no restraint in the direction of loading). The model solved when I ignored the warnings and it looks like I have a real moment result along the beam; however, take a look at the displacements for node 2 and you'll see immediately where the problem is:

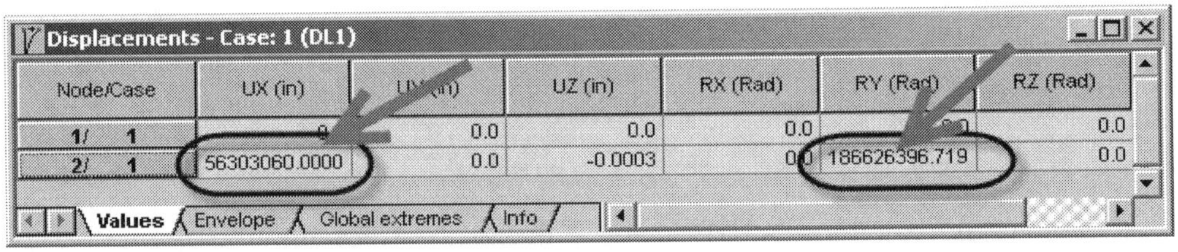
Displacements - Case: 1 (DL1)

Node/Case	UX (in)	UY (in)	UZ (in)	RX (Rad)	RY (Rad)	RZ (Rad)
1/ 1	0	0.0	0.0	0.0	0.0	0.0
2/ 1	56303060.0000	0.0	-0.0003	0.0	186626396.719	0.0

The type of instability may differ, the degree of freedom that Robot identifies as the culprit may vary but you should always do a gut-check of the displacements to be sure that the model is behaving as you expect.

Chapter 4 - Modeling Loads

Creating Load Cases

Now that we have covered most of the basic geometry definition it's time to take a look at applying loads in Robot. Load cases are configured via the Load Types dialog and until we have load cases defined we cannot apply loads associated with those cases.

From the Loads Menu select Load Types… or use the load types button on the right hand toolbar: 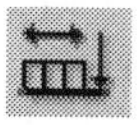 This will open the Load Types Dialog:

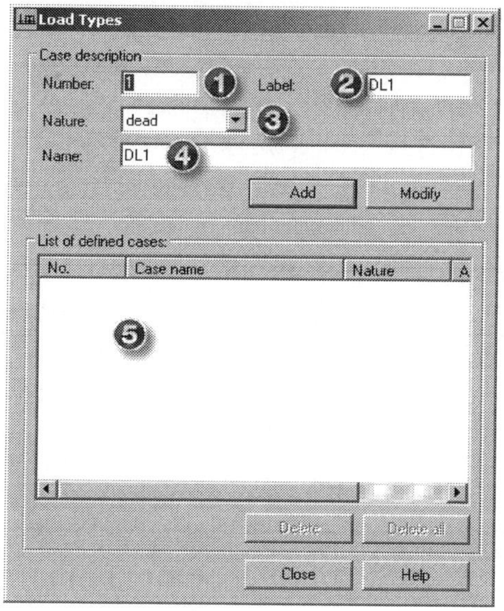

1- This is the number of the load case. Robot will manage this for you so even if you supply a number Robot will automatically replace it with the next in sequence.

2- Label is the name of the load case that will be used in the display of load combinations. Typically this label would be DL1, LL, SL, or something similar.

3- Nature is a setting which will determine how the load case is considered in automatic load combinations based on a code combination. Multiple load cases all with the same nature can all be included in a combination or can be separately included in different combinations. The load nature setting facilitates this. A great example of this might be a bridge crane load where you want to check the load at all support points but do not want them to all be applied at the same time. Robot's automated load combinations can manage this in a very powerful way and we'll look at that in more detail in the load combinations section.

4- Name: This label will be used in most areas of Revit: Load combination assembly, case selection, and in the loads table showing which loads have been applied and to which case they belong. This can be the same as the Label or you can provide a more descriptive name to help you remember the usage.

5- List area: all load cases will be listed here including manually defined load combinations so don't be surprised if you see those in here after you have created some manual load combinations. Included in this list will be an indication of the "Analysis Type" You cannot change that in this dialog but will need to use **Analysis>Analysis Types...** or the Analysis Types button from the main toolbar: 

NOTE: WHEN YOU FIRST CREATE A DEAD LOAD CASE IN ROBOT, ROBOT WILL AUTOMATICALLY CREATE A SELF-WEIGHT LOAD FOR YOU AND APPLY IT TO THE MODEL! THIS MAY OR MAY NOT BE WHAT YOU WANT. IN ORDER TO CHECK, TAKE A LOOK AT THE LOAD TABLE (LOADS>LOAD TABLE) AND SEE IF SELF-WEIGHT HAS BEEN INCLUDED:

	Case	Load type	List				
	1:DL1	self-weight	1to36	Whole structure	-Z	Factor=1.00	MEMO:
*							

IF SO AND YOU WISH TO REMOVE IT, YOU CAN SIMPLY HIGHLIGHT THE LINE IN THE TABLE (MAKE SURE YOU ARE ON THE EDIT TAB AT THE BOTTOM OF THE TABLE. AND PRESS <DELETE>

Exercise 8: Create Load Cases

1. Start a new project with the Building Design UI Configuration

   a. 

2. Open the Load Types dialog from the menu **Loads>Load Types...** or from the toolbar on the right: 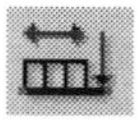

a. 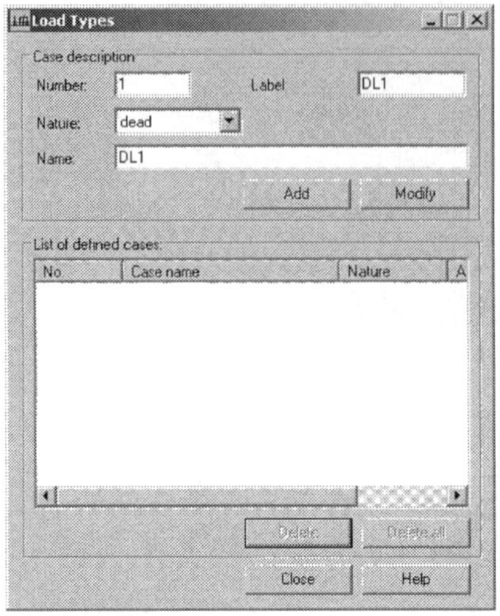

3. Leave nature set to Dead, and Label and Name both set to DL1
4. Press "Add"
5. Now for the next one, change the nature to Live and check to see that Robot has switched the label and name to LL1, if not, type those in yourself.
6. Press "Add"
7. For the next, select wind for the nature and change Robot's Default WIND1 to WindEW for label and Name then Press "Add"
8. Finally add a WindNS case as well.

a. 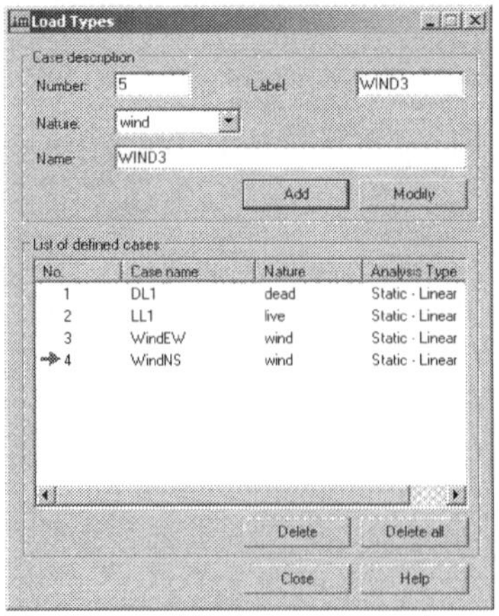

9. Press "Close"
10. Notice that the case selector now has our load cases available for selection:

a. 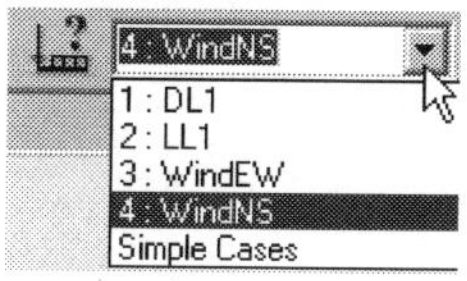

--End of Exercise--

Applying Loads to Elements

Now that we have load cases we can start creating loads to go with them. You can model almost any conceivable type of structural loading in Robot from self-weight to drifted snow to harmonic loading and more. We will cover the basic types of nodal, bar and surface loads and leave advanced types to another text.

Using the Load layout from the layout selector: The Load layout organizes the display to automatically show the load table and the load types dialog. Once you have configured your load cases, you typically will not need to use that dialog again however, it can be used to specify the current load case for load application. That said, the load case selector on the selection bar can also be used so perhaps the most useful feature of the automated Load layout is positioning the load table nicely at the bottom of the screen so that you can watch the entries in the table as you apply loads graphically. You can also simply use the table to manually enter loads in cases where you have the same load to apply to many elements. You could also import loads from a spreadsheet if you have set the spreadsheet up to create information in a format similar to the loads table.

All this said, The Load Layout does not give you anything that simply opening the Load Definition Dialog would not do: the choice is yours.

Using the Load Definition Dialog:

Launch the Load Definition Dialog either from the menu **Loads>Load Definition...** or use the load definition button from the right hand toolbar: 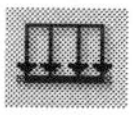

Basic Workflow:

1. Select load case for load you intend to apply (from the load case selector or in the Load Types dialog if you have that open)

2. Select the type of load you wish to define (E.g., Nodal force, Bar uniform load, Surface uniform planar load)

3. Configure the parameters of the load to be defined in the corresponding load configuration dialog which opens after selecting the load type. These parameters include load magnitudes, points of application or change in load value as well as load direction and other parameters.

4. Select "Add" from the load configuration dialog. This queues the load for application to elements. You will then be returned to the Load Definition Dialog where you will select elements to which to apply this queued load.

5. Apply loads to elements:

    a. You may move the cursor directly into the project view and click elements to apply or

    b. You may create or manually type in a selection in the "Apply to" field and then press "Apply" in the Load Definition Dialog.

6. Repeat for each different configuration of load you have

The Load Definition Dialog:

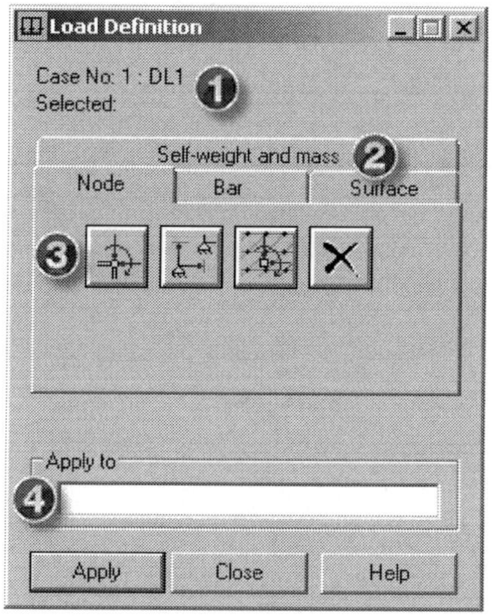

1- This area displays the currently selected load case and as soon as a load type has been configured below (by selecting a load type and adding it. See workflow above) the currently configured load type will display next to "Selected". If the load case displayed here is not the load case to which you would like to add a new load then change it from the case selector on the selection bar or by selecting it in the Load Types dialog.

2- Load types tab-set: Select the tab for the load type you would like to add. Detailed descriptions below.

a. Node: loads or displacements acting at a single point in the structure

b. Bar: Loads on linear members including dilation (elongation or shortening and temperature changes)

c. Surface: Loads on planar surfaces E.g., walls, floors or claddings

d. Self-weight and mass: applying self-weight for whole or part of the structure, lumped masses, body forces and centrifugal forces on elements.

3- Load Type configuration buttons: select one of these to launch a corresponding dialog where you may configure the parameters of the load e.g., magnitudes, distances.

4- Selection list. You may type your selection, copy paste your selection, use the selection dialog, or manually select elements in the project. This edit control is intended for use with the Apply button below.

Load Types:

Nodal Tab	
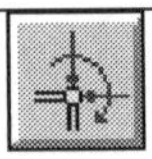 Nodal Force	Nodal forces and moments applied at nodes. May have angle of application specified.
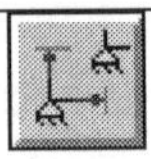 Nodal Displacement	Nodal Displacements: Load is a forced displacement of a specified distance and/or angle
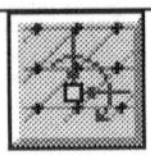 Force in Point	Force In Point: a concentrated force or moment acting at a single point. Intended for use in conjunction with shell and volumetric elements. For concentrated loads on bars use bar force below
Bar Tab	

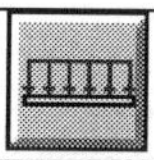 Uniform Load	Uniform Load on a bar element
 Trapezodial Load	Non-Uniform: Allows you to define 2, 3, or 4 points along a beam and corresponding load magnitudes and angles of action. E.g., 2-point is used for triangular load.
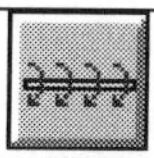 Uniform Moment	Uniform Moment along a bar element
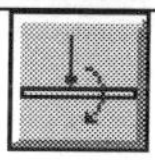 Bar Force	Concentrated force or moment and angle of action at a particular point along a bar. Loads acting eccentrically to the member may also be defined with this tool.
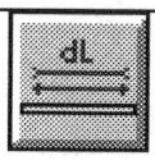 Dilation	Dilation: Load is a lengthening or shortening of a member. May be specified as absolute or relative
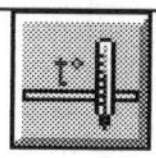 Thermal load	Temperature Load: apply a temperature difference to the member which results in a lengthening or shortening of the element.
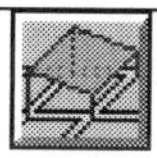 Planar load	Planar load is a method of applying loads to multiple members at once. The preferred method of doing this is to use cladding elements and surface loads applied to the cladding elements.
Surface Tab	

Uniform Planar Load	Uniform Load on the entire surface of specified planar element
Planar Load 3 Point	Planar load applied to an area defined by 3 points. Load magnitude can be specified at each point to create a linearly varying load. For snow drift loads on rectangular areas use the Planar 3p Load on Contour below.
Linear Load 2 Point	Uniform or linearly varying Line load (force and/or moment) on a surface defined by two points or nodes.
Hydrostatic Pressure	Can configure both a uniform plus portion that varies with depth according to a unit weight per unit volume setting. May be oriented with respect to any of the three axes.
Uniform Planar Load on Contour	Define a contour (custom area of action) and load magnitude values for forces in the three axes (global or local)
Planar Load 3 Point on Contour	Define a contour (custom area of action) over which to apply a load which may be uniform or linearly varying by specifying the magnitude of the load at three points in the project. Specified points need not fall within the defined contour. Use this type for applying snow drift loads.
Linear Load on Edges	Uniform linear load on edges of panels
Self-weight and mass Tab	

 Self-weight for Whole Structure	Apply self-weight to the entire structure based on density of materials defined for the elements.
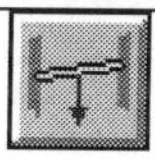 Self-weight for Selected Elements	Apply self-weight only to selected elements based on density of materials defined for the selected elements.
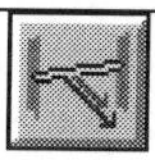 Body Forces	Define static loads using accelerations (a) to create forces based on added masses which is the result of m*a
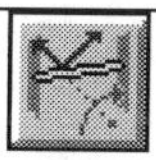 Centrifugal and angular acceleration forces	Allows you to define static loads (radial and tangential) based on a center of rotation and specify an angular velocity and/or acceleration
 Added Masses - Nodes	Define additional mass at nodal locations both mass and rotational mass may be configured.
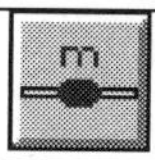 Added Masses - Bars	Define additional mass for linear elements along their length.

Creating and Applying Nodal Loads:

Nodal Force:

Defining the Load Properties: In the Load Definition Dialog select the "Node" tab and then press: 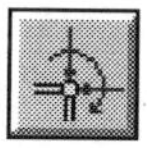 to launch the Nodal Force configuration dialog:

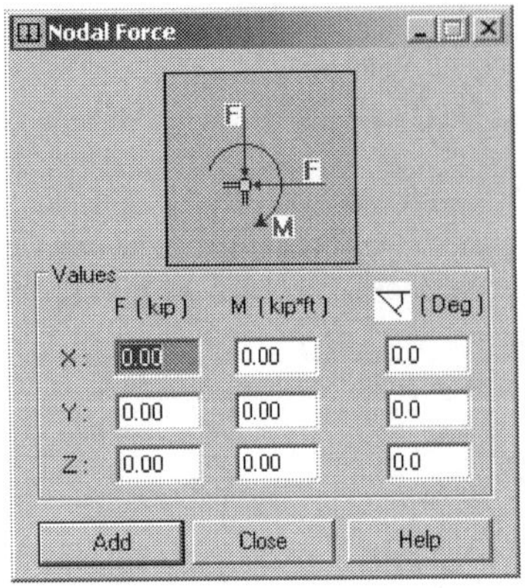

Notice that the current units are displayed at the top of each heading (E.g., kip, kip*ft). Enter the magnitudes of the concentrated nodal force components in each of the three global axes (X, Y, and Z). Next specify the concentrated nodal moments about each of the global axes and last define an angle of rotation about the axis at which the forces act relative to the specified axis. Each angular measurement is an amount of rotation about the specified axis. I.e., if you enter 20 into the direction field on the X row the load will be rotated 20 degrees about the X axis. For example, if you wanted a load at 45 degrees in the horizontal plane (XY plane) you could either figure out the components in Fx and Fy or you could specify Fx and a Z rotation of 45 (rotation about the Z axis of 45 degrees)

Once you have specified the magnitudes and directions for the concentrated force, click the "Add" button which queues this definition for you to apply via the Load Definition Dialog.

Creating the nodal load in the project: Once you have pressed the Add button in the Nodal Force Dialog you will notice that the Load Definition Dialog now shows not only the load case at the top but also the Type of Load currently configured for creation:

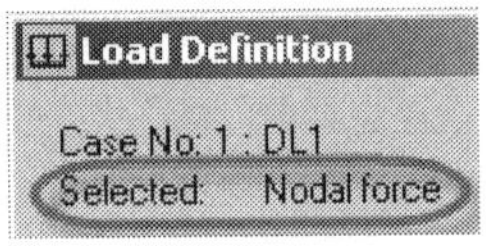

There are now 4 ways in which you can begin applying this load to nodes in the project.

1. **Click to apply**: When you move your cursor out into the project area, the cursor graphic will change to an arrow with a graphic indicting a nodal load.

    a. 

    b. You can simply begin selecting nodes in the project environment by clicking them directly when this cursor is active. The load is applied as you click; there is no need to press "Apply"

2. **Type in selection:** Instead of moving the mouse directly into the project environment, click in the "Apply To" edit control and type in the node numbers to which you would like to apply the nodal load. See "Making Selections" above for keywords you can use in this edit control. Once you have completed the selection you can press "Apply" which will then apply the nodal loads to the selection you have specified.

3. **Create Selection Graphically:** Click in the "Apply To" edit control to indicate that you wish to create a selection and move your cursor out into the view. You will notice that the cursor changes to the selection cursor (a hand) instead of the apply cursor we saw above in "Click to Apply":

    a. 

    b. Once you have done this, the click-to-apply cursor does not like to come back. If you have clicked in the "Apply To" edit control and would like to get back to the click-to-apply cursor you only need to click on the tab of the load you are applying "Node" in this case, and the click-to-apply cursor will return, allowing you to continue in a click-to-apply mode.

4. **Using the Selection Dialog:** You can also use the Selection Dialog to build a selection for applying you loads. Click in the "Apply To" edit control and then click on the node selection button on the selections bar: 

This will launch the Selection Dialog:

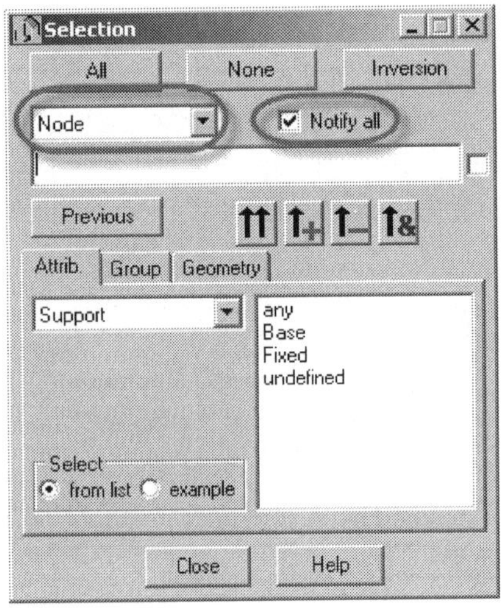

For more information on using the Selection Dialog please refer to the section titled "Making Selections" above. Once you have built a selection in this dialog, simply click back into the "Apply To" edit control in the Load Definition Dialog and the selection you built will be automatically transferred to the "Apply To" edit control for you. Next press "Apply" and the Nodal Load will be applied to all nodes you selected in the Apply To field.

Creating and Applying Bar Loads:

Uniform Bar Loads:

Defining the Load Properties: In the Load Definition Dialog select the "Bar" tab and then press: 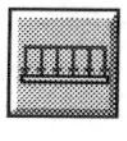 to launch the Uniform Load configuration dialog:

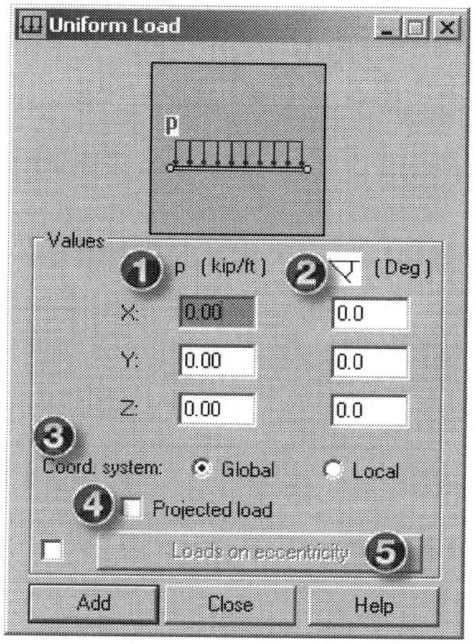

1- Notice that the current units are displayed at the top for each value you must input to use as a reference. In the P column enter the magnitudes of the uniform load in each direction X,Y, and/or Z. These are either global coordinate directions or local coordinate directions depending on which you have selected in the "Coord. System" toggle below (#3). The coordinate system for bars is as shown below: (x is blue, y is green, and z is red)

a. 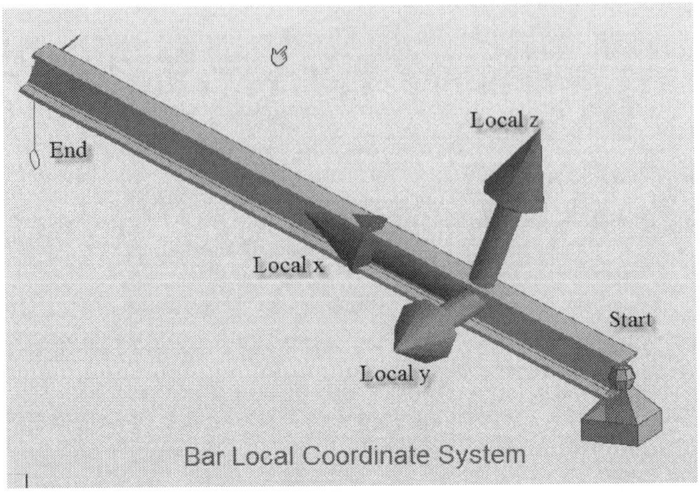

2- In the second column, specify the rotation of the resulting load about each of the specified axes. These values are rotations either about the global coordinate axes or about the local member axes as shown above. Each angular measurement is an amount of rotation about the specified axis. I.e., if you enter 20 into the direction field on the X row the load will be rotated 20 degrees about the X axis.

3- Select whether the load will be applied relative to the global coordinate system or whether it will be applied relative to the bar's local coordinate system. To view the current orientation of the bars' local coordinate systems, use the quick toggle at the bottom of the view window  to toggle on visualization of local coordinate systems.

4- Decide if this load should be projected or not. Projected loads are often used in snow loading for an element. The help files indicate that the load is applied to the length of the member as projected on a plane which is normal to the load's direction. It should be added that this consideration ignores any rotation of the load. If you choose to use rotation of the load, be careful about using projection as the results may not be what you would expect.

5- Loads on eccentricity checkbox enables the Loads on Eccentricities dialog:

a. 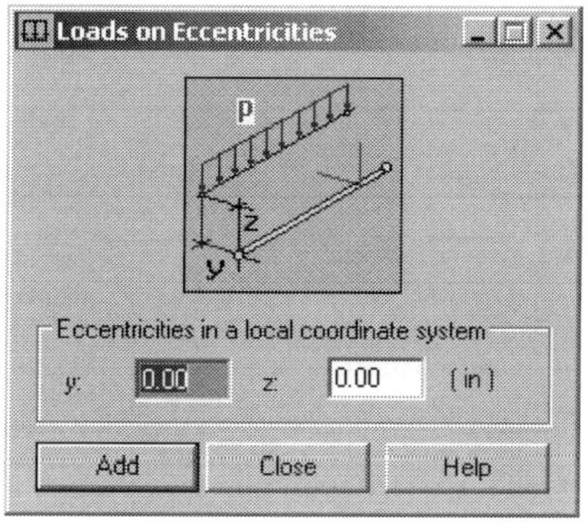

b. Specify offsets from centerline as measured in the member local coordinate system

c. The result of applying an eccentricity is that in the calculation model the uniform load will be applied to the member along with a uniform moments corresponding to the components of the uniform force multiplied by their distances from the bar centerline. Your model, however, will still appear to have a uniform load applied at the eccentric application point.

d. 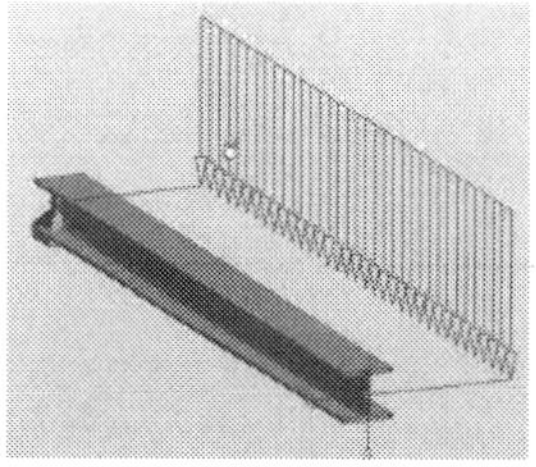

When you have completed configuration of the load press "Add" to queue this load configuration for application.

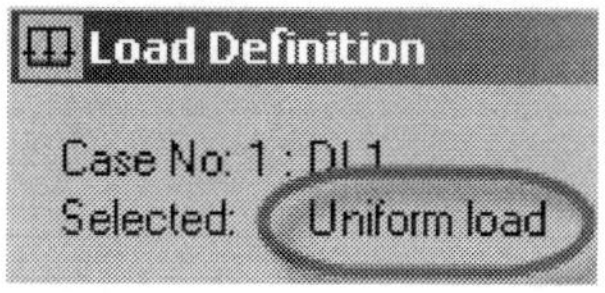

You will notice that "Selection" will now indicate "uniform load".

Creating a Uniform Load in the Project:

1. **Click to apply**: When you move your cursor out into the project area, the cursor graphic will change to an arrow with a graphic indicting a uniform or planar load.

    a. 

    b. You can simply begin applying the load to bars in the project environment by clicking them directly when this cursor is active. The load is applied as you click; there is <u>no need</u> to press "Apply"

2. **Type in selection:** Instead of moving the mouse directly into the project environment, click in the "Apply To" edit control and type in the bar numbers to which you would like to apply the nodal load. See "Making Selections" above for keywords you can use in this edit control. Once you have completed the selection you can press "Apply" which applies the uniform load to the bars you have specified.

3. **Create Selection Graphically:** Click in the "Apply To" edit control to indicate that you wish to create a selection and move your cursor out into the view. You will notice that the cursor changes to the selection cursor (a hand) instead of the apply cursor we saw above in "Click to Apply":

    a. 

    b. Once you have done this, the click-to-apply cursor does not like to come back. If you have clicked in the "Apply To" edit control and would like to get back to the click-to-apply cursor you only need to click on the tab of the load you are applying "Bar" in this case, and the click-to-apply cursor will return, allowing you to continue in a click-to-apply mode.

4. **Using the Selection Dialog:** You can also use the Selection Dialog to build a selection for applying your loads. Click in the "Apply To" edit control and then click on the bars selection button on the selections bar: 

This will launch the Selection Dialog. For more information on using the Selection Dialog please refer to the section titled "Making Selections" above. Once you have built a selection in this dialog, simply click back into the "Apply To" edit control in the Load Definition Dialog and the selection you built will be automatically transferred to the "Apply To" edit control for you. Next press "Apply" and the Uniform Load will be applied to all bars you selected in the Apply To field.

Non-Uniform Bar Load:

Defining the Load Properties: In the Load Definition Dialog select the "Bar" tab and then press:  to launch the Trapezoidal Load configuration dialog:

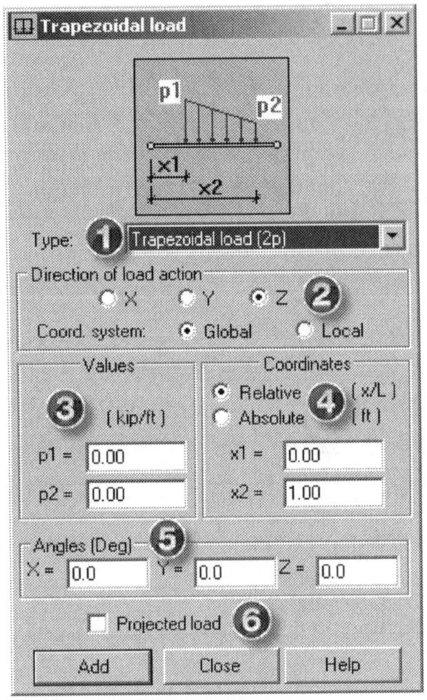

1- Type selector: Trapezoidal load may have from 2 to 4 points where magnitude is defined. The number of edit controls for magnitude and location will adjust as you change this selection.

a. 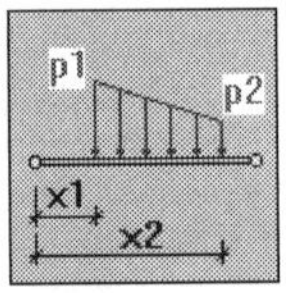

b. 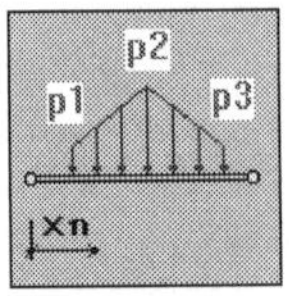

c. 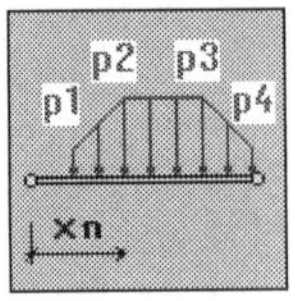

2- Direction selection: The load will be aligned parallel to one of these axes directions. You must decide if X, Y, and Z refer to the global axes of the project or the local axes of the member. (See Uniform Load above for a depiction of the bar local axes) To view the current orientation of the bars' local coordinate systems, use the quick toggle at the bottom of the view window  to toggle on visualization of local coordinate systems.

3- Define the magnitudes of the load at each point of interest. The number of value fields here will directly correspond to the selection made in number 1 above. Current units are displayed for reference at the top of the column.

4- Specify the location of each point of interest along the member length as measured from the member start to the member end. You may specify Relative coordinates (distance along the member/member length) or Absolute coordinates along the member length.

5- Direction allows you to rotate the direction of action of the load. These values are rotations either about the global coordinate axes or about the local member axes as shown above. Each angular measurement is an amount of rotation about the specified axis. I.e., if you enter 20 into the direction field on the X row the load will be rotated 20 degrees about the X axis.

6- Decide if this load should be projected or not. Projected loads are often used in snow loading for an element. The help files indicate that the load is applied to the length of the member as projected on a plane which is normal to the load's direction. It should be added that this consideration ignores any rotation of the load. If you choose to use rotation of the load, be careful about using projection as the results may not be what you would expect.

When you have completed configuration of the load press "Add" to queue this load configuration for application.

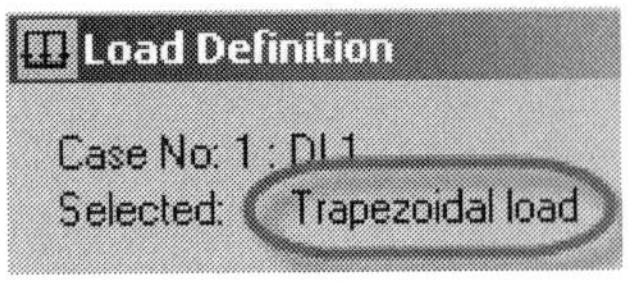

The "Selected" area of the Load Definition Dialog will now indicate "Trapezoidal Load".

Creating a Trapezoidal Load in the Project:

1. **Click to apply:** When you move your cursor out into the project area, the cursor graphic will change to an arrow with a graphic indicting a uniform or planar load.

    a. 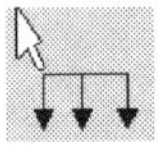 (Note that this icon is exactly the same as the uniform load icon. All bar loads and surface loads will use this same icon)

    b. You can simply begin applying the load to bars in the project environment by clicking them directly when this cursor is active. The load is applied as you click; there is <u>no need</u> to press "Apply"

    c. Notice that hovering near one end of the beam or the other results in two little arrows along the length of the bar. This is indicating the direction of the x axis of the bar for purposes of applying a trapezoidal load. The distances or relative distances you entered in the configuration dialog will be considered starting from the end of the bar away from the little arrows. (ending at the end to which the arrows are pointing.)

d. 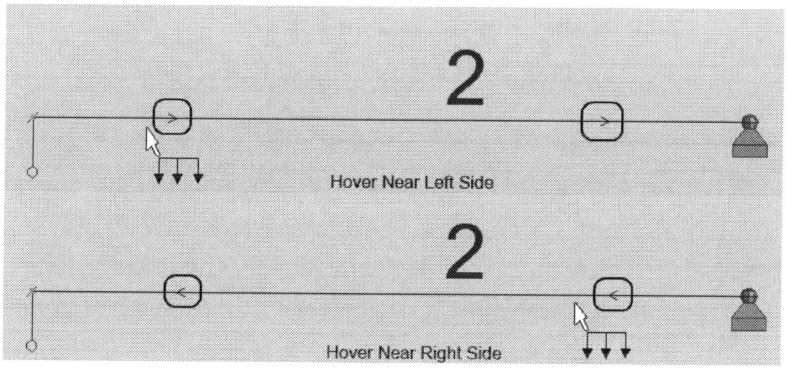

2. Using the Apply To Field:

a. **Type in selection:** Instead of moving the mouse directly into the project environment, click in the "Apply To" edit control and type in the bar numbers to which you would like to apply the nodal load. See "Making Selections" above for keywords you can use in this edit control. Once you have completed the selection you can press "Apply" which will then apply the trapezoidal load to the bars you have specified.

b. **Create Selection Graphically:** Click in the "Apply To" edit control to indicate that you wish to create a selection and move your cursor out into the view. You will notice that the cursor changes to the selection cursor (a hand) instead of the apply cursor we saw above in "Click to Apply":

c. 

d. Once you have done this, the click-to-apply cursor does not like to come back. If you have clicked in the "Apply To" edit control and would like to get back to the click-to-apply cursor you only need to click on the tab of the load you are applying "Bar" in this case, and the click-to-apply cursor will return, allowing you to continue in a click-to-apply mode.

e. **Using the Selection Dialog:** You can also use the Selection Dialog to build a selection for applying your loads. Click in the "Apply To" edit control and then click on the bars selection button on the selections bar:

This will launch the Selection Dialog. For more information on using the Selection Dialog please refer to the section titled "Making Selections" above. Once you have built a selection in this dialog, simply click back into the "Apply To" edit control in the Load Definition Dialog and the selection you built will be automatically transferred to the "Apply To" edit control for you. Next press "Apply" and the Trapezoidal Load will be applied to all bars you selected in the Apply To field.

3. Chain Selection:

a. If you use the "Apply To" field to select a chain of bars, you will have the option to apply the trapezoidal load across all of the bars at once as though they were one long bar. All length and relative measurements can be considered as though they ranged over the entire selection as opposed to a single element.

b. When you press "Apply" with a selection that Robot can determine is ostensibly a "chain" you will automatically be presented with a dialog where you may ask Robot to treat the chain as a whole or treat each element separately.

i. 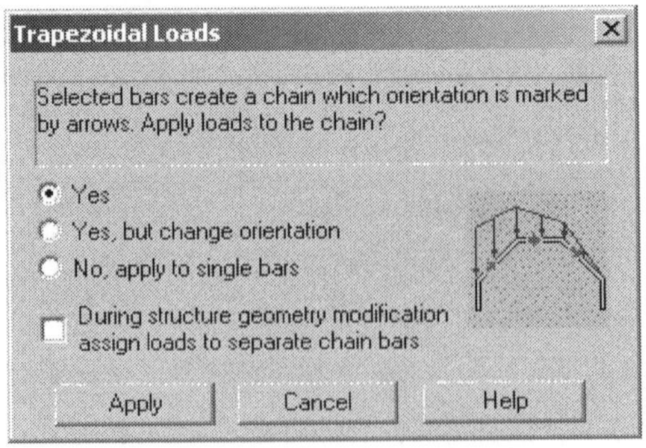

ii. The direction (orientation) will be indicated in the project environment by a set of arrows across the chain of bars to indicate which direction Robot has assumed:

iii. 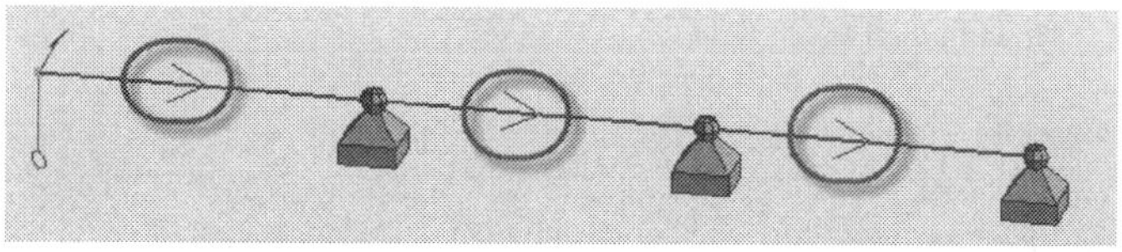

iv. The last checkbox option controls how Robot deals with subsequent changes to geometry. If left unchecked, Robot will continue to think of these bars as a single element where this load is concerned. So if one or more bars grow shorter or longer, the load will adjust itself such that it is re-distributed over the bars as though you had applied it fresh to the chain. If checked, each bar will be treated independently and will maintain the values of load at its ends upon modification. See the Robot Help files for graphics illustrating this.

Concentrated Bar Forces:

Defining the properties of the Concentrated Bar Force: In the Load Definition Dialog select the "Bar" tab and then press: 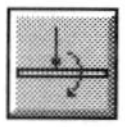 to launch the Bar Force configuration dialog:

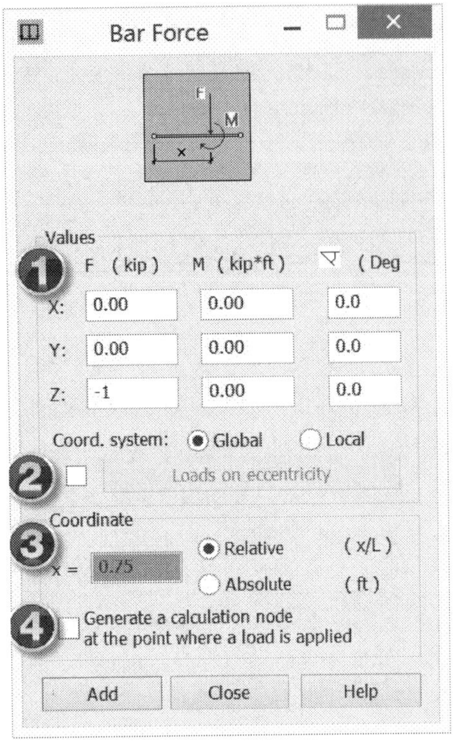

1-3 This Dialog is effectively similar to a combination of the Nodal Force dialog for force and moment component specification, Similar to the Uniform Load dialog for Loads on eccentricities and similar to the trapezoidal load dialog with respect to relative or absolute coordinates of load application. For items 1 through 3 in the above dialog, refer to "Nodal Load", "Uniform Load" and "Trapezoidal Load" above for information on how to use these inputs. Note: Bar Forces cannot be applied to "Chains" of bars as can Trapezoidal Loads.

4: Generate a Calculation Node: The only new thing in this dialog worth mentioning is the option to "Generate a calculation node at the point of load application". This can be incredibly important: Member designs and code checks are performed at only specified locations along the length of a member. It is rare but it *is* possible that the maximum force along the length of a member will occur at or under the bar force location and without a calculation node generated at this location, you may not be testing, or designing, for the worst case moment. We will cover these options in the steel design section but here is an (admittedly contrived) example to illustrate this point:

The default number of calculation points for a bar is 3: (from the Configurations dialog accessible from Steel Design Calculations)

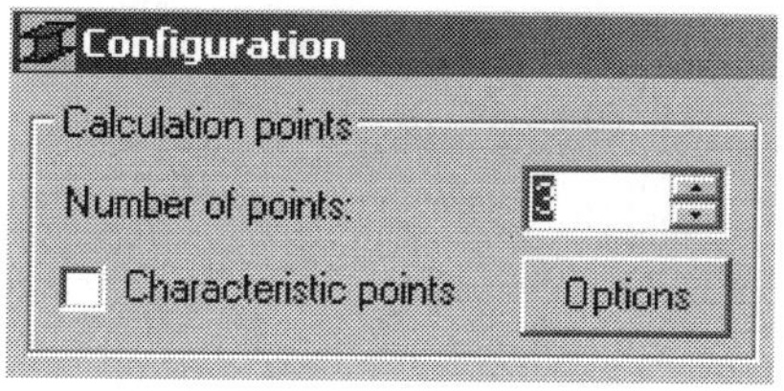

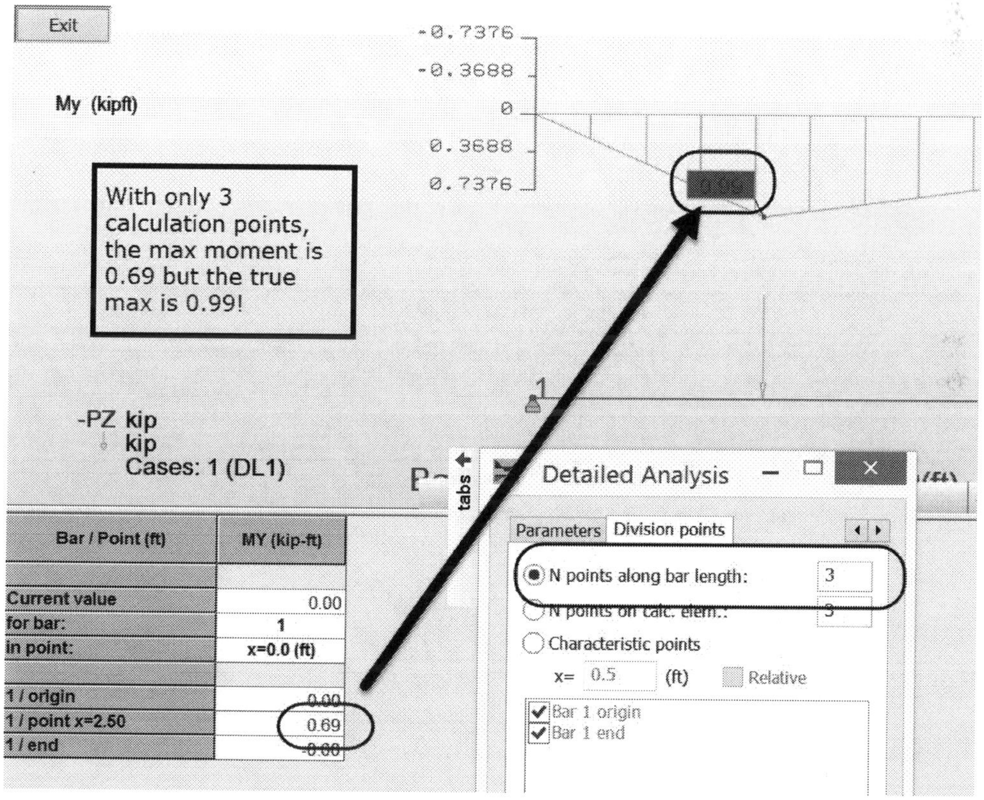

Robot member verifications and design routines are performed at discrete points along the length of the member that you specify called "calculation sections". As you can see from the steel member verification results, the calculated max moment for this member is 0.69!

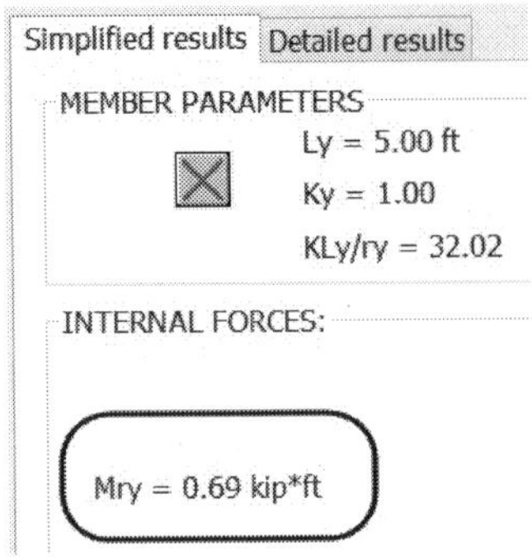

You will increase accuracy by selecting more than 3 points (recommended practice) but even in this case you still may not be getting the results you expect. Here is the same at 10 points:

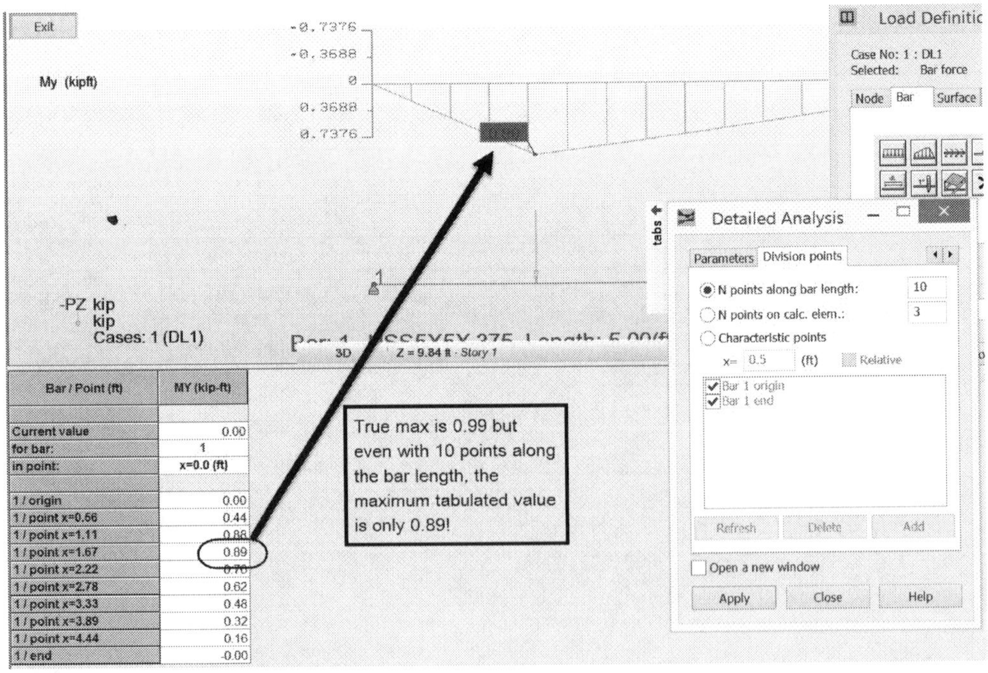

Be aware that generating a calculation node at the location of your concentrated bar force, even with the same setting of 3 calculation points, gives the correct max moment as it occurs under the bar force.

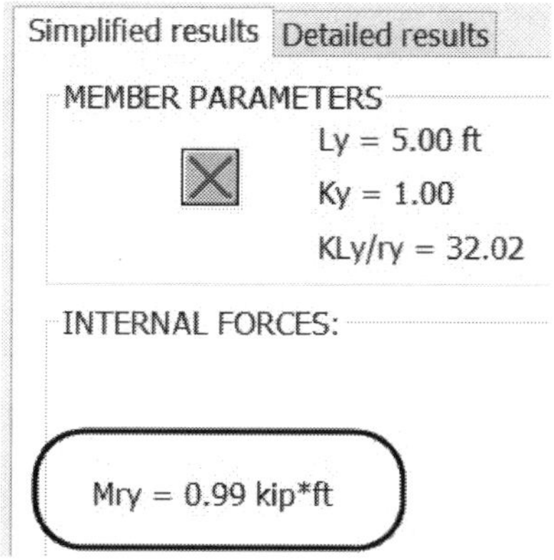

If you are applying bar forces, increasing the number of calculation sections is one way to get closer to this result but generating a calculation node is a better method.

Claddings for Planar Loads

Applying planar loads typically requires some type of planar element upon which to apply the load. As we have not yet covered creating planar deck or wall elements we will discuss the somewhat more straightforward claddings first and then look at how we can use claddings as hosts for planar loads.

The cladding object is purely for load distribution. It will not provide any diaphragm action. Claddings are planar elements which may be created in any orientation. Once created they provide a distribution element which can take planar loads and distribute them to model elements in either a two-way or one-way distribution. Claddings have local coordinate systems which dictates which directions the load is distributed. You can view the local coordinate system for each cladding, modify it and also display the current distribution options selected for the cladding object.

Creating Claddings:

Claddings Dialog: From the menu choose **Geometry>Claddings...** This will open the Claddings Dialog as shown here:

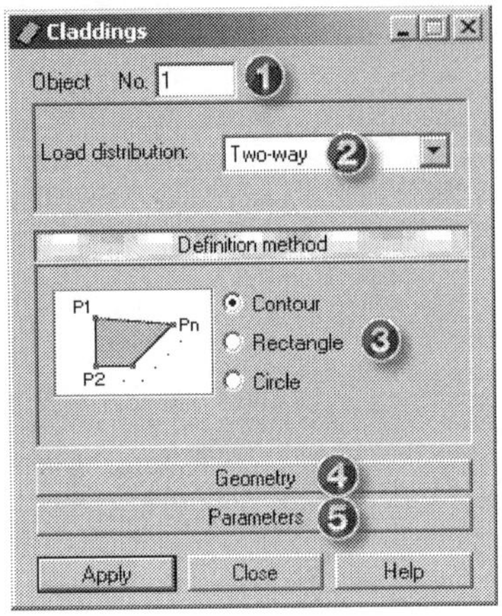

1- Object Number: Robot will automatically pick the next available element number. If you choose to specify one and that element already exists in the project, Robot will ask you if you want to re-define it.

2- Select the initial load distribution direction. Options are:

a. Two-way: loads are distributed with a trapezoidal distribution to members co-planar with the cladding element in both the cladding local X direction and the cladding local Y direction.

b. One-way (X or Y): distribute loads with a rectangular distribution to members in either the cladding local X direction or the cladding local Y direction.

3- Choose the geometric style of cladding you wish to create:

a. Contour: Freeform polygon. Edges may be straight lines or arcs

b. Rectangle: Simplified definition by three points

c. Circle: Simplified definition by three points

4- Geometry section expands to allow configuration of the vertices of the cladding. It will change depending on the style of cladding geometry you have selected in #3 above.

5- Parameters contain advanced settings regarding curve approximation and filleting at corners. Curve approximation will replace any arc you specify with a number of linear segments according to the settings configured in the Parameters section.

a. 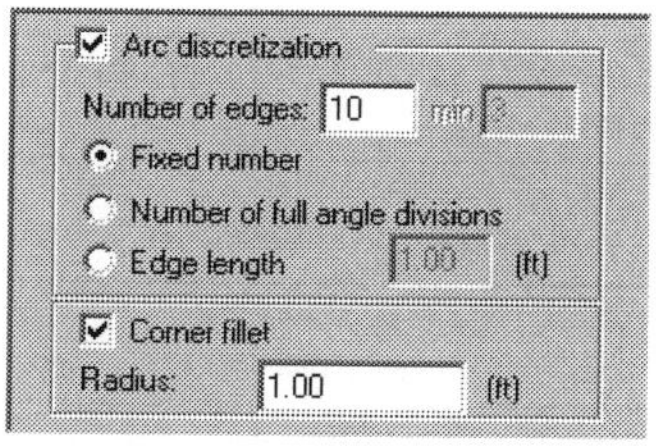

Defining the Geometry: First select the type of cladding you want to create: Contour, Rectangle, or Circle. While it is possible to create a non-planar cladding, doing so will not allow you to apply nor distribute loads.

<u>Contour</u>: The Geometry area of the Claddings dialog will show the contour definition configuration:

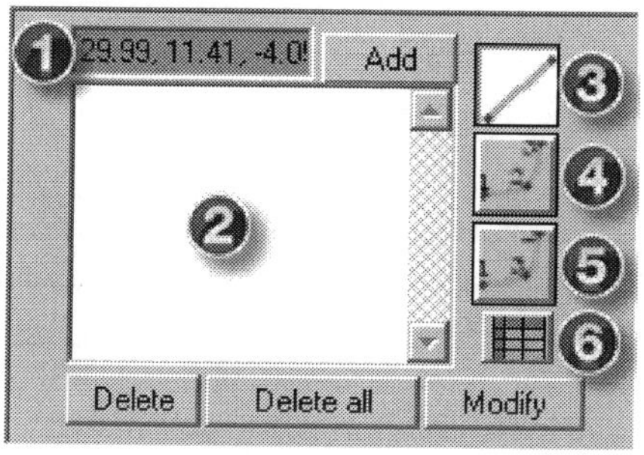

1- Coordinate entry area. You may type coordinates manually and use the "Add" button to add them to the current list of vertices or you may click in the edit control and move your mouse out into the project environment where you may indicate vertices directly by clicking in the project.

2- Current vertex list: As you either manually "Add" points or click them in the project, you will see them populate this list.

3- Create linear segments: Between successive clicks, Robot will create a linear edge for the cladding. For buttons 3-4 as you click in the project environment, you can select back and forth between these options for each edge you wish to create thereby stringing together linear and arc segments in order to specify exactly the cladding configuration you desire.

4- Arc Beginning-Center-End: Create an arc by specifying three points in the project or by manually entering the coordinates

5- Arc Beginning-End-Center: Create an arc segment by specifying the start, the end and a point that the arc should pass though

6- Spreadsheet Input: Opens a spreadsheet interface for indicating cladding vertices in a spreadsheet style. To use this interface, right click in the spreadsheet to insert a new row and configure the parameters. Use "Table Columns..." from the right-click menu to add columns in order to specify arc center points and other parameters.

Rectangle: This is a simplified version of the contour definition above and will likely be the most frequently used. Changing the Definition Method will discard your current vertex Selecting Rectangle will reconfigure the Geometry section as shown here:

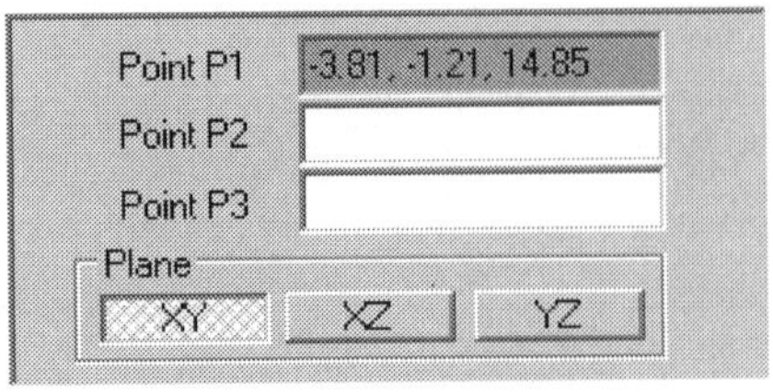

You can type in the coordinates of three points defining the rectangle. The first two points define the first edge and the last point merely defines the distance perpendicular to the first selected line (and potentially the inclination if a z-coordinate magnitude is specified by manually typing in coordinates or snapping to elements in the project).

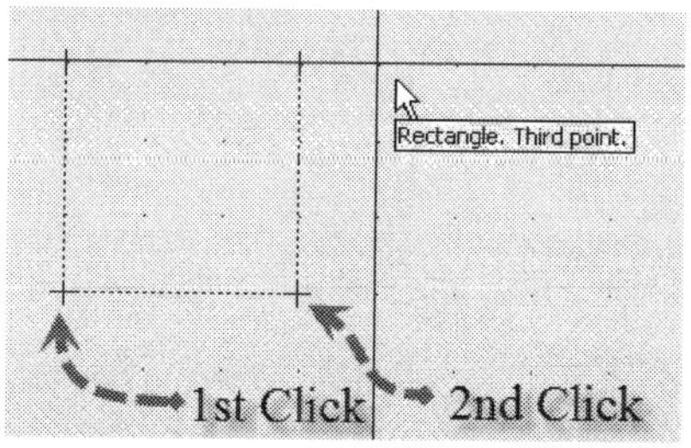

Circle: Requires the definition of three points that lie on the circle. You may manually type in the coordinates or select the points in the project environment. You may create a sloped circle by either manually specifying a z-coordinate or snapping to an elevated piece of geometry in the project.

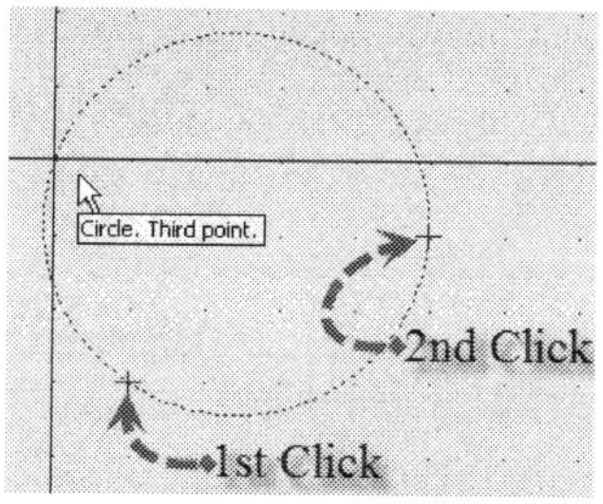

Finishing the cladding definition: If you have manually entered any of the coordinates you will need to use the "Apply" button at the bottom of the Claddings Dialog in order to create the cladding in the project. If you are clicking in the project to define points simply selecting the last point for rectangle or circle will finish the definition and add the cladding. If you are using the contour option, you may either click again on the last vertex to complete the definition to and add the cladding or use the "Apply" button in the Claddings Dialog.

Visibility Options for Claddings:

1. In **View>Display...** under "Panels/FE" you can find the following options:

   a. 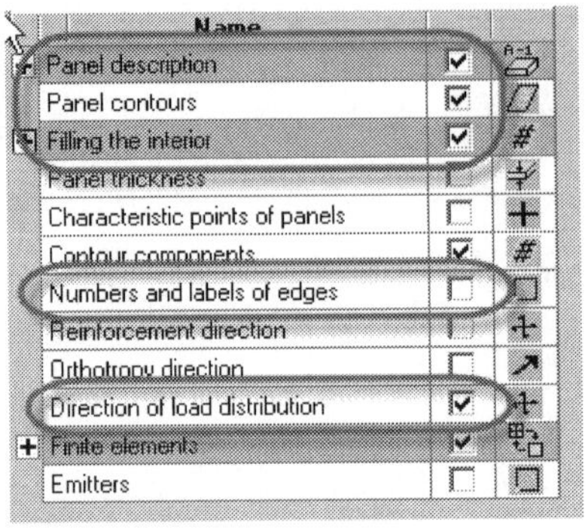

   b. Which correspond to the following visual information: (filling the interior shades the cladding)

   c. 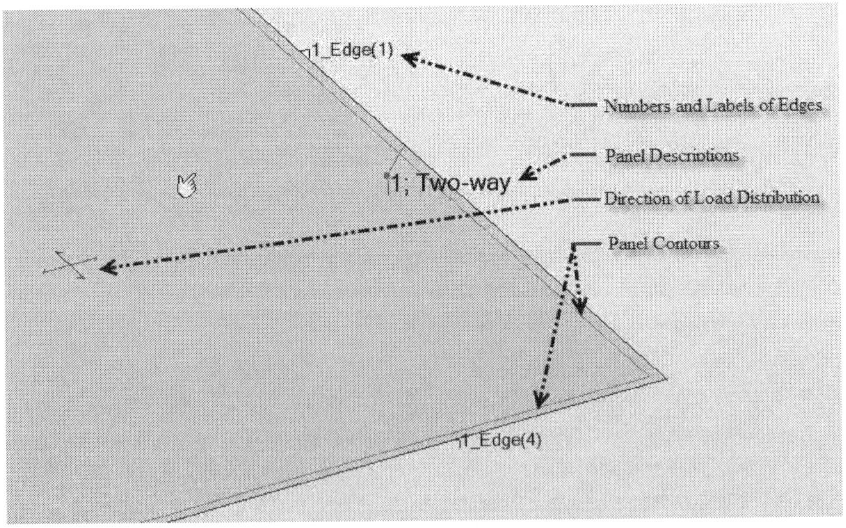

   d. Under the Model Category you can find "Local Systems" which will show the local X, Y and Z orientation for the local system of the cladding. Loads are distributed along the X and Y axes so the orientation is very important information.

e. 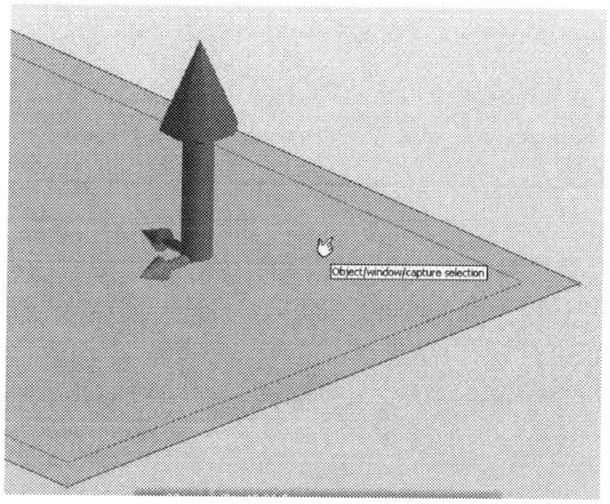

f. Finally under the Loads category, you can visualize the "Load Distribution Regions" and "Forces Generated Automatically" but <u>only</u> after both a load has been applied *and* calculations have been run. An example looks like this:

g. 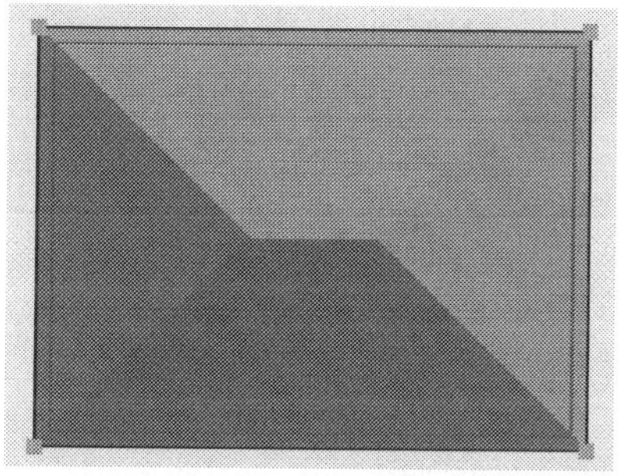

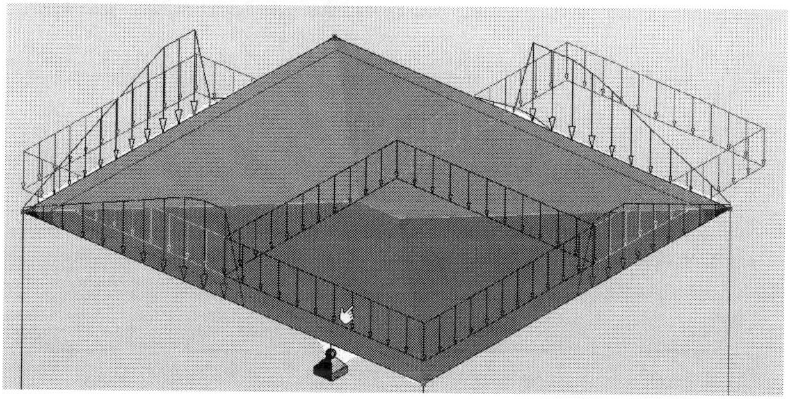

h. Here the two-way distribution cladding has four areas of load distribution to supporting members. (There are only four members at the edges of this cladding) You can also see the automatically generated loads applied to the beam members at the edges of the cladding resulting from the load distribution by the cladding element to the supporting beam elements.

Changing the Local Orientation of a Panel:

In order to adjust the local axes orientation of a cladding object, select **Geometry>Properties>Local Panel Direction...**

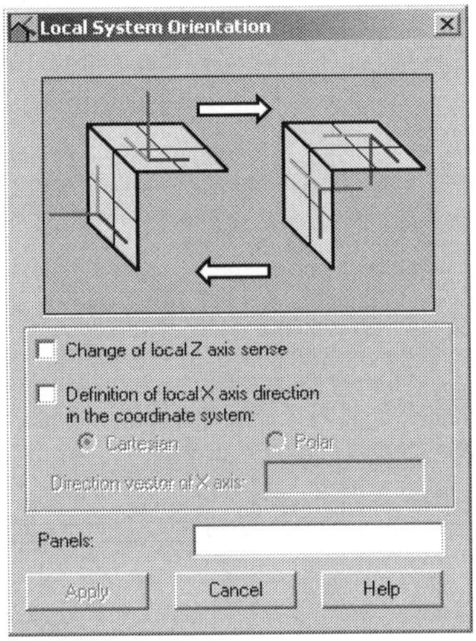

Selecting the "Change of local Z axis sense" will simply flip the Z axis from one side of the panel to the other which will not change the positive sense of the X axis but will change the positive sense of the Y axis. In terms of load distribution, there is effectively no difference between these options as load is distributed in the direction of positive X/Y and Negative X/Y.

Selecting "Definition of local X axis direction" will allow you to indicate a direction vector with which the X axis of the cladding local coordinate system will be aligned. Placing the cursor in the edit control, you can click two points in the project environment to specify this vector or you can manually type in a direction vector referenced to 0,0,0. Robot will align the X axis with this direction vector.

In the "Panels" edit control, enter the panel numbers to which you would like to apply this change in local system orientation and then press "Apply". You may also click once in the "Panels" edit control, then move your mouse into the project environment where the cursor will be the selection cursor and allow you to create a selection of panels. Elements you select in the project environment will populate in the Panels edit control. When you have completed the selection, press "Apply" to apply the change of direction to all selected panels.

Change the Cladding Distribution Method

Loads are distributed to supporting elements by claddings along the local X and/or Y axis directions. Even though you selected a distribution method at the time of creation, you may modify it at any time by selecting **Geometry>Additional Options>Load Distribution - Claddings...** Which will launch the Load Distribution Dialog

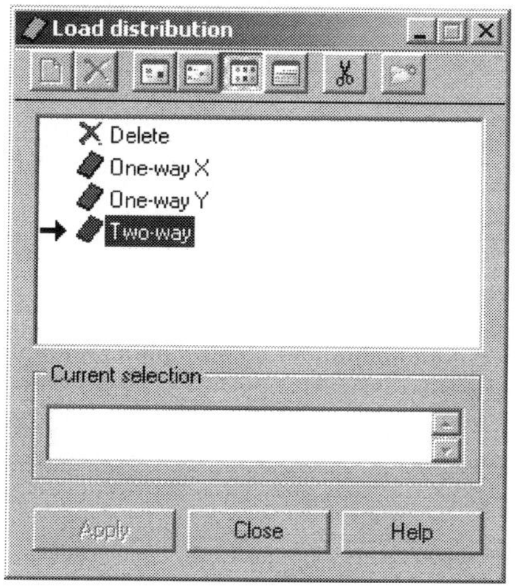

This is another Robot Label dialog: the three available options for distribution are labels which can be applied to cladding elements by selecting the label you want to apply and using one of the following application methods:

1. **Click to apply**: When you move your cursor out into the project area, the cursor graphic will change to an arrow with a graphic indicating a panel.

    a. 

    b. You can simply begin selecting cladding objects in the project environment by clicking them directly when this cursor is active. The change of distribution is applied as you click; there is no need to press "Apply"

2. **Type in selection:** Instead of moving the mouse directly into the project environment, click in the "Current Selection" edit control and type in the panel numbers to which you would like to apply the change of distribution. See "Making Selections" above for keywords you can use in this edit control. Once you have completed the selection you can press "Apply" which will then apply the distribution method label to the selection you have specified.

3. **Create Selection Graphically:** Click in the "Current Selection" edit control to indicate that you wish to create a selection and move your cursor out into the view. You will notice that the cursor changes to the selection cursor (a hand) instead of the apply cursor we saw above in "Click to Apply":

    a. 

    b. Select panel edges to select the panel. Hovering near a panel/cladding edge will highlight the edge.

c. NOTE: Once you have done this, the click-to-apply cursor does not like to come back. If you have clicked in the "Current Selection" edit control and would like to get back to the click-to-apply cursor you need to click on an alternate label, then re-select the label you wish to apply. The click-to-apply cursor will return, allowing you to continue in a click-to-apply mode.

4. **Using the Selection Dialog:** You can also use the Selection Dialog to build a selection for applying the load distribution label. Click in the "Current Selection" edit control and then click on the bar selection button on the selections bar: 

This will launch the Selection Dialog:

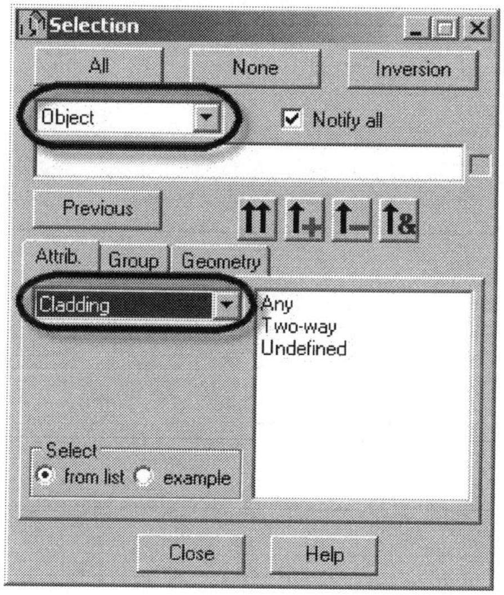

To select Claddings, choose "Object" from the dropdown list at the top, then choose "Cladding" from the Attribute dropdown. You can then use the selection dialog like normal. For more information on using the Selection Dialog please refer to the section titled "Making Selections" above. Once you have built a selection in this dialog, simply click back into the "Current Selection" edit control in the Load Definition Dialog and the selection you built will be automatically transferred to the "Current Selection" edit control for you. Next press "Apply" and the load distribution label will be applied to all claddings you selected in the "Current Selection" field.

Creating and Applying Planar Loads

Uniform Planar Load

From the Load Definition Dialog (**Loads>Load Definition...**) select the "Surface" tab and then the uniform planar load button:  to launch the Uniform Planar Load configuration dialog:

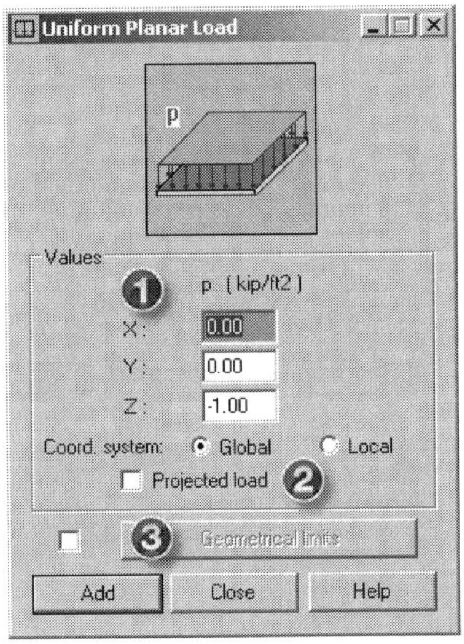

1- Specify the magnitudes of the loads in each of the coordinate directions. These will either be along global or local axes depending on the setting below under Coord. System

2- Determine if you want this load applied in the global or local coordinate system of the element and whether this load should be projected. Projected loads are typically used for applying snow loads to sloped roofs

3- Geometrical limits: This option allows you to select a plane and a direction which will limit or constrain the uniform load application. You can use this in situations where the element you intend to load is partially shielded from load by another element. For instance if you have a slab which is partially exposed as a roof and partially constitutes a floor. You can apply the roof load and the floor load as uniform loads to this panel with a geometrical constraint placed at the exterior wall. You could also simply opt for applying a partial uniform load on a contour. This option, however, will allow the load to somewhat adapt to changes in model geometry whereas a load applied on a contour will need to be updated manually with changes to the model.

Once you have completed configuration of the load properties, press "Add" to queue it for application to the model. You will notice that the Load Definition Dialog now reflects that we are applying a uniform load:

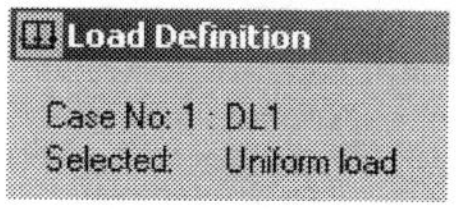

You can now use one of the 4 application methods previously discussed to specify individual cladding elements to which the load will be applied: Click to Apply, Manual Selection, Graphical Selection or Selection Dialog. (See "Uniform Bar Loads:" Page 125 Creating a Uniform Load in the Project above under Creating and Applying Bar Loads)

Linear Load 2 Point

From the Load Definition Dialog (**Loads>Load Definition...**) select the "Surface" tab and then the uniform planar load button:  to launch the Linear Load 2p configuration dialog:

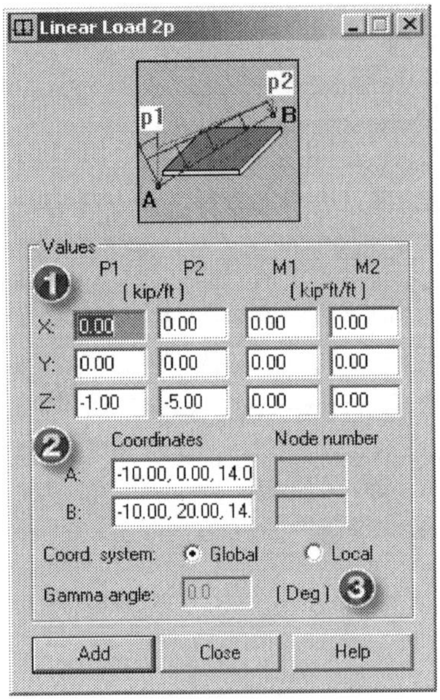

1- You can supply starting and ending values for both force and moment to be used at each end of the linear segment.

2- Coordinates of the end points of the linear segment: These do not need to fall within the structure nor even within the cladding area. They can also terminate within the cladding area. NOTE: Be careful, if you are manually entering these coordinates or clicking locations in the project, if you do not specify points which are co-planar with the cladding or shell element, the load will not be applied to the cladding.

3- Specify whether this load will be oriented relative to the global coordinate system or the cladding local coordinate system. If you select Local you can additionally specify a rotation of the load about the line of action. (Gamma)

Applying a Linear Load to a Surface:

The Linear surface load is different from other load types in terms of its application to the surface/cladding. The major difference is that you do not indicate panels or elements to which this load is applied; it is applied as soon as you select "Add". Notice that the "Apply To" edit control is greyed out after pressing "Add".

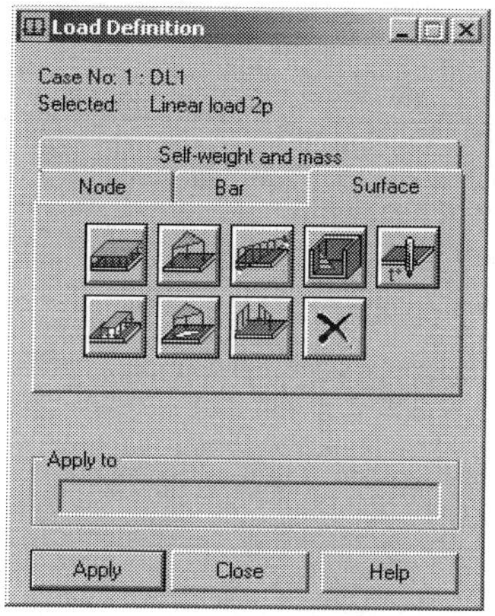

This would usually queue this load type for application but in this case it applies the load instead! After application the load automatically finds coplanar elements (coplanar with the load or that is to say "lie in a plane which includes some portion of the linear line load") and applies itself to them. This means that you can apply a linear load across multiple planar elements and by nature of their co-planarity with the linear surface load, they will be loaded.

Unique Issues for Linear Loads: If by some chance you either manually entered coordinates or selected coordinates in the project which do not lie in the plane of the element, the load might not appear in your project view. It is still very much there, however, which can quickly be verified by looking at the load table (**Loads>Load Table**). If you check, it is possible that you have Story Filters turned on: Check **Geometry>Stories>Filter Stories** to see if it has been turned on. Because the load still exists in the project, if the cladding or shell element subsequently is moved to a location where the defined line and plane are coplanar the load will be applied to the cladding; however, will still not appear in the project view if filter stories is on.

You will also find that linear loads applied to claddings show as "orphaned" by using the Verification routine (**Analysis>Verification...**). Here you will see a warning indicating that "Linear load has not been applied to a panel". Select that warning in the Verification dialog to see the load in question highlighted as shown here: (Note: If the load graphic is not visible, the highlighting still will be visible)

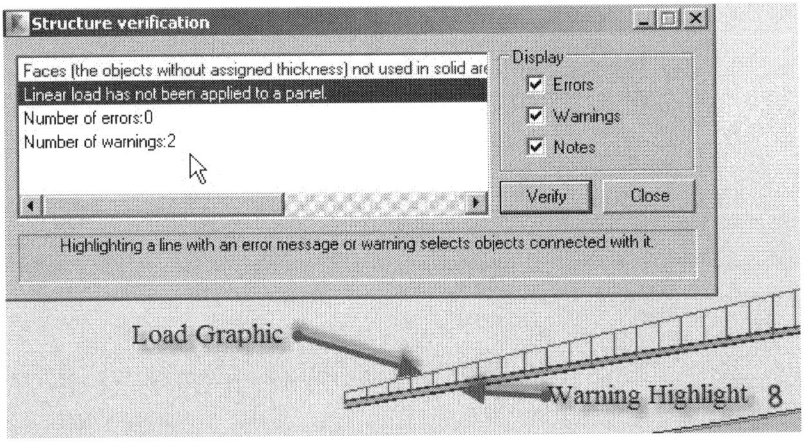

If the linear load is applied to claddings the claddings will distribute it however, the warning dialog will always show this load as not applied to a "panel" which... technical it is not. We think of claddings like panels but they are not in this case so the warning continues even though the load is being distributed by the cladding. To see what is happening and to verify your model is behaving appropriately, turn on "Forces Generated Automatically" in the display properties under the Loads category.

Uniform Planar Load on Contour

Uniform loads on contours have some similarities with linear loads we have just discussed. If they are configured for "Automatic panel selection in the contour plane", then once they are "Added" from the configuration dialog, then they have already been added to the project. However, if you uncheck the automatic option, then you will apply them in a similar way to the uniform planar load and will need to select the elements to which Robot should apply the load. This can be somewhat confusing at first. It will help to think about what type of operation you are intending to perform. The automatic option is best for applying uniform loads to a partial area of a panel where both the panel (cladding) and the load are co-planar. The second option is best for situations where the load needs to be applied to panels which are not necessarily co-planar with the load (E.g., cylindrical shells or sloped planes).

Defining a Uniform Load on Contour: To define a Uniform Load on Contour, open the Load Definition dialog and select the Uniform Planar Load on Contour button:  which will launch the Uniform Planar Load (contour) configuration dialog box:

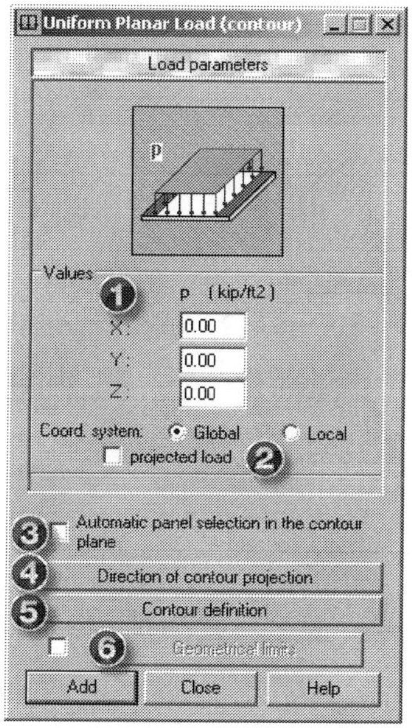

1- Values: Specify the components of the load to be applied with respect to either global or local panel coordinate systems (depends on the setting below in #2)

2- Choose either global axes or local panel axes for which you have entered values above. Also, choose whether this load is intended to be a projected load. Projected loads are most often used for applying snow loads to sloped roofs.

3- When this checkbox is checked, the tool will look for any panels which are co-planar with the defined contour (#5 below) and associate itself with them. If this is left unchecked, then #4 Direction of contour projection becomes enabled see #4 for more information.

4- For typical loading situations leaving the automatic box checked is appropriate; however, if you want to apply a load to elements which are not planar (e.g., a cylinder) you can leave automatic unchecked and expand the Direction of Contour Projection.

a. 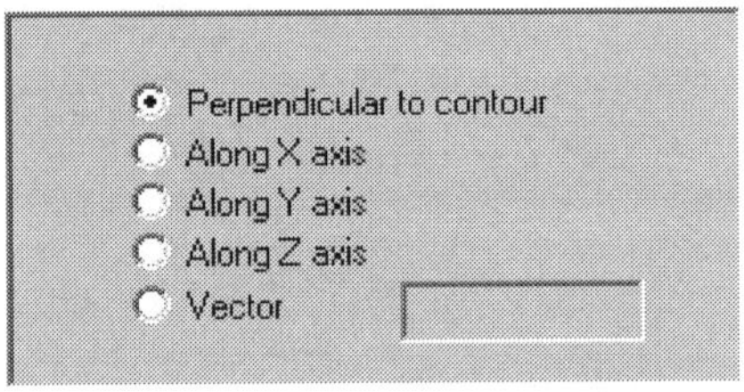

b. These options allow you to specify along exactly which direction you wish the load to be cast. I say cast because it is different than projection in the sense of projected loads. (Uniform loads on contour may also be projected loads!) Casting means that the area defined by the contour (in contour definition section below) will be projected along the direction specified here in order to find the area over which to apply the load. Select the appropriate vector direction or specify your own custom by selecting "Vector" and entering or selecting a direction vector. Here is an example of the contour load applied to the arch below with the Z axis selected as the casting (projection) axis. The contour is applied at an elevation above (could have been below) the arch. The first picture is the applied load, the second picture is with the "Forces generated automatically" turned on in the Display (**View>Display...**.) options under the Loads category (after running calculations). You can see how the contour load has been applied to the individual finite elements meshed into the arched shell.

c. 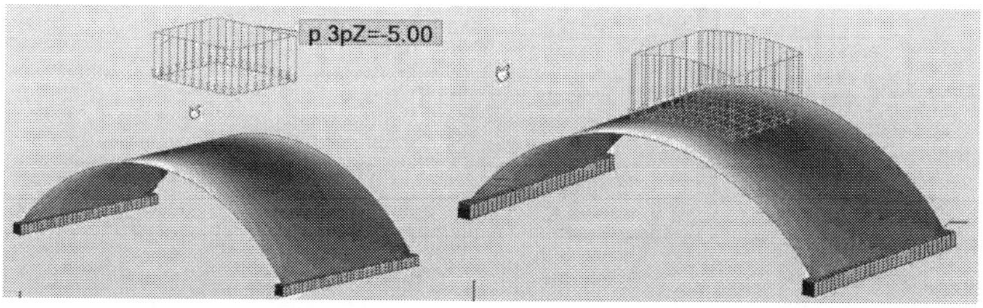

d. Here is another illustration of this principle at work in a more common building configuration. The highlighted load (the one to which the pointer finger is pointing) has been created with the automatic option (#3 above) and because it is above the plane (non-coplanar) it is not affecting the supporting elements. To correct this situation the contour definition for the load must be changed to be co-planar or the load needs to be re-created with the automatic option off with an "along Z axis" projection direction specified.

e. 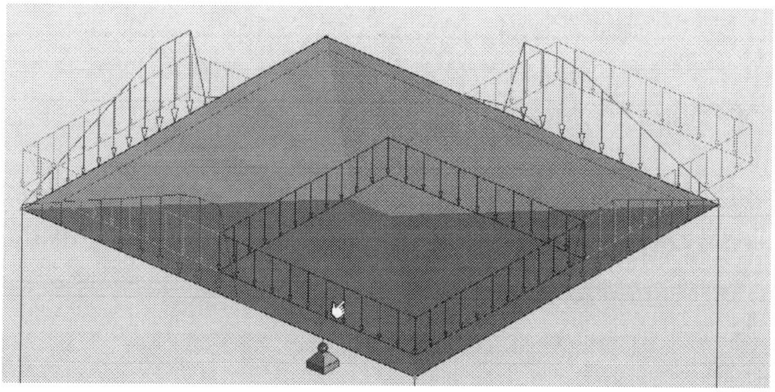

5- Contour Definition: This is where we will specify the vertices of the contour which effectively defines the outline of our load. Expanding the contour definition section reveals:

a. 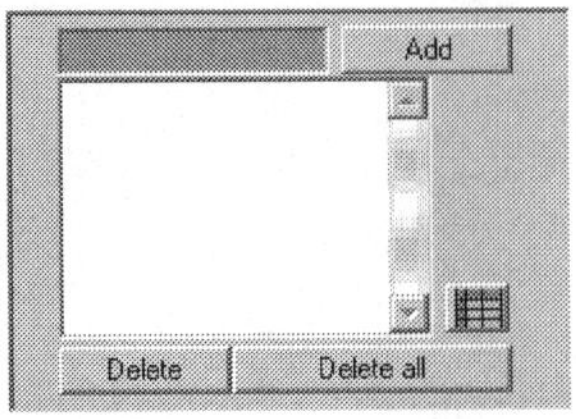

b. Selecting in the edit control next to the Add button you may either select points directly in the project environment (which automatically adds them to the vertex list) or type coordinates manually and add them to the vertex list via the "Add" button in this section. (Note: Not to be confused with the "Add" button at the bottom of the dialog).

c. Spreadsheet Input: 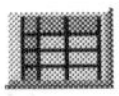 Opens a spreadsheet interface for indicating cladding vertices in a spreadsheet style. To use this interface, right click in the spreadsheet to insert a new row and configure the coordinates. You can use fill special as with any Robot table or cut and paste values from another spreadsheet.

Applying the Uniform Load on Contour:

If you have selected the Automatic option, the "Apply To" field in the Load Definition dialog will be greyed out. This is because the load has already been applied and it will look for any planar elements in the plane of the contour definition to which to apply itself.

If you have not selected the automatic option then you will see the Planar load on contour queued up for application to shells, volumes or claddings.

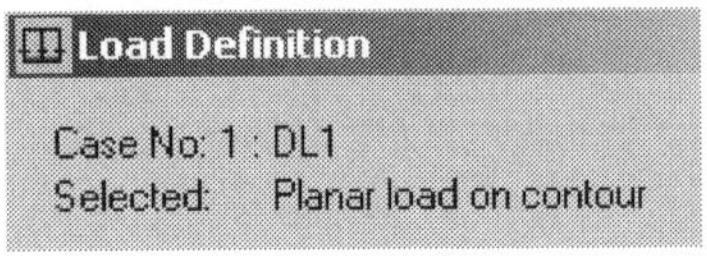

You can now use one of the 4 application methods previously discussed to specify individual cladding elements to which the load will be applied: Click-to-Apply,

Manual Selection, Graphical Selection and/or Selection Dialog. (See "Uniform Bar Loads:" Page 125 Creating a Uniform Load in the Project above under Creating and Applying Bar Loads for details about each of the methods.)

Planar Load 3 Point on Contour

The last planar load type we will cover is important as it is the load type you will use for drift loading or linearly varying wind loads on the windward side of a structure. The general idea is that you define a contour for the loading in almost exactly the same way as we have covered above for planar load on contour. Additionally you will specify three magnitudes and three coordinate points. The magnitudes and coordinates of these points will define a plane and the extents of that plane will be determined by the contour over which the load is applied. (The contour is the "footprint" of the load whereas the three points and magnitudes simply define the plane of the top surface of the load). In order to avoid duplicating information first start by reviewing the information in Uniform Load on Contour above. This will cover use of the "Automatic panel selection in the contour plane" option as well as the Direction of contour projection and also the section on defining the contour.

After selecting the Surface tab in the Load Definition dialog (**Loads>Load Definition...**) pressing the Planar Load 3p on Contour will open the Planar 3P Load (contour) dialog:

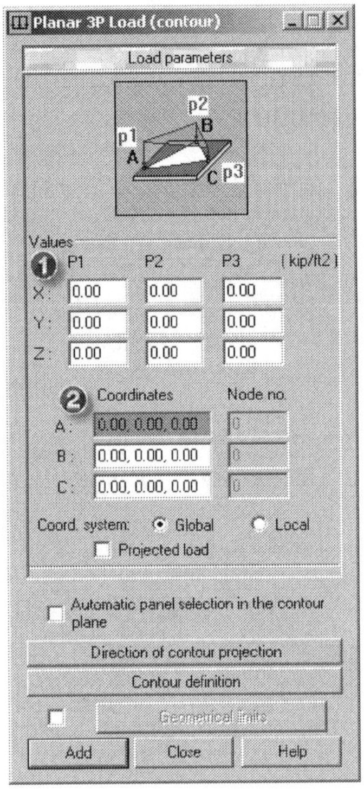

1- This is where you will define the magnitude of the planar load. Specify the X, Y and/or Z components of the load for each of the points. These will define a plane or the top surface of the surface load. The resulting load will be the result of intersecting the contour (footprint) and the plane defined by these magnitudes and the locations defined in Item #2 below.

2- Here you specify the coordinates of the three magnitude points you entered above. Coordinate A corresponds to Point 1, B: Point 2, C: Point 3. You may enter these coordinates manually or if you click in the first edit control then move your mouse out into the project you may specify these points by clicking in the project. The focus will automatically switch to the next edit control as you click to streamline the process.

3- The remainder of the dialog functionality has been covered above in Uniform Load on Contour and Uniform Planar Load. Review those sections for more information on how to configure this type of load as well as define the contour.

Applying the Planar 3P Load (contour):

The application of this load is almost identical to the planar load on contour above. Note: If you select the option to automatically detect panels in the contour plane then the edit control under "Apply To" will be greyed out because it will automatically detect panels for you. If you have selected one of the other options then you will apply the load to panels by indicating them with one of the previously discussed methods above (Click-to-Apply, Type in Selection, Select in Project or using the Selection Dialog).

Example:

At an interior corner of a low roof, you could create two separate Planar 3P Load on contour as shown here:

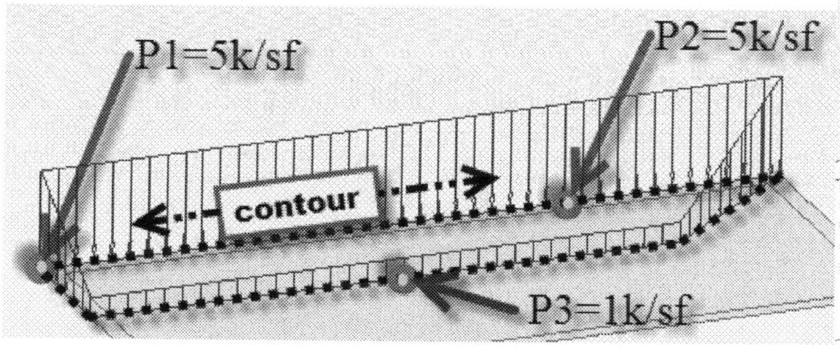

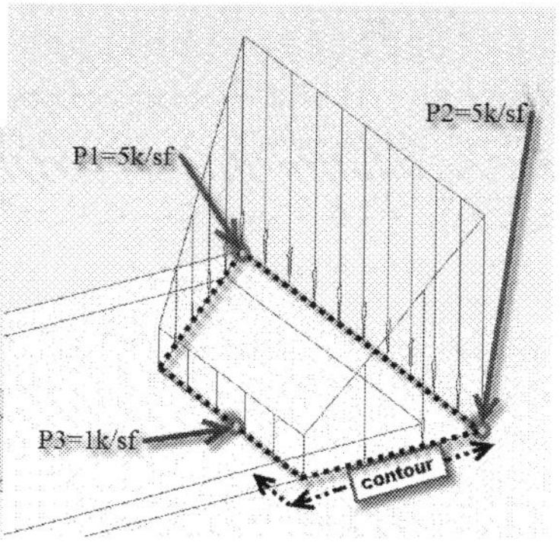

Which will result in the following loading configuration for the low roof interior corner:

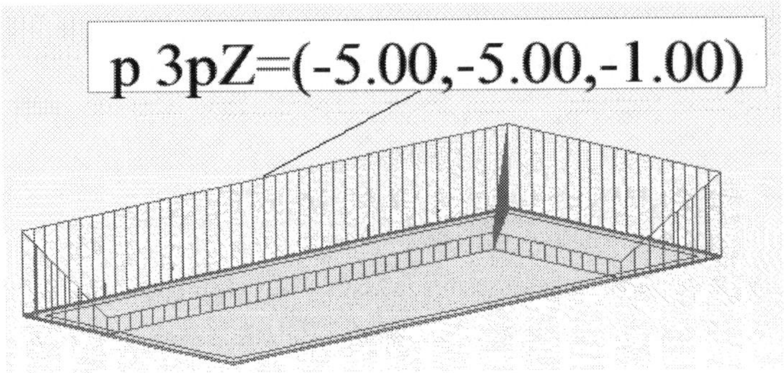

Creating and Applying Self-weight

Robot can consider self-weight of members based on their material densities as defined in the materials dialog (**Tools>Job Preferences**>Materials tab).

Whole Structure Self-Weight

Applying self-weight to the entire structure is a one-click operation. In the Load Definition Dialog, select the Self-weight and Mass Tab and then click the "Self-weight –PZ for the whole structure" button: 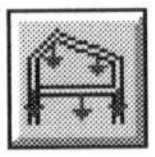

Once you press this button, the self-weight is applied and there are no options to configure. You will see this button greyed out on the self-weight and mass tab until the self-weight load is deleted via the load table. (**View>Tables...**>Loads)

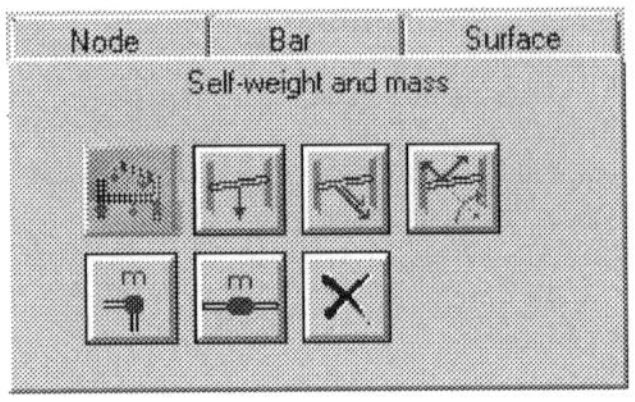

The self-weight load, however, may be modified (or deleted) in the loads table:

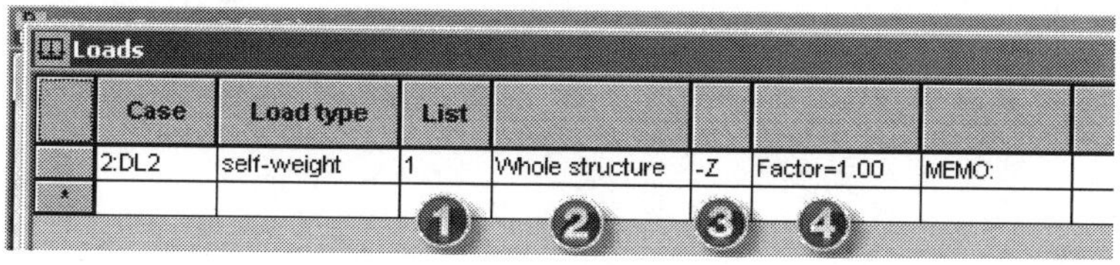

1- List indicates the member numbers to which this load has been applied. Because there was only one element at the time of load application, it only shows one element number. As you add additional members to your model, this list may not update. However, you can verify that it does update upon running the calculations. The "Whole Structure" setting in the next column controls this behavior.

2- The options here are "Whole Structure" or "Part of Structure". When "Whole Structure" is selected, the "List" column will take care of itself. If "Part of Structure" is selected, then only the elements in the List column will receive this self-weight and others will not be loaded by their self-weight.

3- Direction of Load Application: It is possible to change the sense of the self-weight application. Although typically we think of self-weight as acting in the negative Z direction (gravity down), you are not limited to considering only self-weight in this direction. However, in automatically applying self-weight to the entire structure, Robot assumes you mean negative Z and does not give you the option at time of application. This column would allow you to change that setting for this load.

4- Factor is an adjustment to allow you to account for dead-load which is not part of the physically modeled geometry. For instance, paint, fireproofing, other finishes or just bumping up the value of the dead load. This is optional.

Partial Structure Self-Weight

If you wish to only apply self-weight to certain elements, you can select "Self-weight on Selected Elements" with this button: 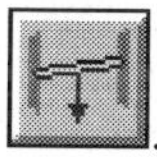. You will then be presented with the Self-weight dialog which allows you to configure the parameters of the self-weight.

(Notice that these are the same parameters which are available in the Load Table above. They are also available for self-weight on "Part of Structure", but may be configured before application to selected elements.

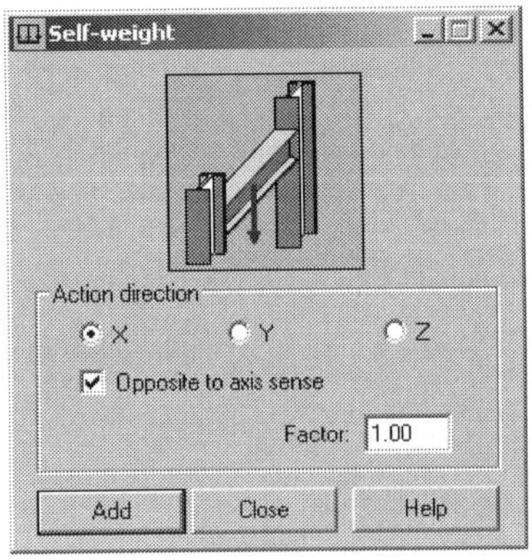

Once you configure these options (typically self-weight would be Z axis and "Opposite to axis sense" since gravity typically acts down. Other options could be used for other types of analysis such as centrifugal or dynamic analysis), press "Add" to queue this load for application. Use the methods we have covered previously for selecting members: Click-to-Apply, Type in Element Numbers, Select in Project or using the Selection Dialog. You can read more about these methods above.

Deleting Loads

When multiple load types have been applied to a single element, using the delete option  available on each tab of the Load Definition dialog is not so straightforward. Robot wants to get an idea of which type of load you intend to delete when clicking or creating a selection in the "Apply To" field of the Load Definition Dialog. To queue a load type for deletion, first queue the load type by launching the corresponding load configuration dialog and pressing "Add" in the load configuration dialog (the settings do not matter), then press the delete button  and begin either clicking elements in the project or use the "Apply to" field of the Load Definitions

Dialog. The concept here is to think of queuing the load type and then queuing the deletion operation. Once you have done this you can use the standard methods of indicating elements as we have covered above (Click-to-Apply, Type in Element Numbers, Select in Project, or using the Selection Dialog. You can read more about these methods above.)

Alternatively you may use the Load Table (**View>Tables...**>Loads) and delete individual loads by highlighting the table row and pressing the delete key.

NOTE: WHEN YOU SELECT AN ELEMENT IN DELETE MODE, ALL OF THE LOADS OF THE QUEUED TYPE APPLIED TO THE MEMBER WILL BE DELETED. MORE REFINED CONTROL IS ONLY AVAILABLE BY EDITING THE LOAD TABLE DIRECTLY.

Exercise 9: Applying Loads

1. Open Exercise_09-ApplyingLoads.rtd
2. This is a simple structure which we will use to add dead, bridge crane, snow, and wind loads to this simple structure using point, line, area, surface, and 3p on contour.
3. 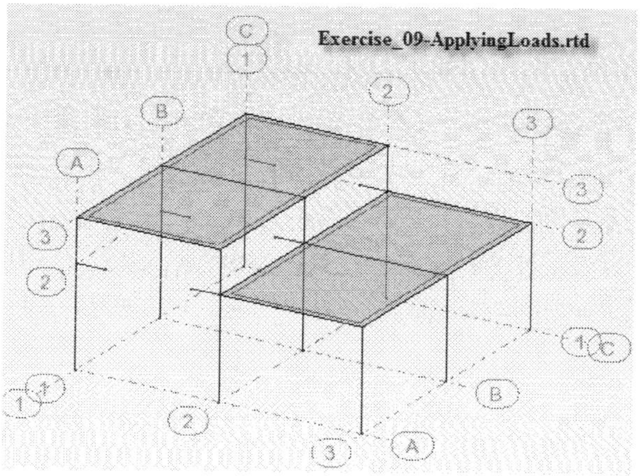

4. Switch to the loads layout from the layout selector on the right hand side of the standard toolbar.

a. 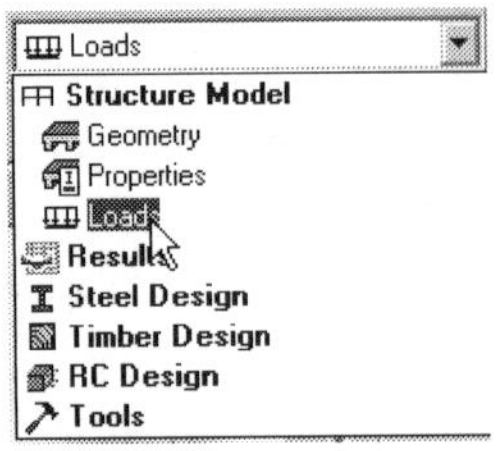

5. Our first task will be to add self-weight to all members in the model. Start by opening the Loads definition dialog **Loads>Load Definition...** and select the "Self-weight and mass" tab:

   a. 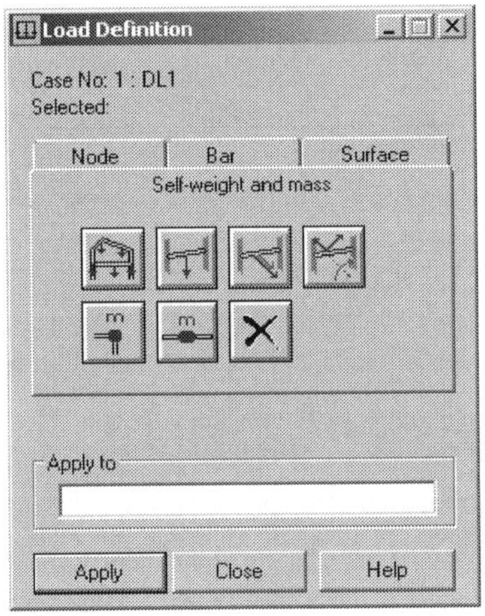

   b. Next, make sure that DL1 is selected in the load case selector: 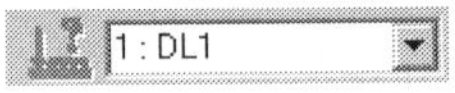

(from the right hand side of the selection toolbar) and verify this at the top of the Load Definition dialog next to the "Case" label.

   c. Next select the "Self-weight –PZ for the whole structure" button

   d. 

   e. As soon as you press this button, it will become greyed out  and self-weight will be applied to the entire structure. This will account for all modeled elements of the structure. We will use surface loads to account for additional dead loads of construction materials. Take a look at the loads table at the bottom of the view to see what Robot has added to the table. This table will continue to build as you add loads to the structure.

i.

Case	Load type	List			
1:DL1	self-weight	1to31	Whole structure	-Z	Factor=1.00

6. Next, let's add surface loads to account for the dead load of other construction materials. In the Load Definition Dialog, switch to the "Surface" tab. Select load case DL2 in the load case selector. Then select the "Uniform Planar Load" button: 

   a. In the Z edit control enter -0.01 as shown here then press "Add":

   b. 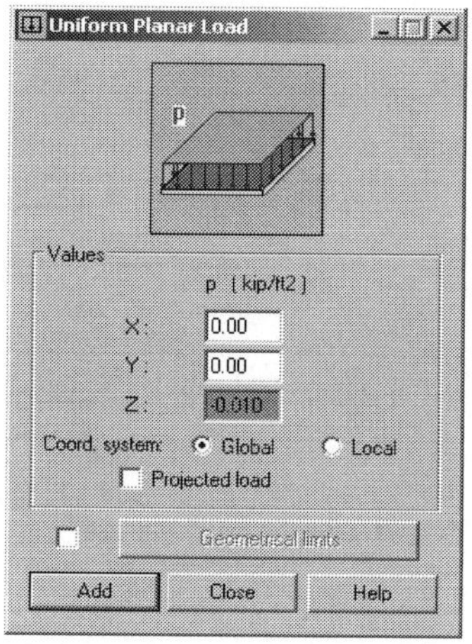

7. This queues the load for application. Next we'll indicate which elements we want to apply that load to. Type in "24, 25" in the "Apply To" field of the Load Definition dialog and press "Apply":

a. 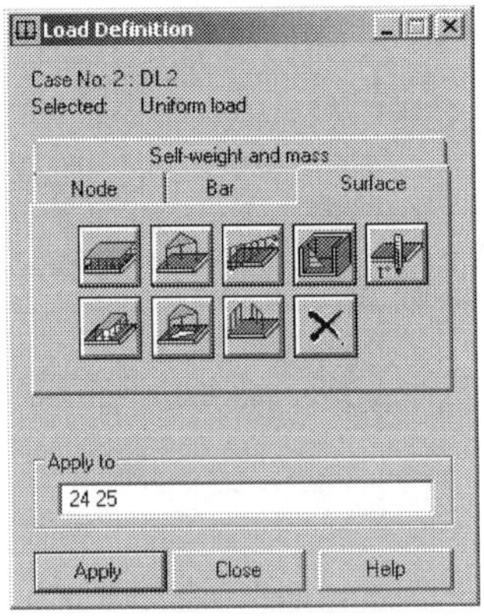

8. Your structure should now show the newly applied surface dead load and there will be a new entry in the Loads Table.

a. 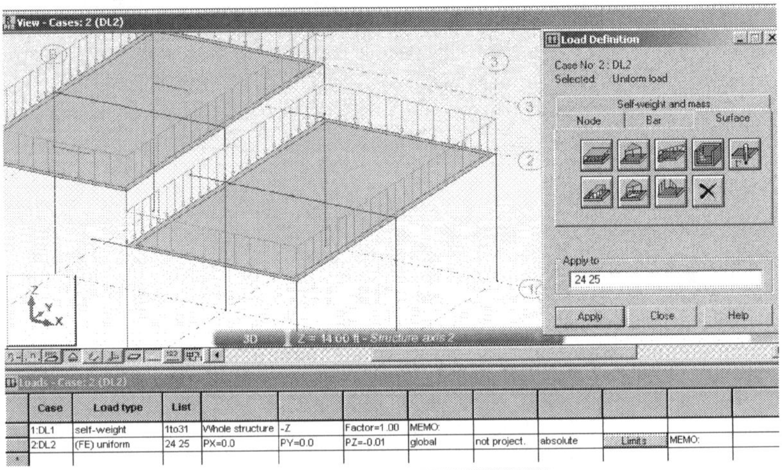

9. Now switch the load case selector to 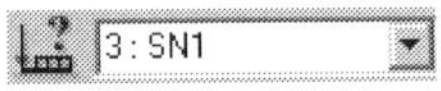

10. On the Load Definition dialog press  Uniform Planar Load on Contour.
11. In the Uniform Planar Load on Contour dialog enter -0.04 for Z then expand the "Contour Definition" section:

a. 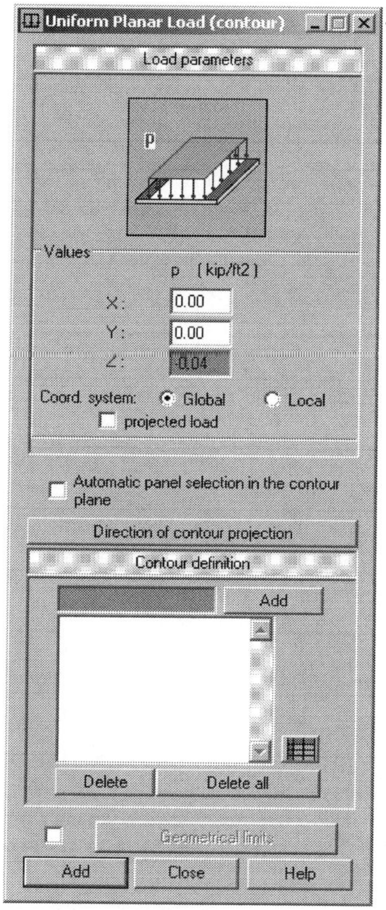

b. First toggle on Node Numbers display and toggle off Members Numbers. Place your mouse in the edit control to the left of the "add" button and then locate nodes 4, 19, 23, and 25. Click each one in turn in a clockwise or counterclockwise fashion.

c. 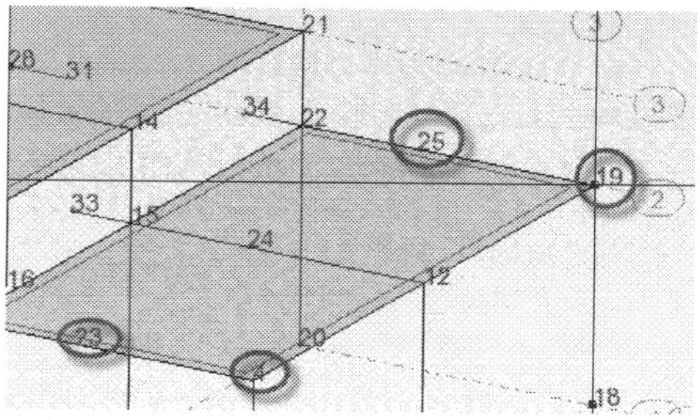

d. Your "Contour Definition" section will now look similar to this:

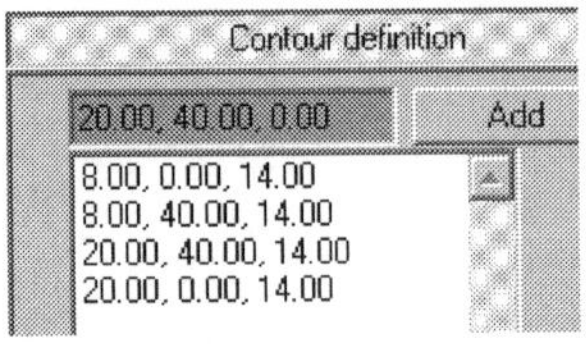

e.

f. Make sure that "Automatic panel selection in the contour plane" is selected: 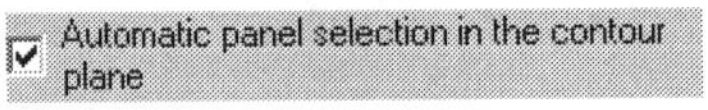

g. Next press "Add" at the bottom of the entire dialog

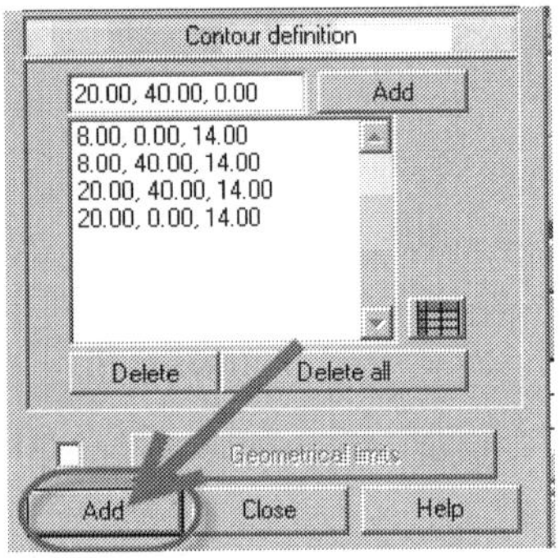

h.

12. Instead of queuing the load for application the "Automatic panel selection in the contour plane" will detect panels which are coplanar with the contour definition. Your project should look like this:

a. 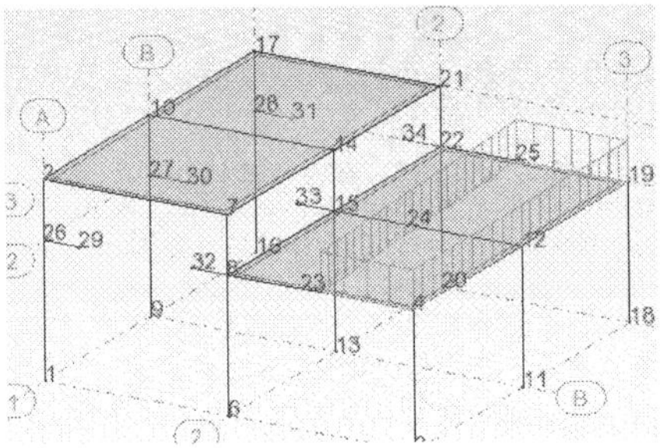

13. Next we'll add the drift load with the "Planar load 3p on contour" tool: . This looks very similar to the Planar load 3p but is distinctly different. Make sure you click the "Planar load 3p on contour" and not just "Planar load 3p" button which right above.

14. In the upper part of the "Planar load 3p (Contour)" dialog start by filling in values for the 3 points we will define. For P1, in the Z direction, enter -0.04, for

P2 enter -0.04, for P3 enter -0.08. Then we will select the coordinates of the points, click in the edit control next to A and select in the project Node 23 for A, Node 25 for B, and Node 15, 22, or 8 for C. (We're defining the plan locations for the magnitudes to define a sloped plane of the drift load). Your dialog should look like this if you chose node 15 for C:

a. 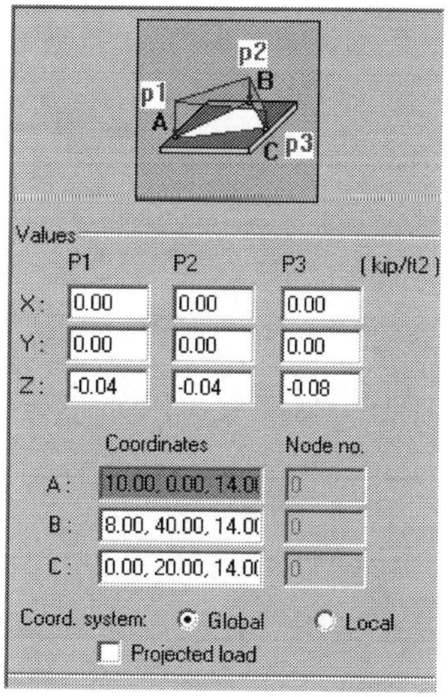

b. Next we'll define the contour so expand the contour definition section and just like before, select nodes 8, 22, 25, 23 which define the boundary of the snow drift.

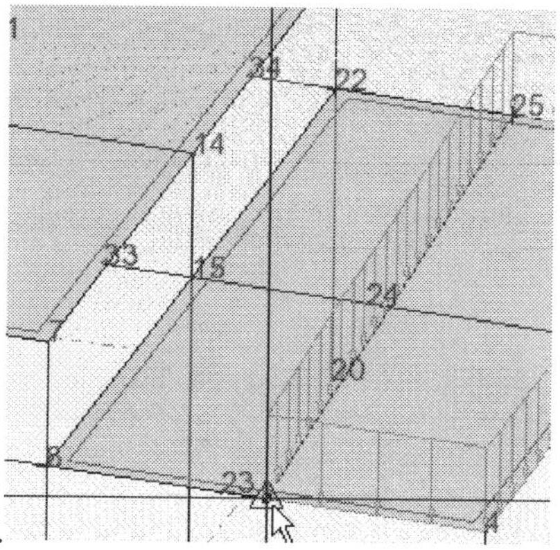

c. Check "Automatic panel selection in the contour plane" and the "Contour definition" section of your dialog should look like this. (Only the numbers in the list matter, the coordinates in the green input control are just where the mouse is currently).

d. 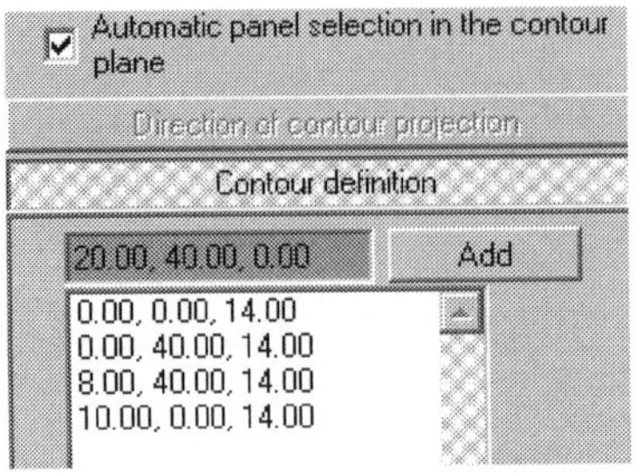

e. Next press "Add" at the very bottom of the dialog (not the one next to the coordinates:

f. 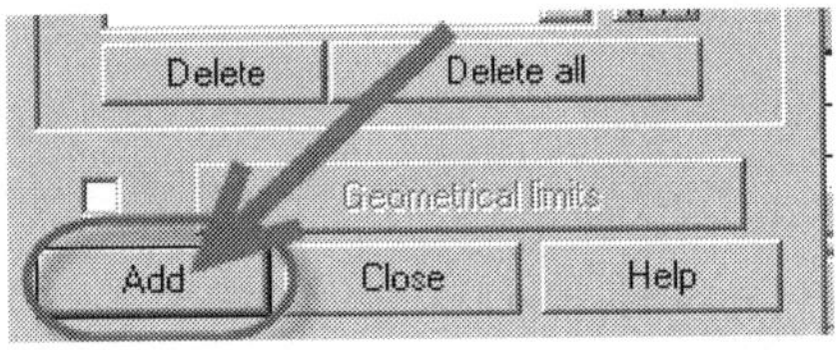

g. You should now have a drift load applied to the low roof:

h. 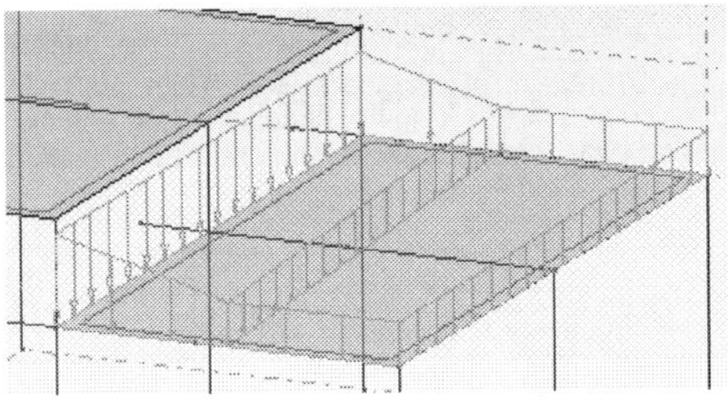

15. Next we'll start adding some wind loads to the columns. We'll start by changing the view to a cross section view of the building. First select on the coordinate axes at the bottom left of the screen: . Then in the "View" dialog press the 2D button, press the XZ button and select 20 grid line B from the dropdown under the 2D button as shown here:

a. 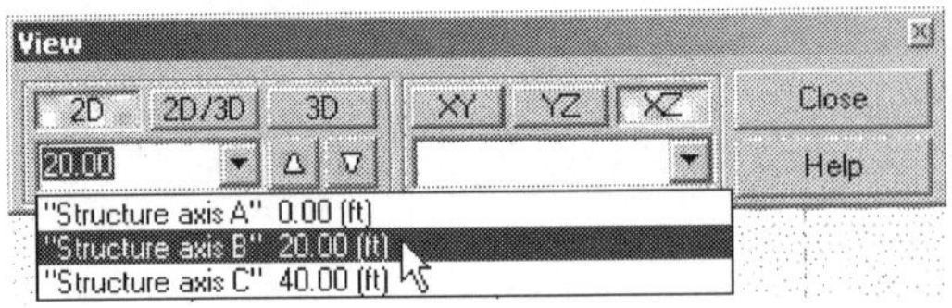

b. Your view should now look like this:

c. 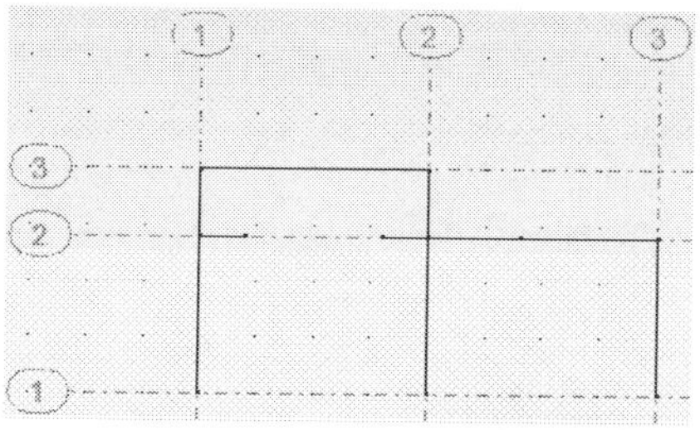

16. Switch to the Wind load case: 

Then select the "Bar" tab on the Load Definition dialog.

17. On the "Bar" tab select the "Uniform Load" button: 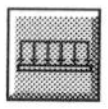 which will bring up the Uniform Load dialog

a. 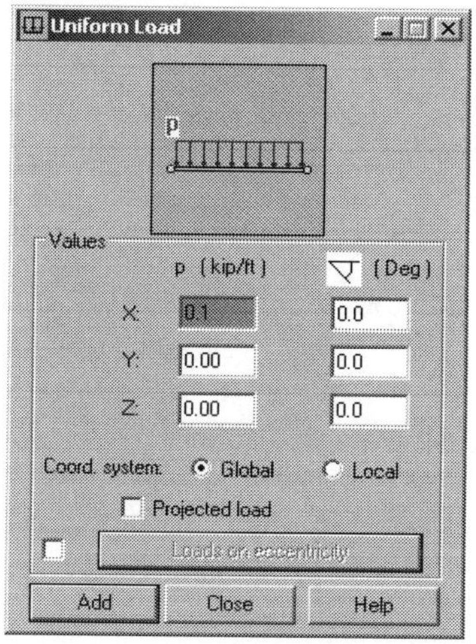

18. Enter 0.1 in the X direction force magnitude and press "Add". Unlike the surface loads on contour, this will only queue the load for application in the Load Definition dialog.
19. Now you will be back in the Load Definition Dialog, simply move your mouse out into the project environment and click elements 10 and 11 to apply the uniform line load to the outside columns. (Toggle on Member Numbers, if you do not currently have them visible in the view.)

a. 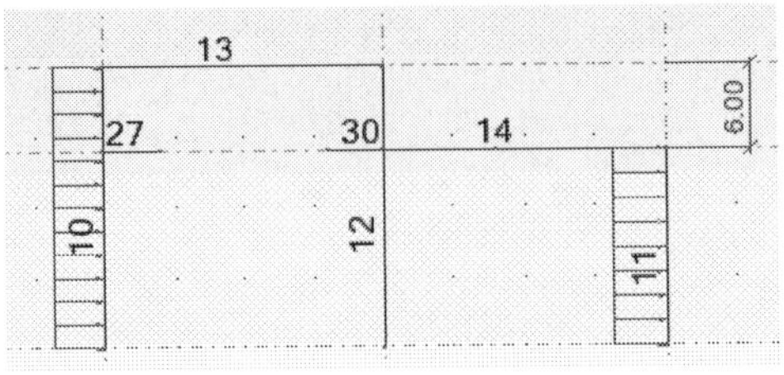

20. Go back to the Load Definition Dialog and this time select the "Trapezoidal Load" button:  which will launch the Trapezoidal Load dialog.
21. In this dialog, for
    a. Type: "Trapezoidal Load (2p)"
    b. Direction of Load Action: Select "X"
    c. Coordinate System: Select "Global"
    d. Values: P1 and P2 enter 0.1
    e. Coordinates: Select "Absolute"
        i. X1: Enter 0.0
        ii. X2: Enter 6.0
22. Next press "Add" which will queue this load for application and return you to the Load Definition Dialog.
23. Move your mouse out into the project environment and hover over element number 12. Notice the direction arrows which are painted on top of the element indicating the direction of the beginning of the element which will be assumed when you click to select the element. If you hover over the lower part of the element these arrows will flip to indicate the opposite sense of the start of the member. This is only for load application and does not affect the defined direction of the member in the database. Once you have the cursor near the top and the direction arrows are as shown below, click to select the member.

a. 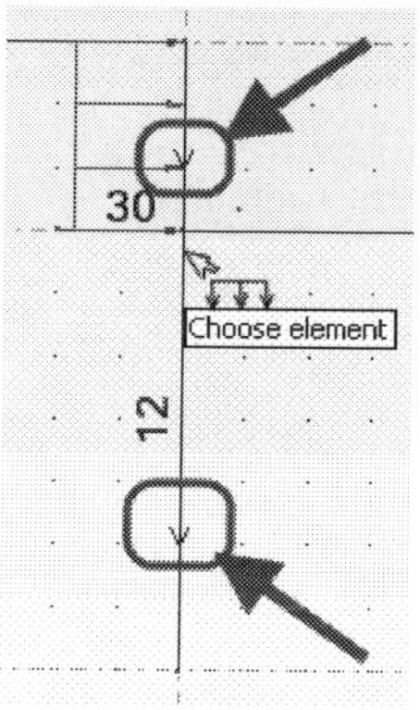

24. Finally, your wind load case should look like this:

a. 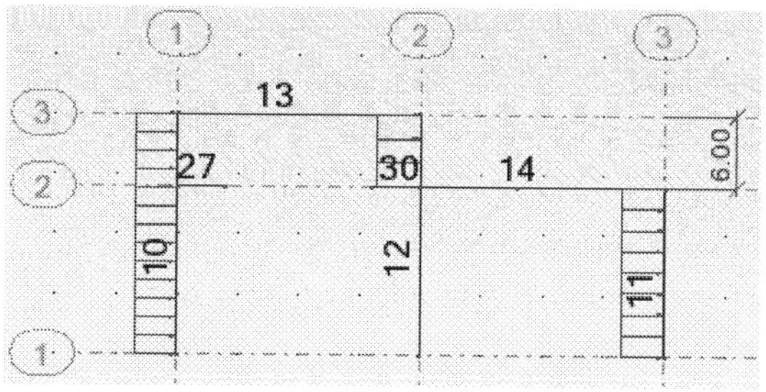

25. Continue working with the loads application dialog to apply nodal loads to the ends of each cantilever.

--End of Exercise—

Wind Load Generator

An exciting new feature in Robot Structural Analysis 2016 which will enable you to take a different look at your wind loading. Better yet, think of using the wind load generator for a wicked quick initial look at lateral wind load on your structure. You could get a very fast approximation of the forces in your lateral system at a very early stage of the project without the pain of calculating static wind loads per the code! Of course you will, ultimately, still need to perform your usual building code wind analysis, manually calculate, and apply code required loading as one set of load cases for your structure: that doesn't change. What does change, however, is the ease with which you can use the wind load simulator to generate some additional load cases which you can check alongside your main code required load cases. This would be especially interesting where you have a somewhat unique structure which doesn't quite fit the requirement for full wind-tunnel analysis but which you might want to gain some additional confidence that you've considered more than just the code requirement. It's also a lot of fun and because it automatically generates the loads and load cases for you, it's almost too easy to not at least try it: so let's take a look.

Configuring Wind Load Generator

Once you have created a model, you can enter the wind loads generator from the loads menu: **Loads>Wind Loads Simulation>Wind Load Generator...** to bring up the Wind Simulation dialog:

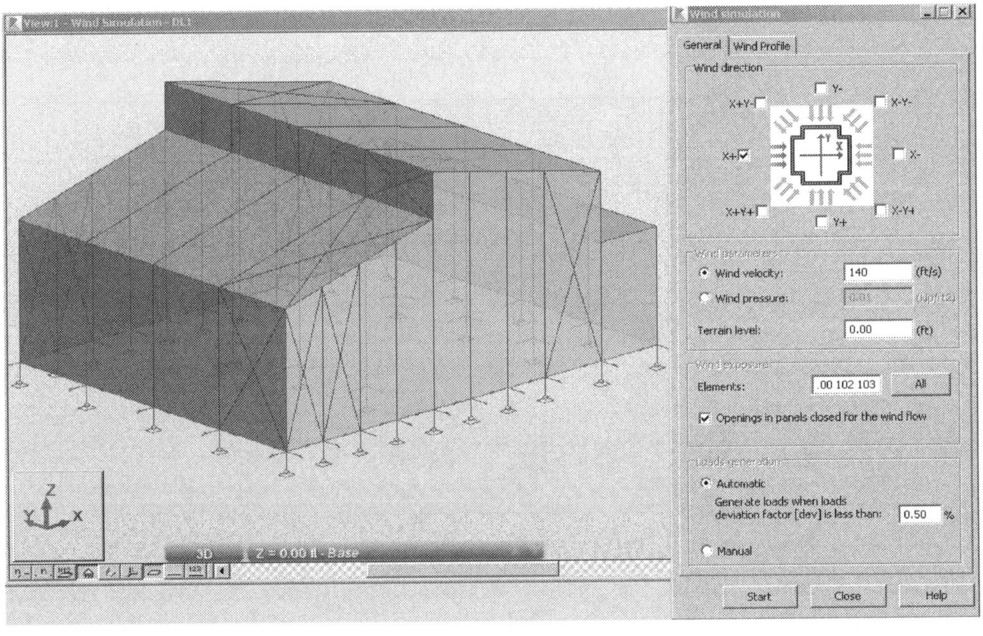

I have a simple yet interesting structure modeled and Robot is showing me the Wind Simulation dialog. The wind load generator works on individual structural elements (beams, columns, panels, floors, and claddings). The simulator will approximate elements as having a thickness even if there is not one assigned. Robot uses the cross section of structural elements approximated as prismatic solid members and uses the thickness of the floor or panel elements assigned. If you use claddings as I have below, you will find that they also gain some sense of thickness for the simulation.

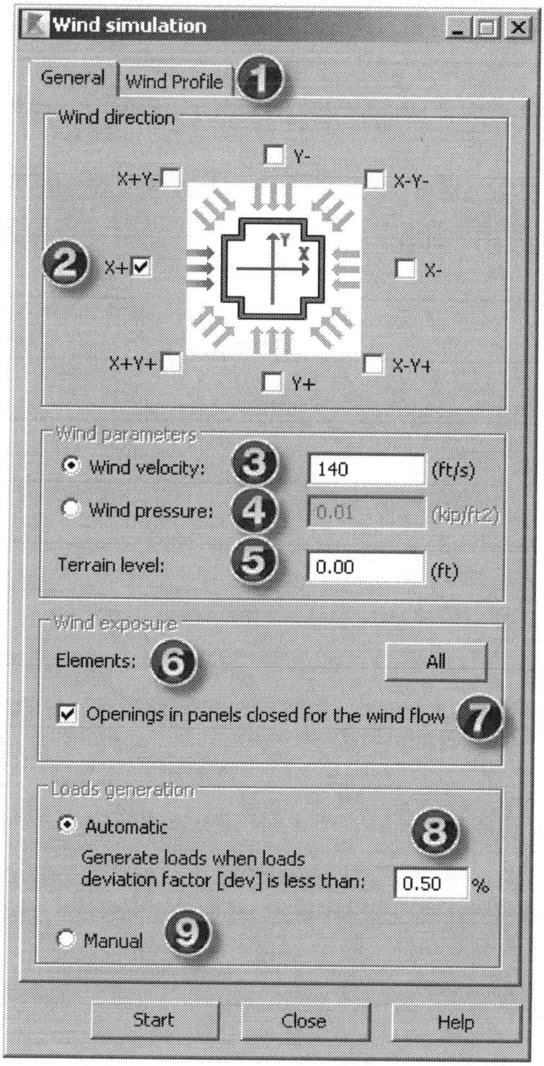

1. General/Wind Profile tab: General settings will be described here. Wind profile settings offer you a chance to adjust the wind profile along the vertical axis of the structure and are described below.

2. Wind Direction: Select the wind directions for which you want to have load cases (and loads) generated. X and Y pertain to the project X and Y axes. Only these selected directions will have loads generated. Each direction selected will result in a new load case. There is no control over the exact angle of the wind to the model, only parallel to the X/Y axes and quartering at 45 degrees to the project XY.

If you have a particular direction, consider rotating your model relative to the project XY axes.

3. Wind velocity: Note that this is in ft/s not miles per hour which you might expect. You will need to convert your applied velocity to ft/s for the simulation. (or m/s if you speak fluent SI)

4. Wind pressure: You can alternatively choose a base wind pressure for the calculation.

5. Terrain level: This allows you to configure the level of ground elevation relative to your model. Specify the Z coordinate of ground level relative to your model and no wind will be generated below this level.

6. Wind Exposure – Elements: Select which elements are going to participate in your simulation. If your building uses cladding or shell elements for the main wind-load application surface, you would select only these elements. If you have beams/columns or other structural elements actually exposed to the wind, you would select those elements. Select the elements that will receive direct wind loading as opposed to the elements which will support the elements receiving direct wind loading. The "All" button is fairly useless in this situation and you'll want to choose the elements to expose to the wind load generation carefully. Use the Selection Dialog to make quick work of selecting claddings or shell elements for wind loading. NOTE: only the elements selected here will be considered in the simulation so if you model "up-wind" structures to understand the contributions to loads on your building, you will need to include the panels from upwind buildings as well.

7. Wind exposure-Openings in panels closed for the wind flow: Robot wind simulation can ignore openings in panels when running the wind load simulation.

There is no optional control for this on a per panel basis, only for the entire model. This can be an interesting way to consider the effects of opening percentage on the structure. Consider using phasing to control which panels have openings at a particular point of interest for your analysis.

8. Load generation – Automatic: Wind load simulation proceeds and measures convergence at each step of the simulation. You can use this option to automatically generate wind loads on elements selected in "Wind Exposure – Elements" configured above. When the simulation convergence reaches the value specified here, the simulation is stopped and loads are automatically generated on the elements. While the simulation is running, you can always manually cause loads to be generated by pressing the "Generate loads now" button on the simulation progress dialog.

9. Load Generation – Manual: Simulation will continue running until the user presses "Stop" or "Generate loads now". You can use the convergence process and Resultant forces information to help inform your decision about when to generate loads.

Next we'll take a look at the Wind Profile tab of the Wind Simulation Dialog. The wind profile tab is a very cool new way of interacting with Robot which is very intuitive and I hope represents a new direction for smoothness and fluidity in interaction for the product. This tab will allow us to configure the wind profile up the building. We can add control points just by clicking on the profile! Once created, simply drag to adjust the factor applied to the load at any point along the vertical profile. There is no way to remove individual control points but the reset button can be used to clear any customization of the wind profile and allow you to start from the beginning. You'll find that this dialog is pleasantly intuitive to use, allowing you to click to add a point and simply hover over and drag existing points to change the value of the multiplier.

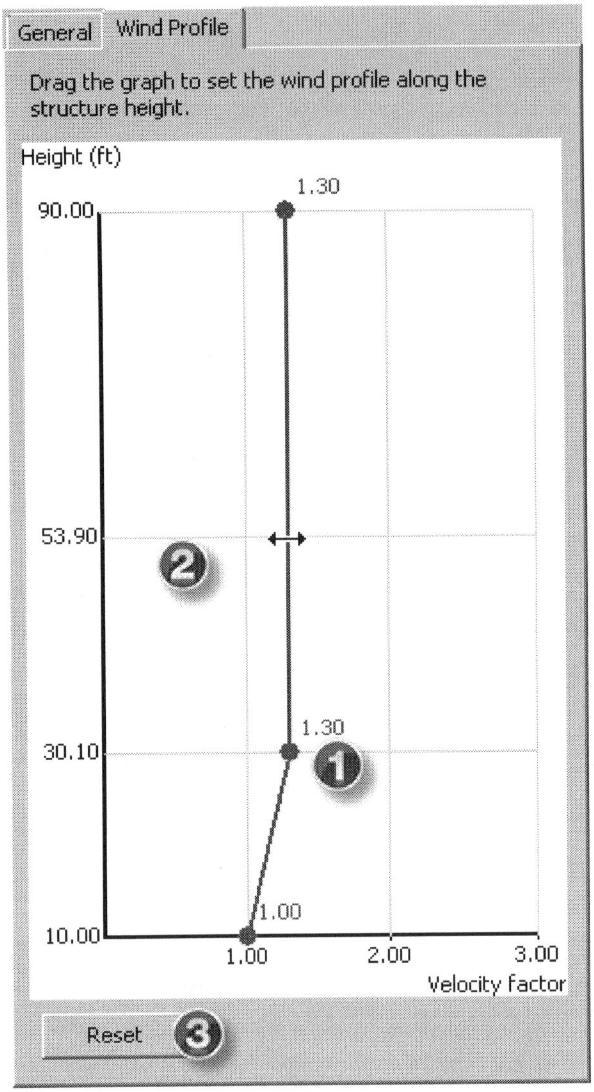

1. Control Point: With either the initial control points at the top and bottom, or with an added control point, hover your mouse over the point and drag left/right to decrease/increase the value of the multiplier respectively.

2. Add a new control point: Hover over a location on the profile line and the dialog offers an indication of the elevation. Click once to place a new control point and then adjust the point as in #1 above.

3. Reset: There is no way to remove a point once it is added. You may simply set the point to the same value of multiplier as other points above/below, or you can use reset to refresh this to a clean state where you can re-apply your customizations. Note that the profile does not necessarily update when the "Reset" button is pressed but only after you move your mouse into the dialog.

Running Wind Simulation

During the simulation, Robot will show you a pressure map on the elements you have selected to analyze and will continuously update it as the simulation progresses. You can orbit the model and watch the simulation take shape as the calculations progress. You will also see a wind direction indicator for quick identification of the simulation direction. It's a bit mesmerizing to watch at first.

Computational Fluid Dynamics (CFD) simulations are an iterative and dynamic process which do not "finish" but "reach convergence" or "reach a steady flow state". Imagine a real wind tunnel in operation, the pressures on the object in the tunnel are only a snapshot in time of the pressures because they tend to fluctuate with turbulence. Allowing the model to reach a steady state is very important to getting results that are reasonable and usable. Robot effectively creates a virtual wind tunnel around your model (the height of which is evidenced in the wind profile tab and the box in the 3D view of the wind simulation) and the simulation runs continuously until a convergence criteria is reached or the user manually stops the analysis. You can watch the convergence progress of the simulation in the dialog shown below. .

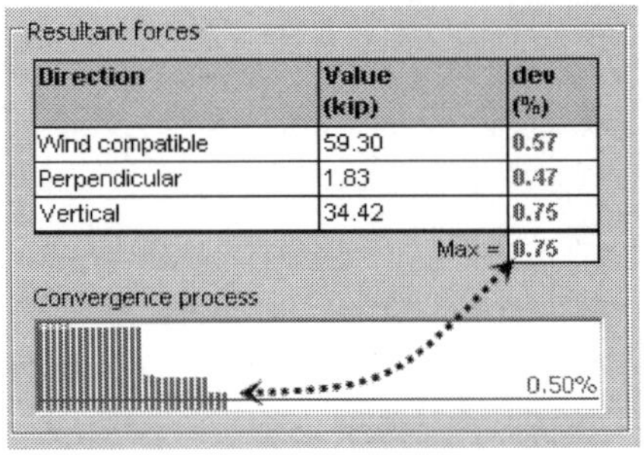

Direction	Value (kip)	dev (%)
Wind compatible	59.30	0.57
Perpendicular	1.83	0.47
Vertical	34.42	0.75
	Max =	0.75

Robot measures three values calculated for the simulation and monitors those values during the simulation

Because the simulation is dynamic, what we are looking for is a "steady" state where the changes in the solution are small and within some tolerance. Robot looks over the past few iterations and determines the maximum deviation in any of the measured values then compares that against the criteria selected.

The suggested value for convergence is 0.5% difference. You can experiment with different values and you can also read more about the simulation in this validation brief found on BIM and Beam blog:

http://bimandbeam.typepad.com/bim_beam/2014/06/autodesk-robot-structural-analysis-professional-wind-simulator-validation-brief-whitepaper.html

It's a very cool white paper and shows good correlation of the simulation with an actual wind tunnel test and thoroughly discusses the effect of simulation duration and convergence tolerance setting on the performance of the model: a great read.

Wind Loads Generation

Once the simulation has converged to within the tolerance specified for automatic generation of loads, or you have pressed "Generate loads now" button, Robot will automatically create a new load case, and apply surface and/or linear loads on the elements selected for that load case. It is important to note that Robot will not assign varying loads on panels, even though the simulation may suggest that the force varies significantly over the panel. Robot takes an average of the pressures on the surface and applies it as a load to the surface. If you want/need finer resolution for wind loading effects, you will need to break up your surface into smaller panels where the average pressure on the panel might be more representative of the pressure in that region of the building. Robot will also create additional linear loads at the edges of panels where there are significant edge conditions. As you can see, this is a very thorough loading:

Case name	Load type	List	Load values
DL1	self-weight	1to65 67to72 74to100 1	PZ Negative Factor=1.00
Wind Simulation X+ 70 ft/s (variable)	(FE) linear on edges	94_EDGE(1)	PZ=0.01(kip/ft) local
Wind Simulation X+ 70 ft/s (variable)	(FE) linear on edges	94_EDGE(4)	PY=0.02(kip/ft) local
Wind Simulation X+ 70 ft/s (variable)	(FE) uniform	94	PZ=0.01(kip/ft2) local
Wind Simulation X+ 70 ft/s (variable)	(FE) linear on edges	95_EDGE(3)	PZ=0.00(kip/ft) local
Wind Simulation X+ 70 ft/s (variable)	(FE) linear on edges	95_EDGE(4)	PY=-0.01(kip/ft) local
Wind Simulation X+ 70 ft/s (variable)	(FE) uniform	95	PZ=0.00(kip/ft2) local
Wind Simulation X+ 70 ft/s (variable)	(FE) linear on edges	96_EDGE(4)	PY=-0.00(kip/ft) local
Wind Simulation X+ 70 ft/s (variable)	(FE) uniform	96	PZ=0.00(kip/ft2) local
Wind Simulation X+ 70 ft/s (variable)	(FE) uniform	97 103	PZ=0.00(kip/ft2) local
Wind Simulation X+ 70 ft/s (variable)	(FE) linear on edges	98_EDGE(2)	PY=-0.00(kip/ft) PZ=0.01(kip/ft) local
Wind Simulation X+ 70 ft/s (variable)	(FE) uniform	98	PZ=0.00(kip/ft2) local
Wind Simulation X+ 70 ft/s (variable)	(FE) linear on edges	99_EDGE(1)	PZ=0.01(kip/ft) local
Wind Simulation X+ 70 ft/s (variable)	(FE) linear on edges	99_EDGE(3)	PZ=-0.00(kip/ft) local
Wind Simulation X+ 70 ft/s (variable)	(FE) uniform	99	PZ=0.01(kip/ft2) local
Wind Simulation X+ 70 ft/s (variable)	(FE) linear on edges	100_EDGE(2)	PY=0.00(kip/ft) PZ=-0.01(kip/ft) local
Wind Simulation X+ 70 ft/s (variable)	(FE) linear on edges	100_EDGE(3)	PY=-0.00(kip/ft) PZ=-0.00(kip/ft) local
Wind Simulation X+ 70 ft/s (variable)	(FE) linear on edges	100_EDGE(4)	PY=0.00(kip/ft) PZ=0.00(kip/ft) local
Wind Simulation X+ 70 ft/s (variable)	(FE) uniform	100	PZ=-0.00(kip/ft2) local
Wind Simulation X+ 70 ft/s (variable)	(FE) linear on edges	103_EDGE(1)	PY=-0.01(kip/ft) local
Wind Simulation X+ 70 ft/s (variable)	(FE) linear on edges	103_EDGE(2)	PY=0.00(kip/ft) local
Wind Simulation X+ 70 ft/s (variable)	(FE) linear on edges	103_EDGE(5)	PY=0.01(kip/ft) local
Wind Simulation X+ 70 ft/s (variable)	(FE) linear on edges	103_EDGE(6)	PY=-0.01(kip/ft) local
Wind Simulation X+ 70 ft/s (variable)	(FE) linear on edges	103_EDGE(7)	PY=0.01(kip/ft) local
Wind Simulation X+ 70 ft/s (variable)	(FE) linear on edges	103_EDGE(8)	PY=-0.01(kip/ft) local
Wind Simulation X+ 70 ft/s (variable)	(FE) linear on edges	102_EDGE(1)	PY=-0.01(kip/ft) local
Wind Simulation X+ 70 ft/s (variable)	(FE) linear on edges	102_EDGE(2)	PY=-0.01(kip/ft) local
Wind Simulation X+ 70 ft/s (variable)	(FE) linear on edges	102_EDGE(3)	PY=-0.01(kip/ft) local
Wind Simulation X+ 70 ft/s (variable)	(FE) linear on edges	102_EDGE(7)	PY=0.01(kip/ft) local
Wind Simulation X+ 70 ft/s (variable)	(FE) linear on edges	102_EDGE(8)	PY=-0.00(kip/ft) local
Wind Simulation X+ 70 ft/s (variable)	(FE) uniform	102	PZ=-0.00(kip/ft2) local

Remember that the goal of wind load simulation is to give you a fun, informative, easy-to-use tool for quickly looking at wind effects on your building and enabling you to get a very rapid early-stage approximation of loading on your lateral system. With Revit and Robot together, you can make much more informed decisions much earlier in the project without a lot of up-front effort. Have fun with this tool, it's crazy to see the effects on your own structures!

Remember to set up your wind load cases as groups exclusively or'd together or create relations between them which exclusively OR them together so that you can include wind simulation as well as your code required wind load analysis in your design. More on load combinations in chapter 5!

Chapter 5 - Preliminary Structural Analysis

At this point we have discussed all of the major components of creating a basic model including defining bars, supports, creating load cases and applying loads to these elements. We could simply go ahead and start running some calculations as Robot will consider each load case and perform calculations for each. We could then begin exploring the results for each load case but before we do, we'll take a few pages to discuss load combinations and how they are managed in Robot.

Setting up Load Combinations

Robot has three main methods of creating and managing load combinations. The first is to manually create each one, the second is to use an automated "composed" combination management and the last is to use automation tools to automatically generate individual manual load cases. The middle one (automatic composed combinations) is particularly interesting as Robot has a very unique method for generating and managing linear combinations. All load combinations are generated as "components" of one main combination effectively collapsing them down to a single load combination called "ULS" or "SLS" (Ultimate Load State or Service Load State). . This allows you to deal with massive amounts of load combinations as though they were only one combination even though you can still drill in and look at the effects of each component. As long as you are only considering linear static load cases (non P-Delta), you can take advantage of this really cool functionality for dealing with huge numbers of load combinations; however, if you need to consider

P-Delta or non-linearity, then you will need to generate individual load combinations either manually or by automatic generation. The reason being that static linear load cases can, using the principle of super-position, be superposed to determine the effects of the load combination on the structure. Non-linear or P-delta cases, however, must be considered each in turn as it participates in the load combination. I think a significant amount of confusion arises simply out of the fact that Robot has some awesome capabilities for dealing with the sheer number of load combinations in a very facile manner in contrast to the fact that any p-delta analysis cannot take advantage of this awesome functionality…

First we will discuss creating manual load combinations which is great for considering just a few combinations, then we will discuss the automated ULS and SLS load combinations and working with the components. Finally we'll discuss automatically generating manual combinations which can be used with non-linear and p-delta analyses.

Manual Load Combinations

Manually creating combinations can be tedious and time-consuming. If you only have a few combinations to manage or you simply need to adjust the parameters of a previously generated manual combination, launch manual combination definition from **Loads>Manual Combinations...** If you have not previously created any load combination, you will be presented with the Combination Definition/Modification Dialog:

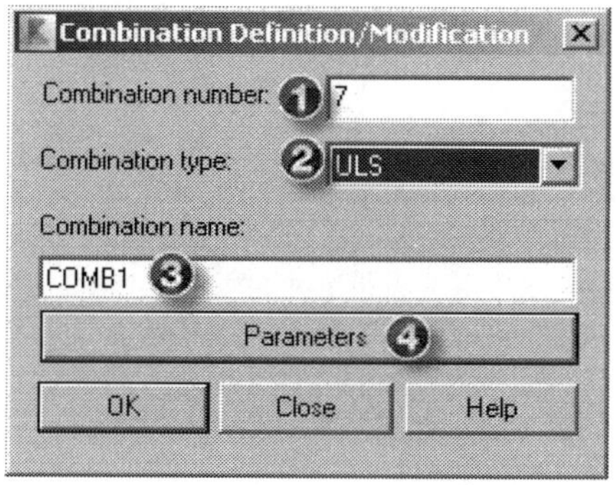

1- Combination Number: Robot will offer the next free number in the list of combinations. You may change this if you wish but it is usually best to allow Robot to manage this number.

2- Combination Type: Select the type of combination.

a. ULS = Ultimate Load State. A super-grouping of load combinations for which you intend to apply load factors and will be used in design and member verification routines for material strength

b. SLS – Service Load State: A super-grouping of load combinations for which you intend to combine load effects for the purpose of evaluating the serviceability of the structure.

c. ACC – Accidental Combination: These are more commonly used in European codes where accidental loading scenarios are more commonly considered in design codes.

3- Name of the load combination

4- Parameters opens the Combination Parameters dialog:

a. 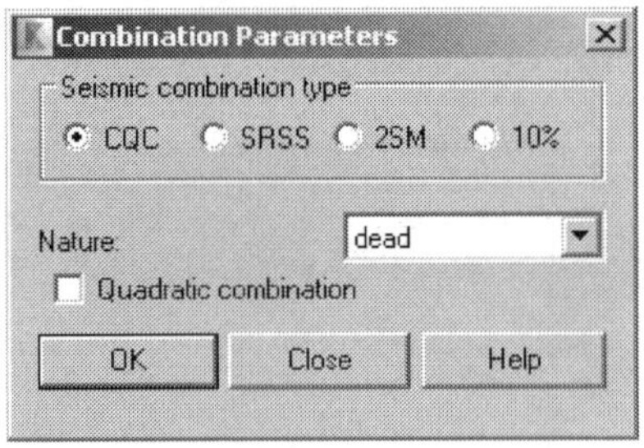

b. Seismic combination type is a specification for managing the modal contributions for seismic loads.

c. Nature allows you to configure the nature of the load combination which can be used to combine this load combination as though it were a load case when automatically generating manual combinations.

d. Quadratic combination will use the Square Root of the Sum of the Squares to determine the effects on the structure. $Rcomb = \sqrt{R_1^2 + R_2^2 + \cdots + R_n^2}$

After Configuring parameters and the name of your new load combination you will enter the Combinations Dialog:

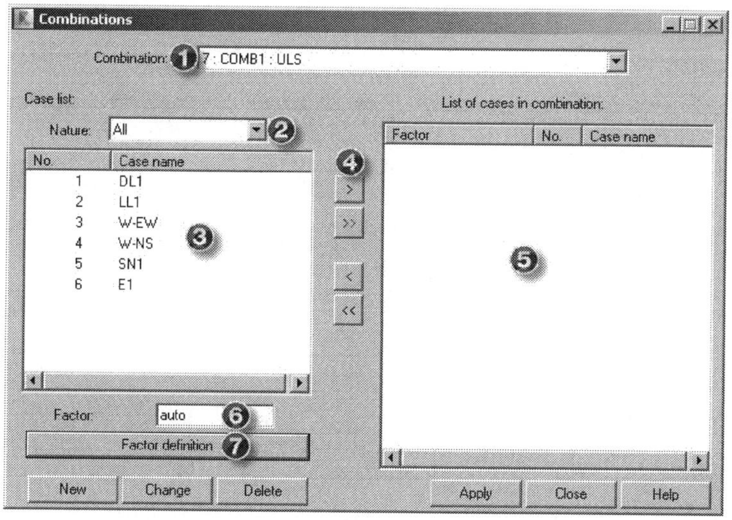

1- Combination Selector: If you are creating a new combination the name you provided in the Combination Definition/Modification Dialog above will appear here. If you want to edit a different combination you can select from the list of previously defined combinations.

2- Nature filter: Use this to filter the case list below (#3) by nature. If you have dozens of cases this can be quite useful although in the example here there is not much use (unless you have forgotten which cases are which nature or you want to double check that your cases are defined properly)

3- Case List: This will contain not only all defined load cases but also all previously defined load combinations. All cases and combinations in this list may be used to build load combinations.

4- Add/Remove buttons: After selecting a load case or combination you wish to add (or remove) to (from) the current combination use these buttons to move it into or out of the combination case-list on the right. (check the factor #6 before you add it to the list)

5- Participating cases and factors list: Here all cases and combinations currently participating in this combination (this combination is in the dropdown list #1 above) and their factors are listed here.

6- Factor control: When you select a case or combination you would like to add to the current combination, you can specify the factor which should be applied to the case/combination. Set this before pressing the "Add to combination" buttons (#4). If you leave this set to "Auto" which is the default, the factor will be applied per the nature. You can view and modify automatic factors in #7 below.

7- Automated factors definition. One factor can be specified for each nature to be used automatically. This is of somewhat limited functionality as factors for each nature vary considerably with each different type of combination. It seems best to simply type in the factor manually for each case. It is tedious but probably no more tedious than manually creating your load combinations in the first place. Comparatively speaking the automatic methods below will seem much more efficient.

You can use "Apply" and "Close" to finish this combination or use the "New" button on the left to create a new combination. "Change" will launch the Combination Definition/Modification dialog to allow you to adjust the parameters and combination name.

How Robot does Automatic Code Combination Generation

Here is the general workflow for creating load combinations automatically in Robot Structural Analysis:

1. Start the Automatic Combinations dialog from the loads menu: **Loads>Automatic Combinations**

a. 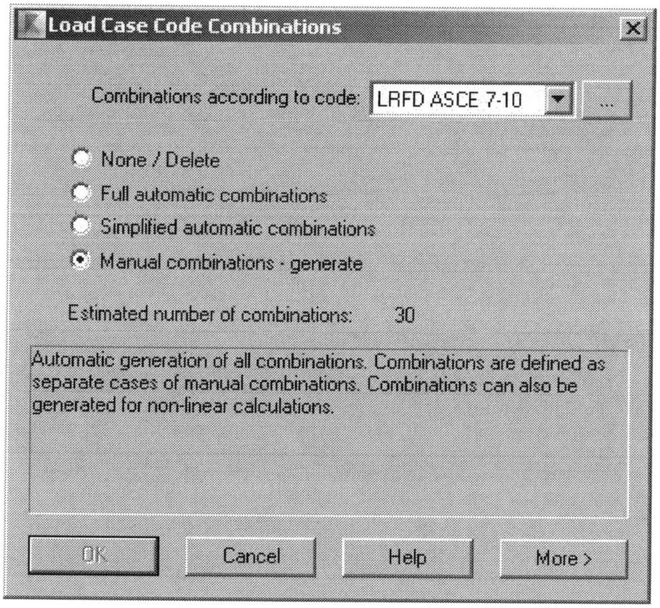

2. Verify the selection of code combination regulation in "Combinations according to code:" dropdown. Make a different selection if it is not correct or exit this dialog and adjust settings in Job Preferences from the Tools menu: **Tools>Job Preferences...**>Design Codes

3. Select Full Automatic or Generate Manual (see below for more information)

4. Choose "More" and check Cases, Combinations, Groups and Relations (see below for more information)

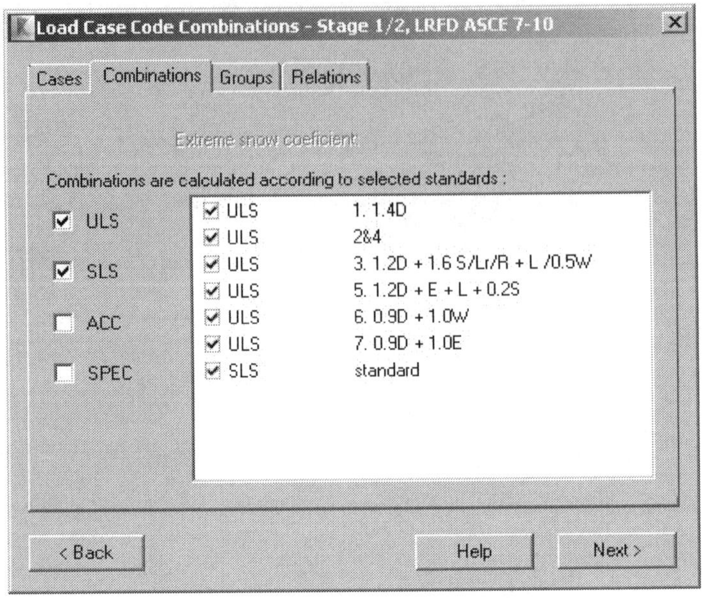

a.

5. If automatically generating manual combinations, choose "Next" and review the list of combinations Robot proposes to generate:

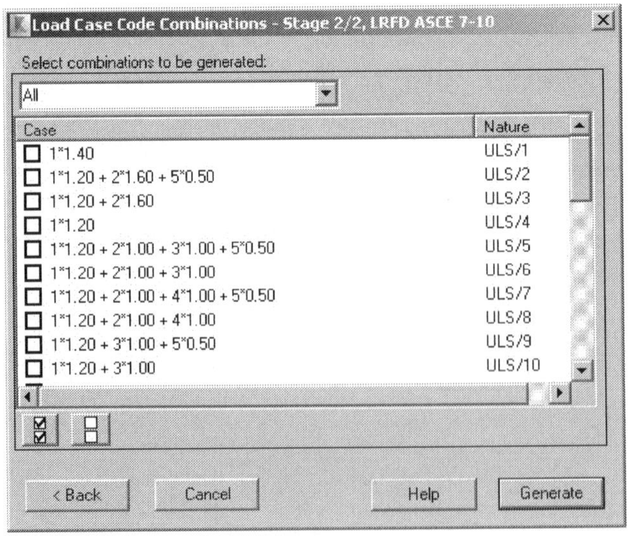

a.

6. Select all combinations you wish to generate or select all or none with the buttons below the checklist.

7. Finally choose "Generate" in the last dialog.

THE LOAD CASE COMBINATIONS DIALOG:

Robot uses a customizable prescription for generating load combinations which can be further controlled by selection of participating cases and through the use of case groups and case relations. This provides very powerful control over the exact results of the combination generation allowing you to create combinations for many many different loading situations in a flexible manner.

First let's look at the basic combination mechanism and then we'll look at case selection, groups and relations. From the Loads menu select **Loads>Automatic Combinations...** you will be presented with the Load Case Code Combinations Dialog:

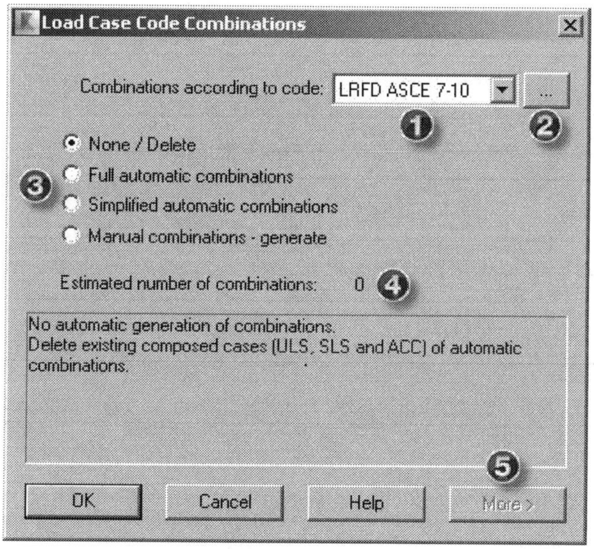

1- Governing code selector: This dropdown contains the list of codes you have configured for your project under **Tools>Job Preferences...**>Design Codes. See Essential Settings above for more information about how to add other codes to this list.

2- Code Combinations Regulation editor: This launches an editor where you can modify the behavior of the specified code combination. See below for more information.

3- Generation Option:

   a. None/Delete: Remove generated composed cases (ULS/SLS/ACC)

   b. Full Automatic: Generate composed cases (ULS/SLS/ACC). All combinations will be generated as "components" of the three combination types (ULS, SLS, ACC). These combinations cannot be used with non-linear or p-delta analyses. Use "Manual Combinations – Generate" for non-linear and/or p-delta analyses.

   c. Simplified Automatic Combinations: Allows you to generate only combinations which produce maximums in various member forces, deflections, and/or reactions. This option is not covered in this text.

   d. Manual Combinations – Generate: Automatically generate "manual" combinations. Does not produce composed cases and results in a significant number of combinations but, can be used with p-delta analysis.

4- Estimated number of Combinations: Based on the current settings this is an estimation of the number of combinations that will be created.

5- More> Access advanced configuration options to control which cases will participate in the combinations, which combinations will be generated and to create and manage groups and relations between cases.

CODE REGULATION EDITOR:

The settings for typical code combination have already been prepared for you and you will not find much need to adjust these settings. However, I wanted you to be aware of their origin. If you press the ellipsis button next to the selected code:

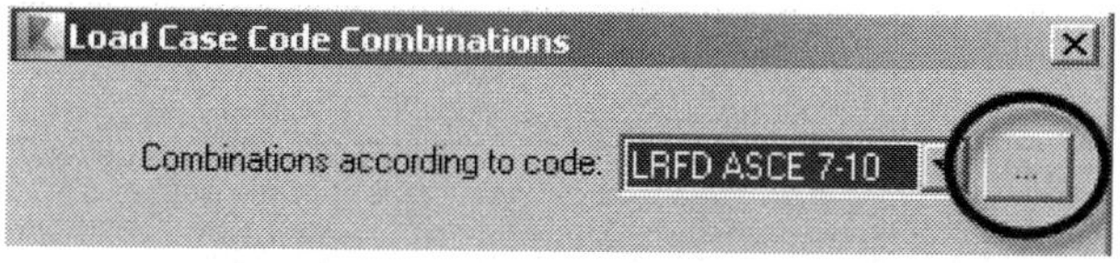

This will launch the code regulation editor:

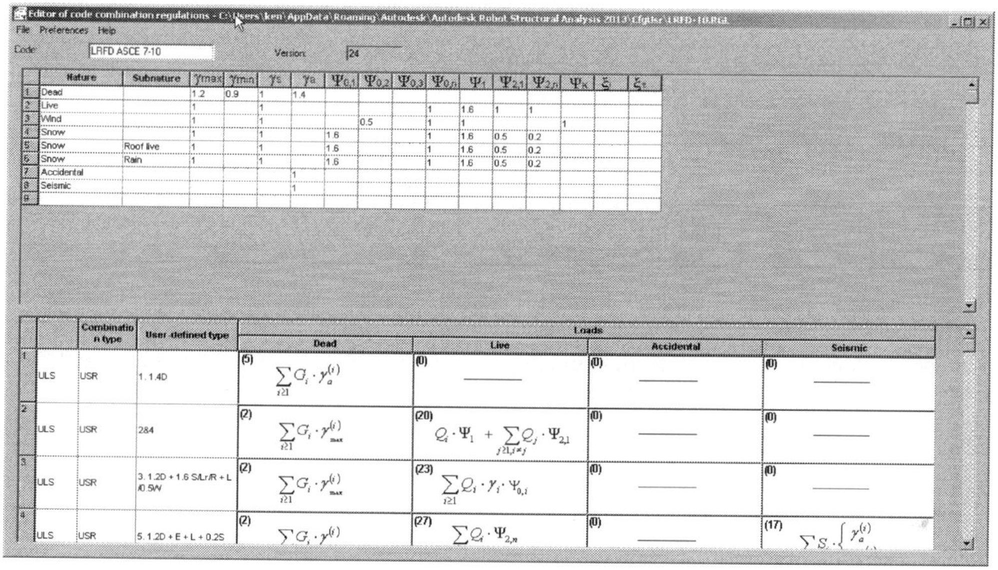

This combination regulation editor was developed with Eurocode combination as the basis. Typically the Eurocodes operate on a principle of combining all variable loads (referred to generically by the letter Q), called variable actions, in one large group meaning that wind, live, snow are all considered "variable actions" and are simply treated in turn in the combination. Dead loads (referred to generically by the letter G) are considered "permanent actions" and are treated separately. This is why there are only 4 columns: Dead, Live, Accidental and Seismic. Anyone familiar with Eurocode style load combination will immediately recognize the summation equations under the action columns as well as the factors Gamma(γ) and Psi (Ψ) and Xi(ξ). This system has been adapted for use with the US codes (LRFD, ACI, ASCE-7, etc.) So while you will see the familiar factors of 1.2, 1.4, 1.6 for dead and live loads in the factors section, they will be under the Eurocode style Gamma (γ) and Psi (Ψ) factor columns.

At the bottom of this regulation editor, the first column indicates which group the combination falls in (ULS, SRS, ACC or SEIS). You will see these as super-groupings in the next dialog. The User-defined type column is the name of the combination that we'll also see in the next dialog. Then under each loading type column is the summation equation which will be applied to the load cases based on their natures to algorithmically create the load combination.

We will dig into this editor in more depth in an advanced text. For now, just be aware that it is here and its job is to provide the definitions of the combinations. You can edit this information but, make sure you make a copy of any regulation files you modify as there is no undo functionality for these edits.

INTO THE LAND OF MORE> (GROUPS AND RELATIONS)

At first you may be thrown for a loop by the advanced capabilities of Robot load combination generation and then in the end, you will likely be amazed at the power of the combinations generator.

With either the Full Automatic Combinations or Manual Combinations – Generate selected, press the "More>" button at the bottom of the Load Case Code Combinations dialog:

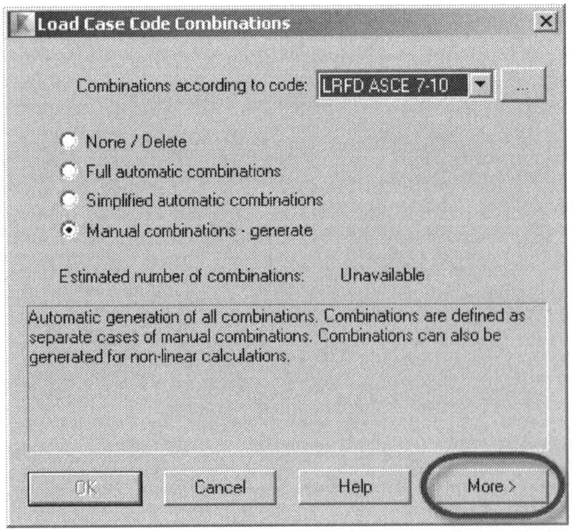

This will take you to the first stage of configuration options for load combination generation where you can select which load cases will participate in load combination generation, which combinations defined in the regulation you wish to generate, the ability to create groups of load cases and finally a place to define relationships between those load case groups.

Here is the Stage 1 part of the dialog:

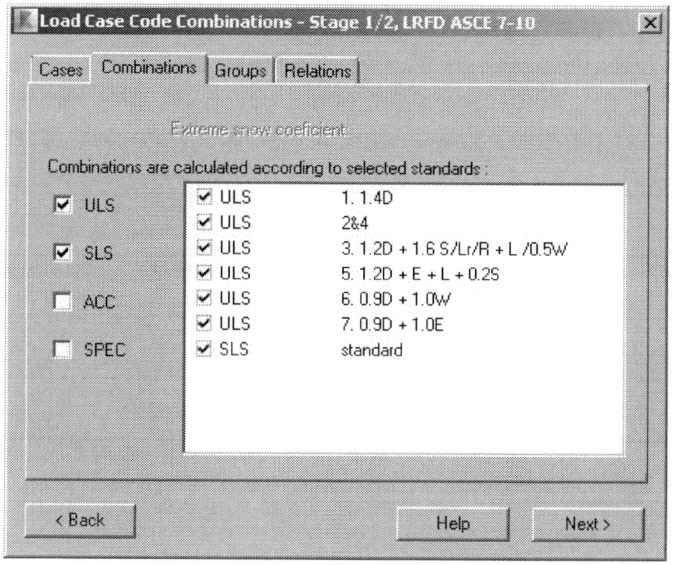

It starts on the Combinations tab but contains 4 tabs: Cases, Combinations, Groups, and Relations.

The Cases Tab:

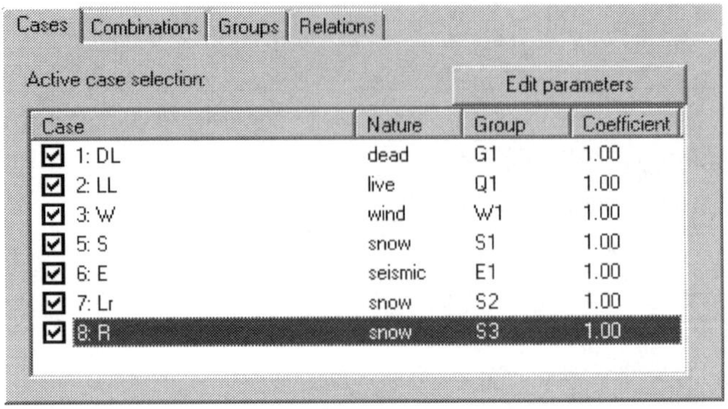

Here you may select which load cases defined in your project will participate in generation of load combinations. You can also review the defined nature of each load case (**Loads>Load Types...**) as well as the current group defined for each. Note: You will have control over how they are grouped but not the naming convention of the group. In the last column you will see the current coefficient. You may apply a coefficient to particular load cases by using the "Edit Parameters" dialog to adjust the coefficient for individual load cases:

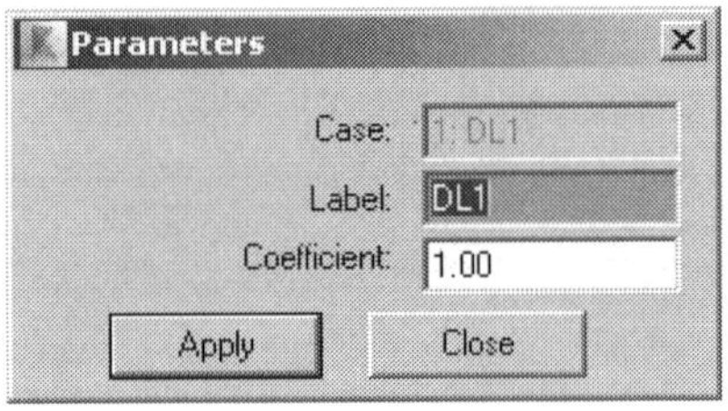

The Combination Tab:

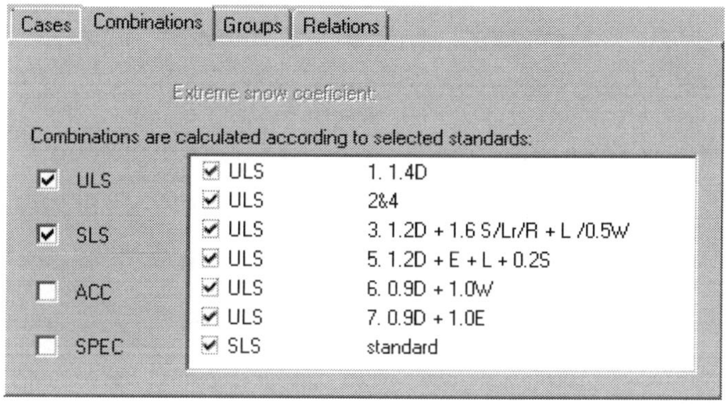

On the left hand side, these are the super-groups I mentioned previously in the regulations editor. Checking or unchecking will check or uncheck all of the corresponding combinations in the list to the right. In the ASCE7 code combinations there are no ACC or SPEC combinations so there is nothing for ACC and SPEC checkboxes to do. You can think of the checkboxes at the left as short-cuts to checking or unchecking several load combination definitions at once. On the right hand side, you will see checkboxes allowing you control which defined combinations will be generated. These are not the final combinations: They only allow you to select which combination algorithms will be used. The actual number of individual combinations will likely be much, much larger than this. That is what it means by "Combinations are calculated according to selected standards" at the top of this list. These each correspond to a line in the regulations editor we briefly touched on above:

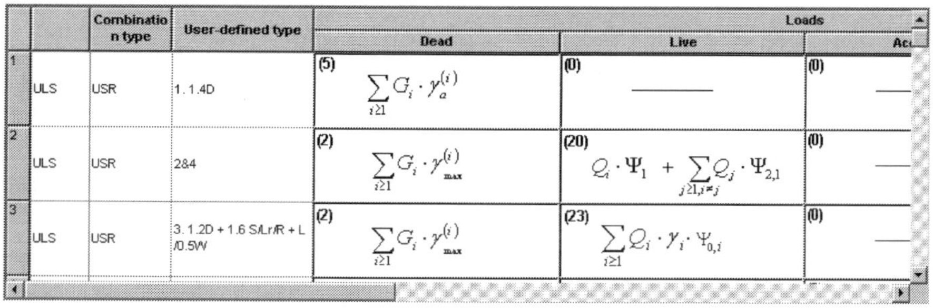

		Combination type	User-defined type	Loads: Dead	Live	Ac…
1	ULS	USR	1. 1.4D	(5) $\sum_{i\ge1} G_i \cdot \gamma_a^{(i)}$	(0) ———	(0) ———
2	ULS	USR	2&4	(2) $\sum_{i\ge1} G_i \cdot \gamma_{max}^{(i)}$	(20) $Q_i \cdot \Psi_1 + \sum_{j\ge1, i\ne j} Q_j \cdot \Psi_{2,1}$	(0) ———
3	ULS	USR	3. 1.2D + 1.6 S/Lr/R + L /0.5W	(2) $\sum_{i\ge1} G_i \cdot \gamma_{max}^{(i)}$	(23) $\sum_{i\ge1} Q_i \cdot \gamma_i \cdot \Psi_{0,i}$	(0) ———

These are the "formulas" or "algorithms" by which the load combinations will be generated. You use the combinations tab to select which algorithms you would like to be employed in the generation of the final load combinations. You would typically leave all selected.

The Groups Tab:

NOTE: FOR MOST NORMAL SITUATIONS BOTH THE GROUPS AND RELATIONS TABS WILL ALREADY BE FILLED OUT FOR YOU. FOR ANY LOAD COMBINATION GENERATION, EACH CASE MUST BE ASSIGNED TO A GROUP AND EACH GROUP MUST PARTICIPATE IN A RELATION (EVEN IT IF ONLY "RELATES" TO ITSELF). NOT HAVING GROUPS CREATED WILL LEAD TO AN ERROR WHEREAS NOT HAVING RELATIONS CREATED WILL SIMPLY LEAD TO NO LOAD COMBINATIONS BEING GENERATED.

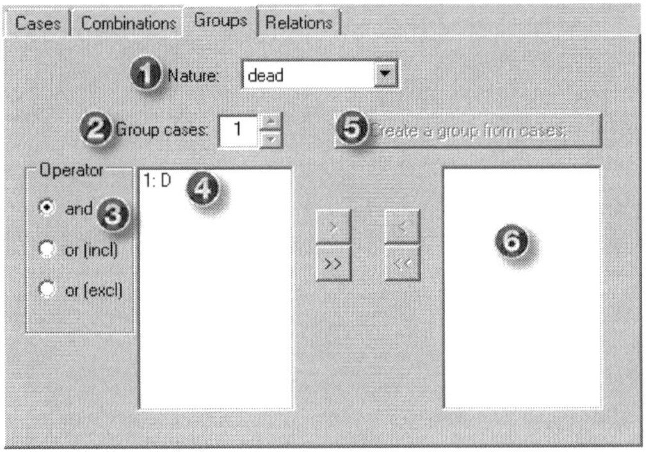

1- Nature selector: Groups are configured on a per-nature basis. Use this selector to choose the nature for which you would like to create or modify the group assignments.

2- Group Cases: Groups are named with an indicator and a number. Dead load groups will be G1, G2, …Gn, Live Load groups Q1, Q2,…Qn, and wind W1,

W2,...Wn and so on. This Group cases selector allow you to view the case groupings for each. The number shown here corresponds to the suffix of the group name e.g. for dead loads 1 means G1 whereas if wind were selected for the nature above, this would be W1. Cycle up and down through the "Group Cases" by number to view the currently defined case groups under the currently selected nature.

3- Operator: You have 3 options for how individual cases will be combined into the group:

a. And: All cases will be combined and applied simultaneously. Dead loads or permanent action are typically combined this way. You want all cases assigned to this group to be applied at the same time.

b. Or(incl): Or the cases in this group together in an inclusive way. This will create all possible combinations of the cases in this group. E.g., if you have cases LL1, LL2, LL3 and add them to a group with or(incl) you will get load combinations which include LL1 or LL2 or LL3 or LL1+LL2 or LL1+LL3 or LL2+LL3 or LL1+LL2+LL3. That means that for each load combination involving Live loads, you will finally have 7 load combinations where the live load portion is each of the above combinations in turn.

c. Or(excl): Or the cases in this group together in a mutually exclusive way. This will only create combinations where the load cases in this group are applied one at a time. E.g., if you have cases LL1, LL2, and LL3 and add them to a group with or(excl) you will only get load combinations which include LL1 or LL2 or LL3.

4- Group Cases List: All load cases which are currently assigned to this group (this group refers to the number next to "Group cases:" just above this list control. (See #2 above for more information) The operator on the left applies only to the cases included in one group.

5- Create group from cases button: Use this button to create a new case group from the cases currently listed in the available case list on the right (see #6 below). Note that this button works on all cases currently in the available case list. In order to break multiple cases into multiple case groups you will need to repeatedly use this button to create a new case group or remove unwanted cases after each use (returning unwanted ones to the available case list). A case may not participate in more than one case group. Robot automatically manages case group designations; you will not be able to create a G2 and not have a G1 for instance.

6- Available Case List: If there are any cases which have not yet been added to a case group they will appear in this list.

NOTE: IF YOU REMOVE ALL CASES OF A NATURE FROM THE LAST GROUP FOR THAT NATURE (GROUP CASES: GROUP NUMBER 1) THEN THE NATURE SELECTOR MAY BECOME GREYED OUT. YOU WILL NEED TO RESOLVE THIS SITUATION BY ADDING AT LEAST ONE LOAD CASE TO GROUP 1 OF THIS NATURE BEFORE YOU CAN MOVE ON.

The Relations Tab:

The relations tab allows you to create logical relationships between the case groups that were configured on the groups tab. So there are potentially two levels of logical operations, one at the group level acting on individual cases and one at the relations level acting on groups of cases. If there are no relations defined there will not be any load combinations generated. So there must be groups *and* relations defined in order for Robot to generate load combinations for you.

Here is the relations tab:

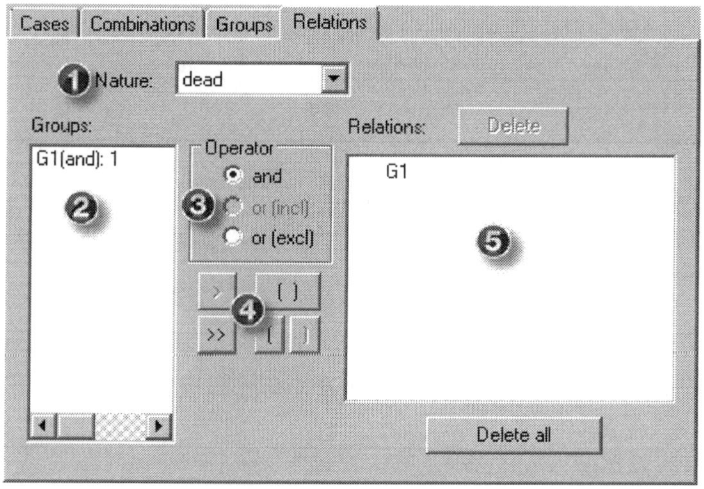

1- Nature selector: Relations are set up on a per-nature basis. The Groups and Relations lists will populate according to the current selection in the nature selector.

2- Groups list: Here all groups created on the groups tab (whether created explicitly by you or automatically created) will show here. First is the group name (G1 in this example), then the logical relationship applied to the cases (and) and then a list of the cases by number. You can review the case numbers on the cases tab and you can review the groups on the group settings on the groups tab.

3- Logical operators: After selecting a group for which you would like to create a relation you can specify how this group should be related to groups previously added to the relations list (#5)

4- Add Groups to relation and parenthesis buttons: In order to allow you to create advanced logical relationships between the groups, the parenthesis buttons give you more flexibility and control over how the relations are combined. For instance, if you had three live load groups Q1, Q2 and Q3 and you wanted Q2 and Q3 always applied together and you wanted either Q1 or (Q2 and Q3) you could use the parentheses to effect this relationship.

5- Relations List: The currently configured relations are shown here. In the picture above, there is only one relation and it is G1, there are no other case groups for the dead load nature hence it only has a relation with itself. However, in order for this case group to participate in combinations, it must have at least this minimum relation.

Combinations Selection:

If you are automatically generating manual combinations you will further need to select which combinations you wish to have generated. Once you are satisfied with the settings on the cases, combinations, groups, and relations tabs press the "next" button to move to the Combination Selection dialog:

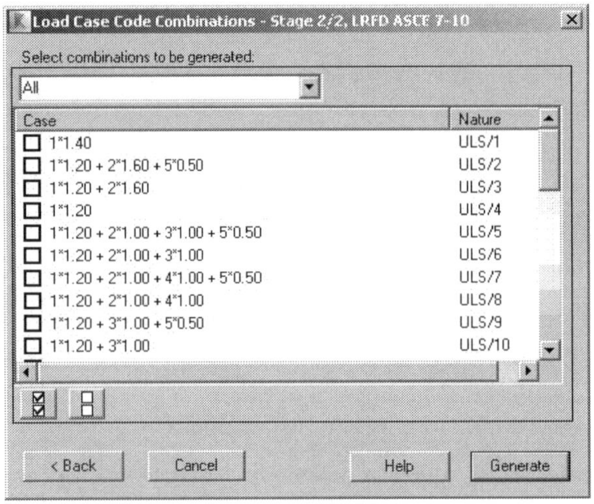

Here you are presented with all possible combinations based on you settings in the previous dialog. Now you must decide which combinations you actually intend for Robot to generate. You may check each checkbox individually or you may use these buttons to either check all or check none , then make adjustments individually after that.

When you are satisfied with the combinations Robot will generate, press "Generate"

Using the Automatic Combinations

We did a bit of this backwards, in the sense that the selection of Full Automated calculations or Generation of manual combinations really should have come first. However, now that we have been through the process of configuring combination options, which apply to both generation types, we can focus more closely on the differences between Full Automatic combinations and Manual Combinations – Generate Options

AUTOMATIC GENERATION - COMPOSED COMBINATIONS

When you first start automated load combinations from the loads menu: **Loads>Automatic Combinations...** Select the "Full Automatic Combinations" radio button.

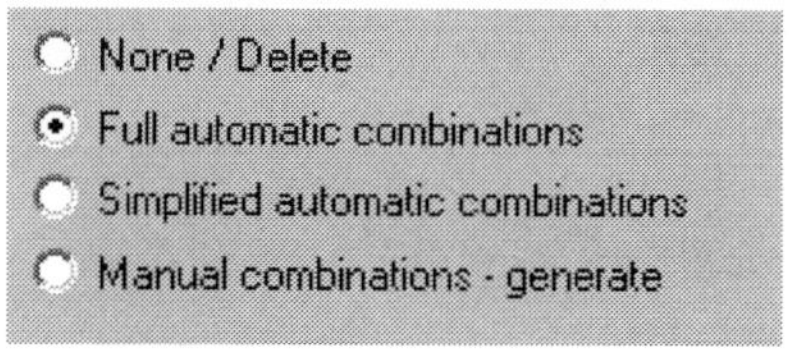

Then select "more" (at the bottom of the Automatic Combinations dialog) and configure the options for cases, combinations, groups and relations as necessary. (covered in the previous section). After you have configured options and selected combinations to generate in the following dialogs, pressing the generate button will commit your changes.

NOTE: WHEN YOU COMMIT YOUR CHANGES, IT MAY NOT SEEM LIKE ANYTHING HAS HAPPENED! IN FACT, IF YOU GO TO THE COMBINATIONS TABLE FROM THE LOADS MENU LOADS>COMBINATIONS TABLE, IT MIGHT IN FACT BE EMPTY. USING FULL AUTOMATED COMBINATIONS REQUIRES SOME AMOUNT OF THE CALCULATIONS TO HAVE BEEN RUN. IN OTHER WORDS THEY ARE NOT NECESSARILY GENERATED UNTIL NECESSARY. THIS CAN MAKE IT SOMEWHAT CHALLENGING TO UNDERSTAND WHAT HAS HAPPENED ONCE YOU GENERATE LOAD COMBINATIONS THIS WAY. EITHER RUN CALCULATIONS PRELIMINARILY OR MAKE A CHANGE WHICH REQUIRES PREPARATION OF THE CALCULATION MODEL (ADJUSTING THE TABLE COLUMNS IN THE COMBINATION TABLE SEEMS TO WORK)

What Robot has done for you is created several additional load cases called "composed cases". All of the combinations we have generated are effectively components of those composed cases. The composed cases will be named according to the type of combination (e.g., ULS, SLS). Take a look in the load case selector dropdown. Notice that 6 new cases have been added.

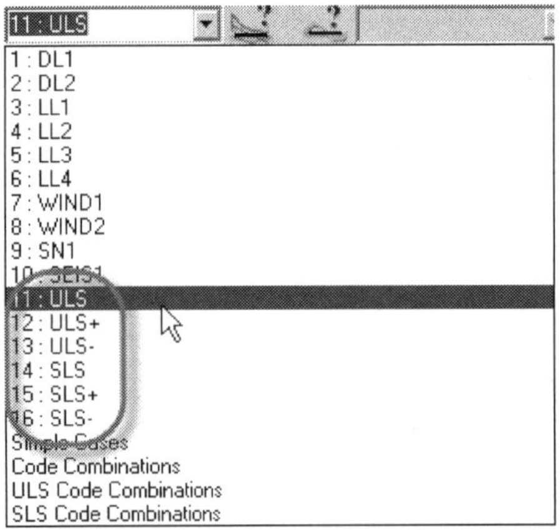

All of the ULS combinations are collapsed under the ULS load case and all of the service load combinations are collapsed under the SLS load case. To understand this more clearly, let's select the ULS case (Number 11 in the above image) and then press the component selector button to the right. For review, the load case selection bar looks like this: 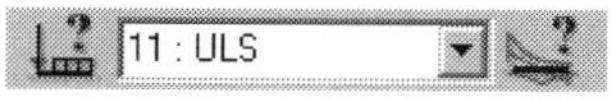

 This is the load case selector which launches the selection dialog with cases preselected in the selection type dropdown. See the section on Making Selections, to review how this dialog works.

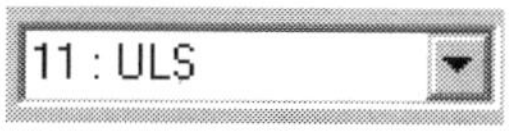

This is the load case selector dropdown

 This is the load combination component selector. This will only work for envelope cases and for composed cases (and for moving load cases which we will cover in an advanced text)

Selecting the ULS case and pressing the component selector launches the Case Component selector dialog:

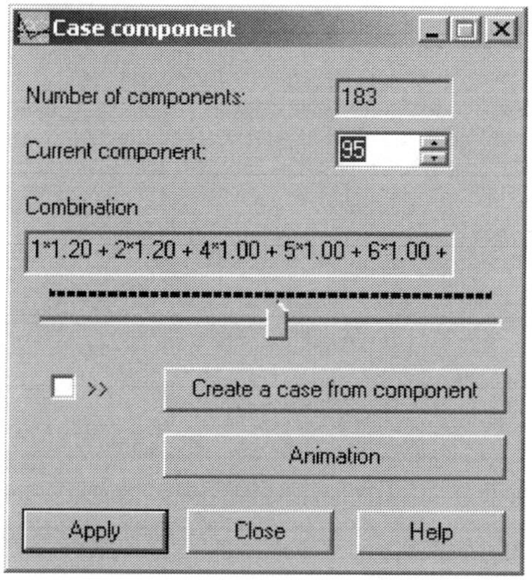

While viewing results and/or loads, you can choose which load combination component you wish to view by using this selector, either type a component into the current component edit control or use the up/down arrows to cycle through or use the slider to choose a component. The combination associated with the component will show in the Combination edit control (greyed out). Other parts of this dialog allow you to do things like automatically create a standalone (manual) load combination from the currently selected component. The animation button will allow you to "play through" some or all of the case components. This is most useful for viewing the effects of moving loads, but can be interesting for watching through all load combinations composing this case to see which ones stand out as potential problems.

Briefly, on animations: the settings are not complicated but once you press start, it can be a bit confusing. Here are the settings. Choose which components you want to see animated (cycled) and the frame rate:

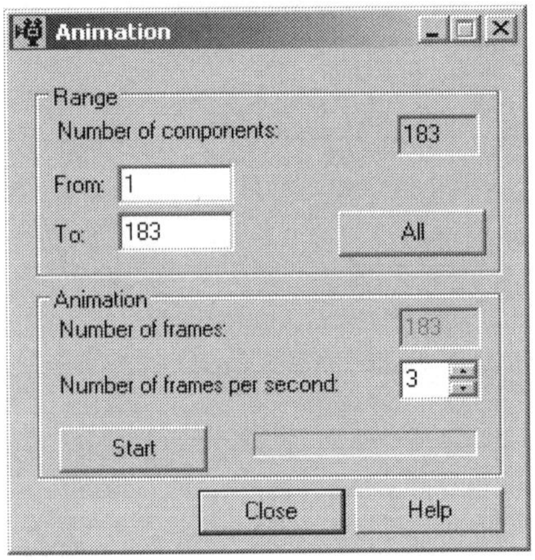

After you press start, you will see the display cycle through each component of the composed case in turn. (Most often you would want to do this in a results view to see the effects of the different combinations.) Notice that somewhere on the screen you will see the animation toolbar:

As long as this toolbar is visible, you are in animation play mode. You will not be able to zoom/pan etc. From this bar, you can open previously saved animations or save the current animation. There are also the standard play/pause/stop/etc. buttons for controlling the current playback. In order to get back to your model, simply close this animation control dialog.

Using the Load Combinations Table with Full Automatic Combinations:

If you open the load combinations table you might see case components listed or you might only see manually added load combinations.

Here is the combination table displaying manual combinations. Notice that the column heading for the left most column is "Combinations" and that it lists case/combination numbers.

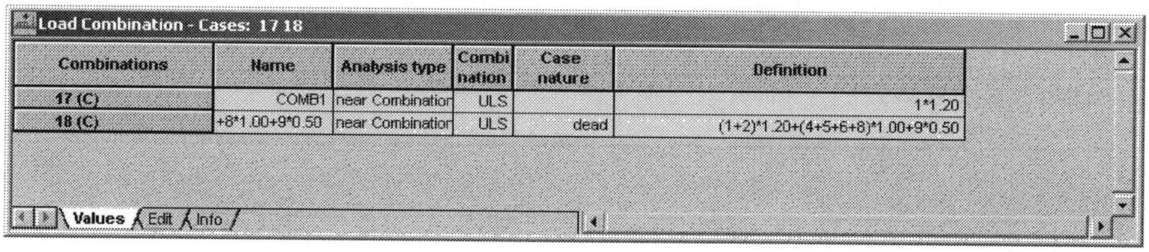

Load Combination - Cases: 17 18

Combinations	Name	Analysis type	Combination	Case nature	Definition
17 (C)	COMB1	near Combinatior	ULS		1*1.20
18 (C)	+8*1.00+9*0.50	near Combinatior	ULS	dead	(1+2)*1.20+(4+5+6+8)*1.00+9*0.50

This is the same table but with case components shown. Notice that the left most column is titled "Combinations/Comp." This is defining the information in that column as first the combination followed by a slash "/" and then Comp or Component. In this image you can see the Combination is ULS and several of its components are shown.

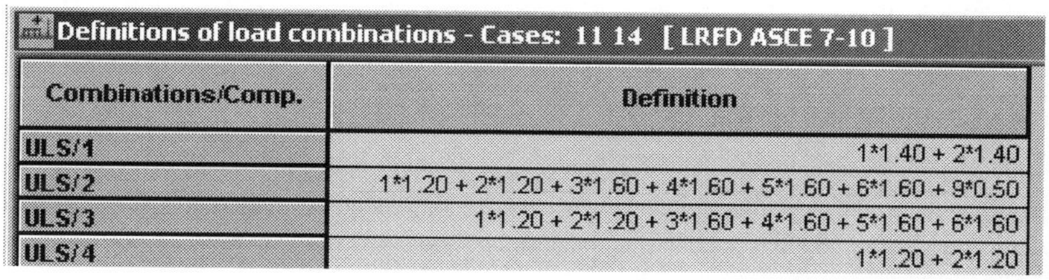

Definitions of load combinations - Cases: 11 14 [LRFD ASCE 7-10]

Combinations/Comp.	Definition
ULS/1	1*1.40 + 2*1.40
ULS/2	1*1.20 + 2*1.20 + 3*1.60 + 4*1.60 + 5*1.60 + 6*1.60 + 9*0.50
ULS/3	1*1.20 + 2*1.20 + 3*1.60 + 4*1.60 + 5*1.60 + 6*1.60
ULS/4	1*1.20 + 2*1.20

The key difference in the functioning of this table lies in the table columns setting (Right-Click>Table Columns…)

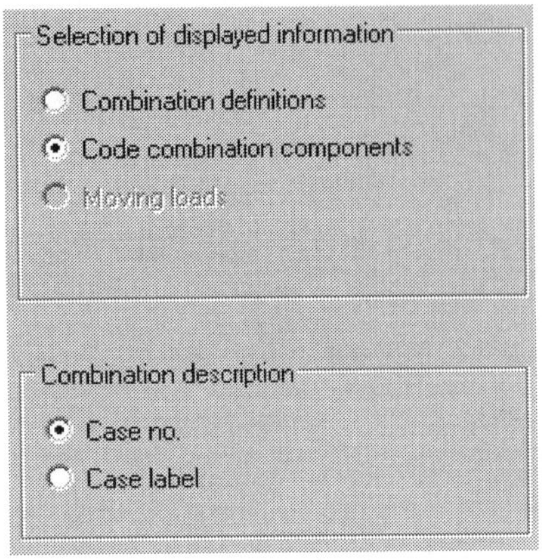

Under "Selection of displayed information" you can switch back and forth between Combination definitions and Code Combination components.

ULS+ and ULS- et al

In addition to the composed case ULS, Robot also generates two additional cases for each composed case ULS+ and ULS-. There are corresponding cases generated for SLS and others. These cases are used to represent envelope values in result tables: ULS+ maximum positive value and ULS- maximum negative value. Take a look at this result table excerpt which includes both manually created load combinations (17 and 18) as well as full automatic ULS and SLS composed combinations. You can see that the maximum and minimum magnitudes of FX, FY, and FZ are presented for Bar1/Node1 for ULS, SLS and then for the two manually created load combinations (17 and 18)

Internal Forces - Cases: 12 13 15to18

Bar/Node/Case	FX (kip)	FY (kip)	FZ (kip)
1/ 1/ ULS+	66.34	0.02	14.03
1/ 1/ ULS-	-1.35	-0.00	-13.96
1/ 1/ SLS+	48.04	0.02	5.78
1/ 1/ SLS-	0.94	-0.00	-5.72
1/ 1/ 17 (C)	13.93	0.00	0.61
1/ 1/ 18 (C)	18.95	0.01	-5.68

In result views, however, selecting ULS+ or ULS- will display the envelope of the results being displayed. For example, in this simple portal frame, selecting ULS+ or ULS- will show an envelope of max/min forces (My in this image):

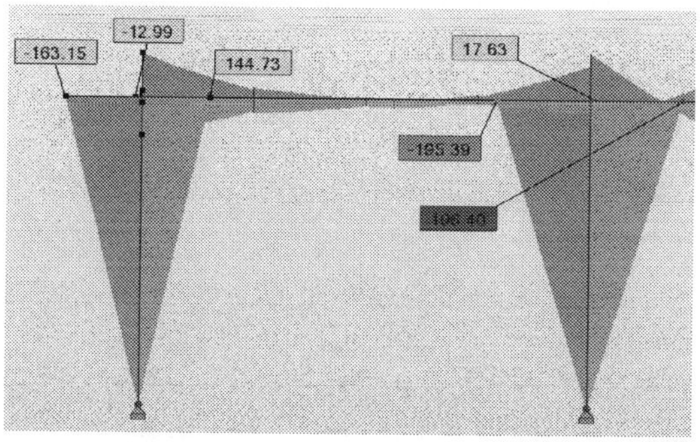

One last note about using full automatic code combinations is that you will also get additional selectable cases called "ULS Combinations" and "SLS Combinations" (assuming you generated both ULS and SLS full automatic combinations.) These can be used in results exploration to view envelope diagrams as well.

So what you have in full automatic combinations is an incredibly powerful way to deal with the sheer magnitude of combinations by letting Robot "componentize" them under ULS and SLS (or ACC and SEI if your building codes make use of those combination types). This can make it a breeze to work with so many combinations at once, essentially distilling all values down to maxima and minima for the mass of combinations. However, you can always drill down into individual load combination components to investigate each and every single combination. The only drawback is that full automatic load combinations cannot be used with non-linear or p-delta type analyses. This is because non-linear and p-delta type analyses require loads to be applied concurrently in order to produce the resulting load-displacement effects on the structure. They cannot be simply added up via superposition as elastic cases can. To use non-linear and p-delta, use the manual cases-generate option below.

AUTOMATIC GENERATION - MANUAL COMBINATIONS

When you want to use non-linear and/or p-delta type analyses you will need to generate manual combinations. You can either create them manually as we have discussed previously or you can use the code regulation to automatically generate all of the combinations you will need. The process is similar in almost all respects except for the selection of "manual combinations – generate" in the automatic combinations dialog (**Loads>Automatic Combinations...**)

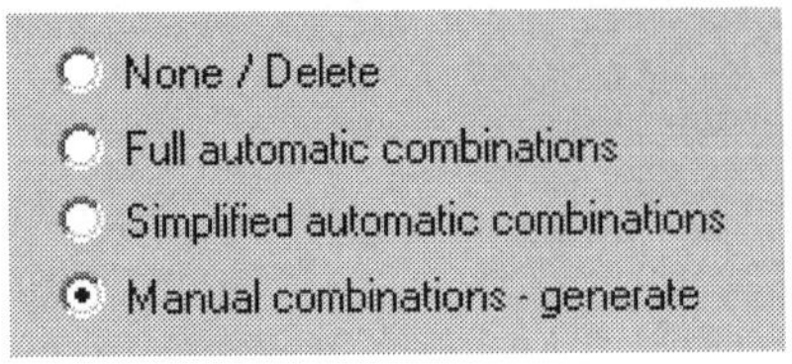

Then select "more" (at the bottom of the Automatic Combinations dialog) and configure the options for cases, combinations, groups and relations as necessary. (covered in a prior section). After you have configured options and selected combinations to generate in the following dialogs, pressing the generate button will commit your changes. A quick look in the combinations table (**Loads>Combinations Table**) will allow you to admire your handiwork:

Load Combination - Cases: 11to293

Combinations	Name	Analysis type
11 (C)	ULS/1=1*1.40 + 2*1.40	Linear Combination
12 (C)	ULS/2=1*1.20+2*1.20+3*1.60+4*1.60+5*1.60+6*1.60+9*0.50	Linear Combination
13 (C)	ULS/3=1*1.20 + 2*1.20 + 3*1.60 + 4*1.60 + 5*1.60 + 6*1.60	Linear Combination
14 (C)	ULS/4=1*1.20 + 2*1.20	Linear Combination
15 (C)	ULS/5=1*1.20 + 2*1.20 + 3*1.60 + 9*0.50	Linear Combination
16 (C)	ULS/6=1*1.20 + 2*1.20 + 3*1.60	Linear Combination
17 (C)	ULS/7=1*1.20 + 2*1.20 + 4*1.60 + 9*0.50	Linear Combination

Unless you have also created either moving loads or full automatic combinations, in tables columns (Right-click>Table Columns...) you will see component and moving load options greyed out:

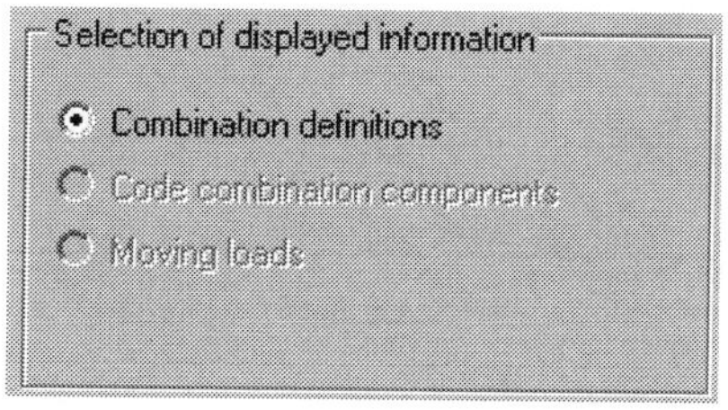

(This is because we do not have any moving loads nor fully automated composed combination cases.) There is nothing prohibiting you from using both composed combinations together with either manually created combinations or automatically generated manual combinations. The only slightly confusing part is that fully automated combinations collapse components into a single case.

Your load case selector will as well, contain all of your recently generated combinations in addition to your simple cases.

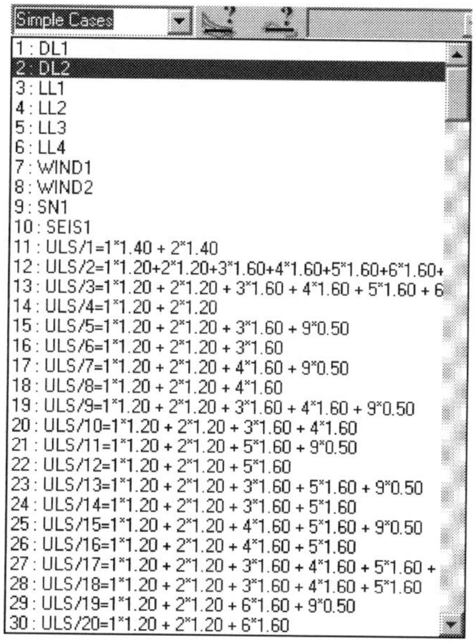

In order to view envelopes of results for your cases, you will find both "Simple Cases" and "Combinations" at the bottom of the case list:

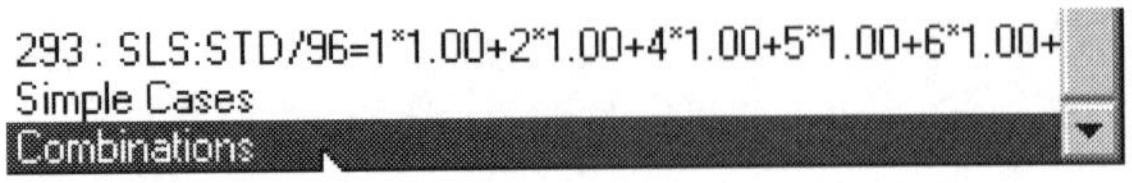

Selecting either while in results exploration will show an envelope of results for all manual combinations.

Exercise 10: Generating Automatic Combinations

1. Start a new project with the Building Design UI Configuration

   a. 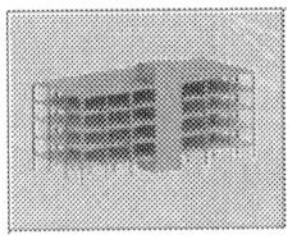

2. Open the Load Types dialog from the menu **Loads>Load Types...** or from the toolbar on the right side of the screen: 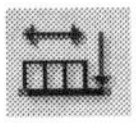

   a. 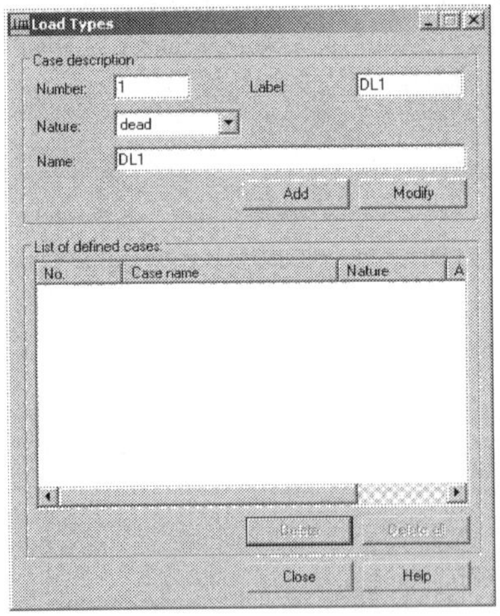

3. Go to **Tools>Job Preferences...** and expand “Design Codes”, then select “Loads” and for “Code Combinations” choose “LRFD ASCE 7-10”. Finally press “Close”.

4. Add the following load cases using what you have learned in the previous chapter:

No.	Name/Label	Nature	Sub-Nature
1	DL1	dead	
2	DL2	dead	
3	LL1	live	
4	LL2	live	
5	LL3	live	
6	LL4	live	
7	WIND1	wind	
8	WIND2	wind	
9	SN1	Snow	
10	Lr1	Snow	Roof Live

5. Your Load Types dialog should look like this:

a. 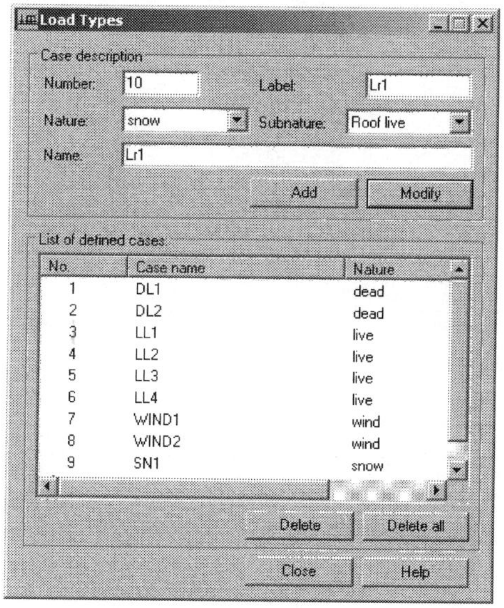

6. Next we will create a couple manual combinations
7. Go to **Loads>Manual Combinations...** and enter "1.4DL+1.7LL" in the Combination Name edit control then press "OK"

a. 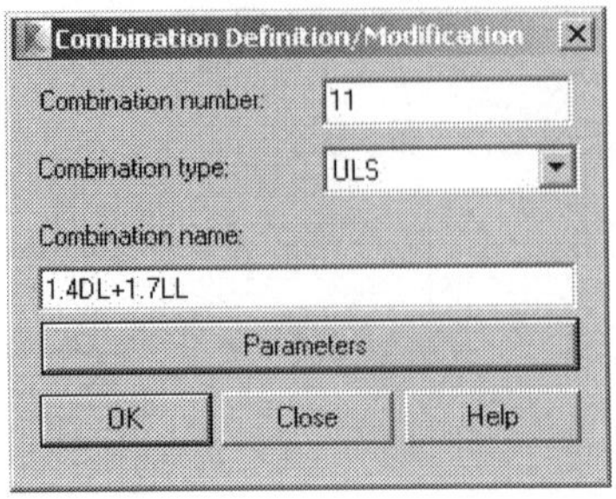

8. In the next dialog (Combinations dialog) select case 1 and enter 1.4 in the factor edit control then press the right arrow to move that case into the combination.

a. 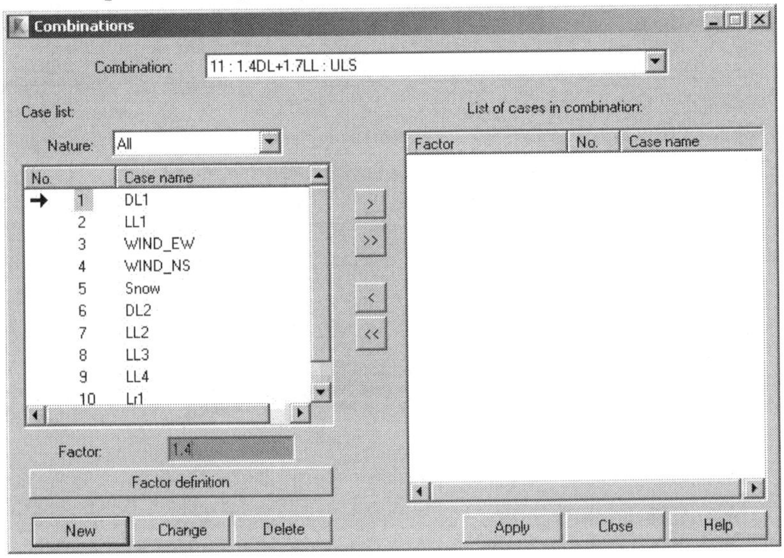

9. Repeat this process for LL1 with a factor of 1.7 Your dialog should now look like this:

a. 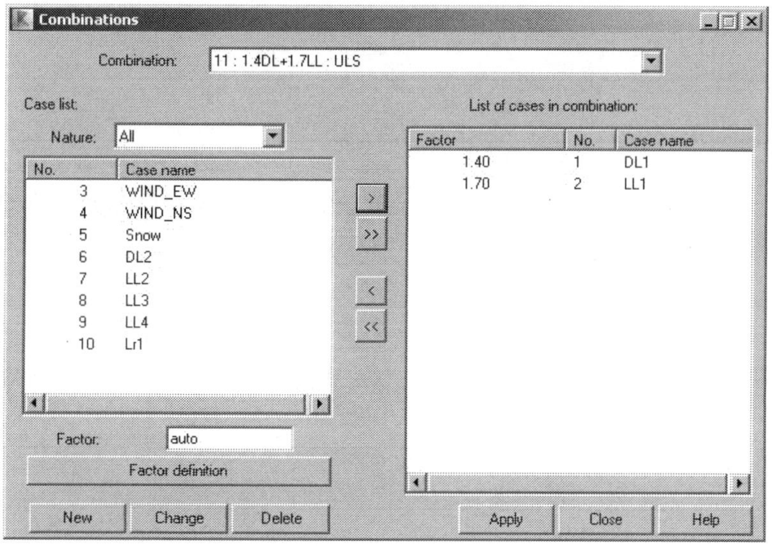

10. Press "Apply" and then "Close"
11. Take a look at our load cases dropdown and you will see our newly created manual load combination:

a. 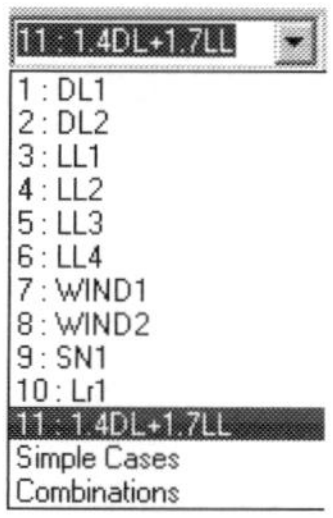

12. Let's also take a quick look at the combinations table. Go to **Loads>Combination Table** and you will see our newly created combination here.

a. Load Combination - Case: 11 (1.4DL+1.7LL)

Combinations	Name	Analysis type	Combination	Case nature	Definition
11 (C)	1.4DL+1.7LL	Linear Combination	ULS		1*1.40+3*1.70

13. Right-click on this table and select "Table Columns..." and in the table columns dialog select case label. Then press "Ok":

a. 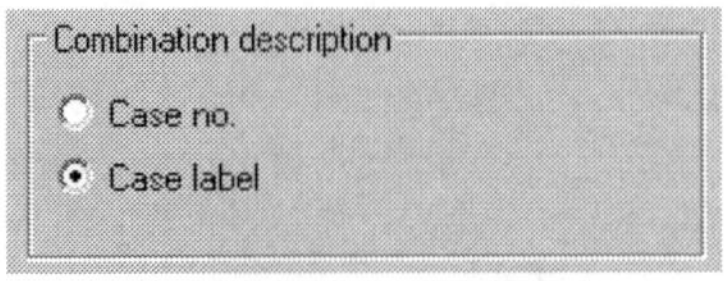

14. The definition in the table will now use case labels instead of case numbers:

a. 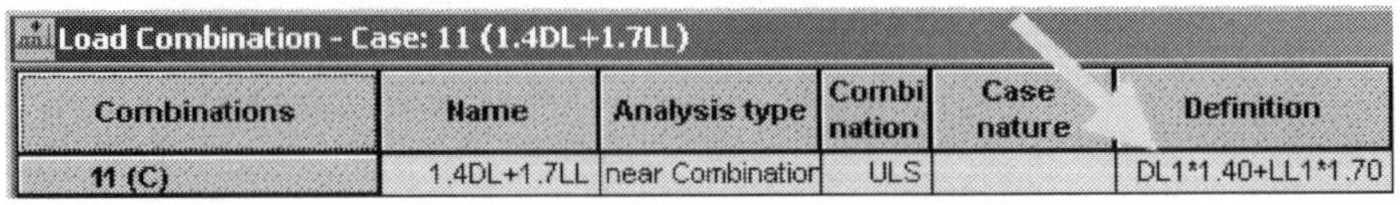

Load Combination - Case: 11 (1.4DL+1.7LL)

Combinations	Name	Analysis type	Combination	Case nature	Definition
11 (C)	1.4DL+1.7LL	near Combinatior	ULS		DL1*1.40+LL1*1.70

15. Close the combinations table and let's proceed with fully automated combinations. Go to **Loads>Automatic Combinations...** Check to see that LRFD ASCE 7-10 is selected in the "Combinations According to Code" dropdown and that "Full Automatic Combinations" radio button is selected.

a. 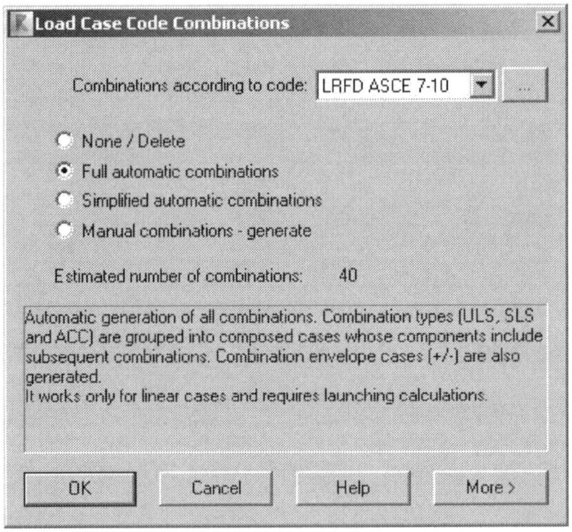

16. Next press "more" and switch to the cases tab: Ensure that all load cases are checked so that they will participate in load combination generation:

a. 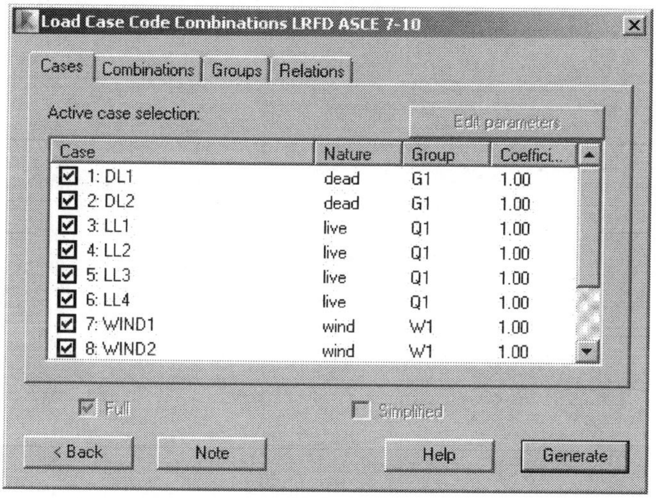

b. Take a quick look at the Groups (on the Groups tab) to which each load case has already been assigned. Remember that Robot automatically assigns cases to groups and creates relations for you. They typically work well for most situations, but we will be modifying them for this example.

17. Switch to the Combinations tab and unselect combinations 5 and 7 which both involve earthquake loads. As we do not currently have any seismic load cases, we will not generate these load combinations

a. 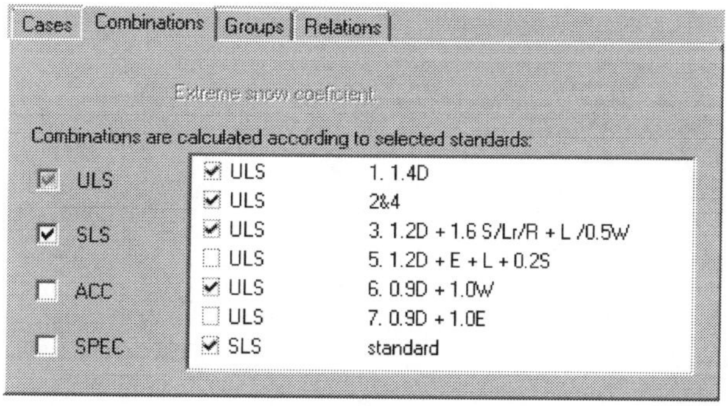

18. There is also a "note" button at the bottom of the dialog which will give you a nice printout of the current settings for cases, combinations, groups and relations. Go ahead and check that out.

19. Next switch to the groups tab. It will initially be on dead loads. Go ahead and cycle through the natures to see what the current groups configurations are. Lastly switch to the live nature so that we can reconfigure this grouping.

a. 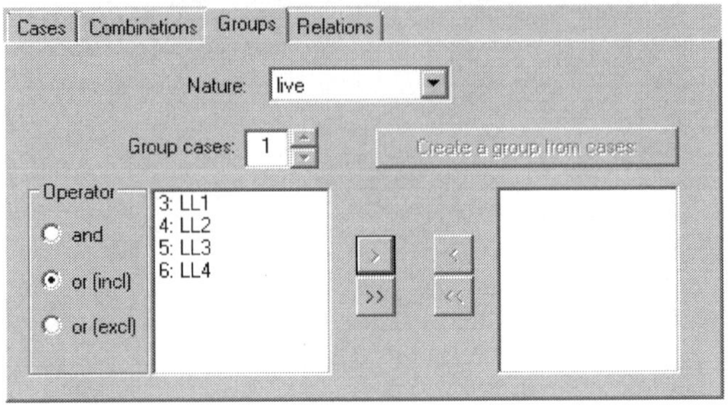

b. Currently there is one live load group with four participating load cases: LL1, LL2, LL3, and LL4. LL1 is a uniform live load but LL2-LL4 are worst-case positions of another variable load. We want to have LL1 applied all the time but LL2-4 only applied each in turn as the variable load can only adopt one position at a time. To do this we will create two groups for which we can create logical relations to give us the combination we desire.

c. One by one, select LL2 – LL4 and press the right arrow button to move these three cases out of the current group and into the available cases list:

d. 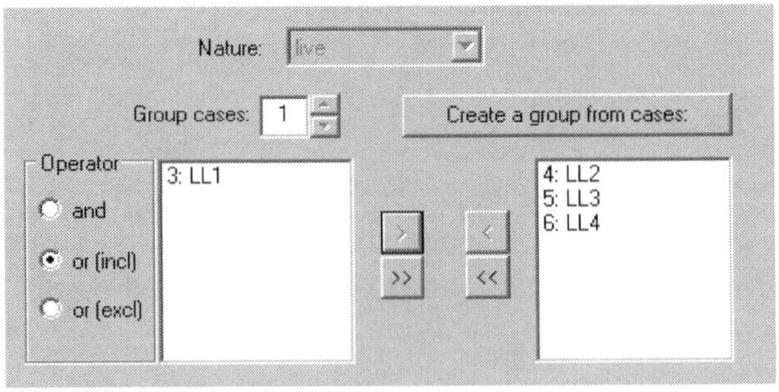

e. Next press the "Create a group from cases:" button:

f. 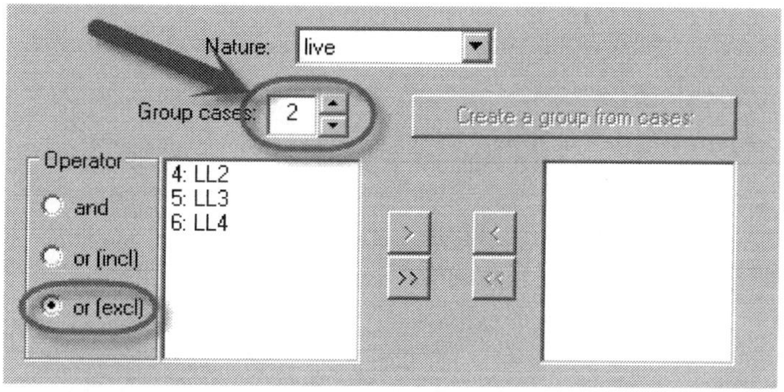

g. Notice that Group Cases is now incremented to "2" because we've just created a new group. Make sure "or(excl)" is selected for this group. Use the up/down arrows to cycle back and forth between live load groups 1 and 2 to get a feel for them.

h. Now that you have the hang of this, check out the snow load nature: notice that there are two groups already created for you?

20. Now that we have set up our groups, let's switch to the relations tab. Take a few minutes to select the different natures in the drop down and look at the currently configured relations:

a. 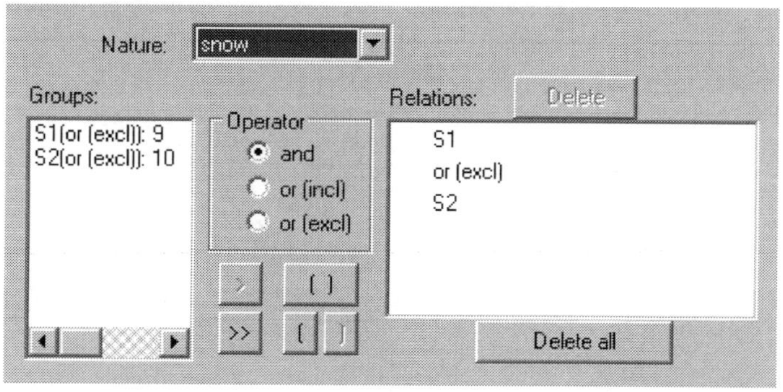

b. Take a look at snow. Remember that Snow had automatically had two groups created? (one for case 9: Snow and one for case 10: Live Roof) Here in the relations tab, there is already a relationship set up for group S1 and group S2 which is mutually exclusive. In other words, either snow OR live roof load will participate in each combination but never both at the same time.

c. Switch to the live load Nature from the dropdown so that we can set up our new relationship for these live load cases:

d. 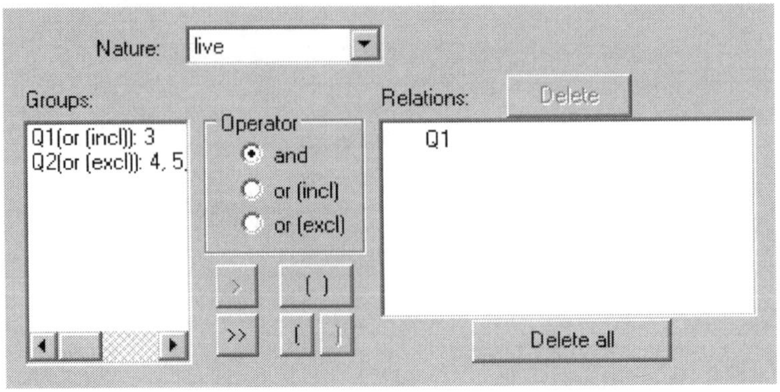

e. Currently only Q1 is participating in the relationship. We have two case groups though. Q1 includes case 3 (LL1) and Q2 includes cases 4,5, and 6 (LL2 – LL4) with an "Or(excl)" or exclusive or relationship. (You can read this information in the "groups" list).

f. What we want to do is add group Q2 to the relationship with an "And" relationship to Q1. Select Q2 then select the "and" radio button and press the right arrow to add Q2 to the relationship:

g. 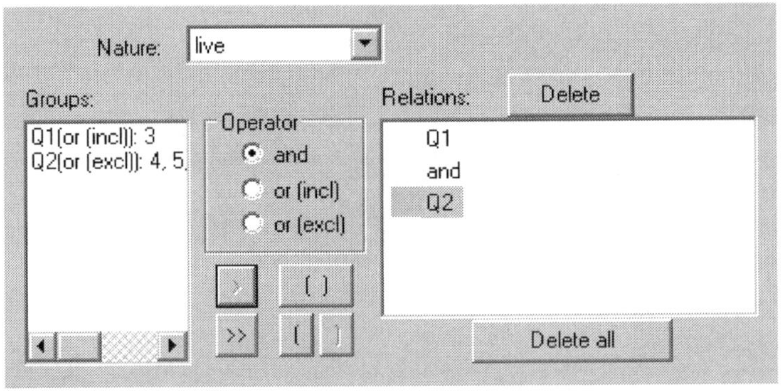

h. Now you will have a relationship that should look exactly like the one above. You will notice that it is similar to the relationship set up for roof loads (snow Nature) but uses the "and" logical operator.

i. In this configuration, every time Live load participates in a load combination, we will have three combinations generated: LL1 + LL2 or LL1 + LL3 or LL1 + LL4

21. Press "generate". Then create some geometry for the model, this can be as simple as a single beam or column. You don't even need supports but you do

need some geometry in order for Robot to run calculations. As we have mentioned before, calculations are required for generation of load case components. After you have created a piece of geometry, press the calculations button, and then open the combinations table from **Loads>Combinations Table**. Note: you may need to add materials for members and/or press escape to ignore warnings during calculations. We only need the combinations generated so ignoring these warnings is ok.

22. Once the combinations table is open, right-click and select Table Columns... then choose the following options:
    a. 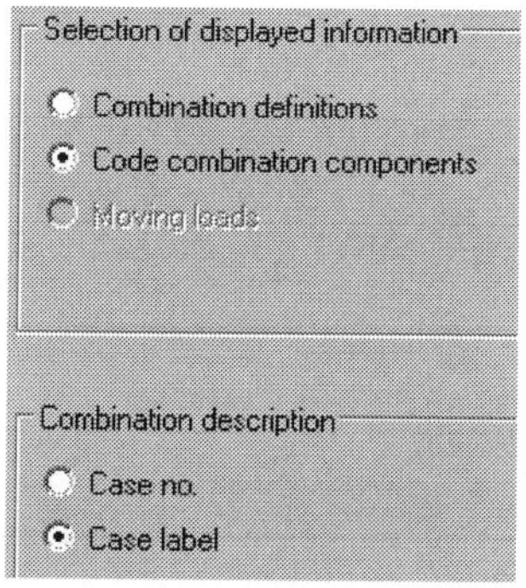

23. Your combination table will then display case components for the fully automatic combinations and use case labels instead of case numbers:
    a.

Combinations/Comp.	Definition
ULS/1	DL1*1.40 + DL2*1.40
ULS/2	DL1*1.20 + DL2*1.20 + LL1*1.60 + LL2*1.60 + SN1*0.50
ULS/3	DL1*1.20 + DL2*1.20 + LL1*1.60 + LL2*1.60
ULS/4	DL1*1.20 + DL2*1.20 + LL1*1.60 + LL3*1.60 + SN1*0.50
ULS/5	DL1*1.20 + DL2*1.20 + LL1*1.60 + LL3*1.60
ULS/6	DL1*1.20 + DL2*1.20 + LL1*1.60 + LL4*1.60 + SN1*0.50
ULS/7	DL1*1.20 + DL2*1.20 + LL1*1.60 + LL4*1.60
ULS/8	DL1*1.20 + DL2*1.20
ULS/9	DL1*1.20 + DL2*1.20 + LL1*1.60 + LL2*1.60 + Lr1*0.50
ULS/10	DL1*1.20 + DL2*1.20 + LL1*1.60 + LL3*1.60 + Lr1*0.50
ULS/11	DL1*1.20 + DL2*1.20 + LL1*1.60 + LL4*1.60 + Lr1*0.50
ULS/12	DL1*1.20 + DL2*1.20 + LL1*1.00 + LL2*1.00 + WIND1*1.00 + SN1*0.50
ULS/13	DL1*1.20 + DL2*1.20 + LL1*1.00 + LL2*1.00 + WIND1*1.00
ULS/14	DL1*1.20 + DL2*1.20 + LL1*1.00 + LL2*1.00 + WIND2*1.00 + SN1*0.50
ULS/15	DL1*1.20 + DL2*1.20 + LL1*1.00 + LL2*1.00 + WIND2*1.00
ULS/16	DL1*1.20 + DL2*1.20 + LL1*1.00 + LL3*1.00 + WIND1*1.00 + SN1*0.50
ULS/17	DL1*1.20 + DL2*1.20 + LL1*1.00 + LL3*1.00 + WIND1*1.00
ULS/18	DL1*1.20 + DL2*1.20 + LL1*1.00 + LL3*1.00 + WIND2*1.00 + SN1*0.50
ULS/19	DL1*1.20 + DL2*1.20 + LL1*1.00 + LL3*1.00 + WIND2*1.00

    b. The case is before the slash and the component is after the slash then under definition you will see the load combination for that component.

Notice that LL1 is applied with LL2, or LL3, or LL4 but LL2-4 are never applied together. This is because of the relation that we set up.

24. Start Automatic Combinations again (**Loads>Automatic Combinations...**) and this time choose "Manual Combinations – Generate"
25. Repeat the configuration options as above or try to exercise what you have learned by setting up 4 different live load groups and creating an advanced live load relation as shown here. Creating 4 groups will be a bit of a challenge as you'll have to first dismantle the existing 2 groups and create new ones by moving each case over and using "create group from cases" button repeatedly. Here in the Relations Tab, notice the use of the "( )" button. You can select a logical relation then press the "( )" button to effect this relationship. (Select "and" and then press "()" will create an "and" relationship between Q1 and everything in the parenthesis). Then select each group Q2-Q4 in turn and add them to the relation using the "Or(excl)" option.

a. 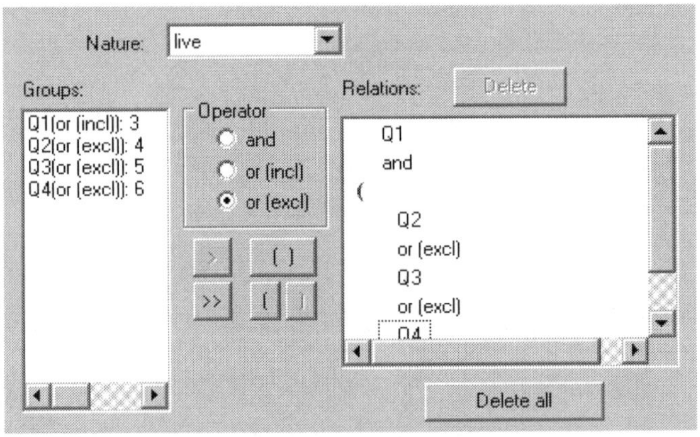

b. This will give you exactly the same effect as the previous configuration, but gives you a chance to practice setting up groups and relations in a different way. Because there are logical relations applied to cases in groups and between groups you can operate on them in many different ways.

c. Open the combinations table again, right-click and choose Table Columns... then select "Combination Definition" and "Case Label" to review the automatically generated manual combinations:

d.

Combinations	Name	Analysis type	Combination	Case nature	Definition
11 (C)	1.4DL+1.7LL	near Combination	ULS		DL1*1.40+LL1*1.70
18 (C)	1*1.40 + 2*1.40	near Combination		dead	(DL1+DL2)*1.40
19 (C)	4*1.60 + 9*0.50	near Combination		dead	(DL1+DL2)*1.20+(LL1+LL2)*1.60+SN1*0.50
20 (C)	3*1.60 + 4*1.60	near Combination		dead	(DL1+DL2)*1.20+(LL1+LL2)*1.60
21 (C)	1*1.20 + 2*1.20	near Combination		dead	(DL1+DL2)*1.20
22 (C)	*1.60 + 10*0.50	near Combination		dead	(DL1+DL2)*1.20+(LL1+LL2)*1.60+Lr1*0.50
23 (C)	5*1.60 + 9*0.50	near Combination		dead	(DL1+DL2)*1.20+(LL1+LL3)*1.60+SN1*0.50
24 (C)	3*1.60 + 5*1.60	near Combination		dead	(DL1+DL2)*1.20+(LL1+LL3)*1.60
25 (C)	*1.60 + 10*0.50	near Combination		dead	(DL1+DL2)*1.20+(LL1+LL3)*1.60+Lr1*0.50
26 (C)	6*1.60 + 9*0.50	near Combination		dead	(DL1+DL2)*1.20+(LL1+LL4)*1.60+SN1*0.50
27 (C)	*1.60 + 6*1.60	near Combination		dead	(DL1+DL2)*1.20+(LL1+LL4)*1.60
28 (C)	1.60 + 10*0.50	near Combination		dead	(DL1+DL2)*1.20+(LL1+LL4)*1.60+Lr1*0.50
29 (C)	*1.00 + 9*0.50	near Combination		dead	(DL1+DL2)*1.20+(LL1+LL2+WIND1)*1.00+SN1*0.50
30 (C)	*1.00 + 7*1.00	near Combination		dead	(DL1+DL2)*1.20+(LL1+LL2+WIND1)*1.00
31 (C)	*1.00 + 9*0.50	near Combination		dead	(DL1+DL2)*1.20+(LL1+LL2+WIND2)*1.00+SN1*0.50
32 (C)	*1.00 + 8*1.00	near Combination		dead	(DL1+DL2)*1.20+(LL1+LL2+WIND2)*1.00
33 (C)	1.00 + 10*0.50	near Combination		dead	(DL1+DL2)*1.20+(LL1+LL2+WIND1)*1.00+Lr1*0.50

--End of Exercise--

Model Verification

After you have completed a modeling exercise, go to the Analysis menu and choose Verification... (**Analysis>Verification...**) This will launch a check of the model for fitness to run calculations. Your report may contain some errors or warnings which will be presented in a dialog box similar to this:

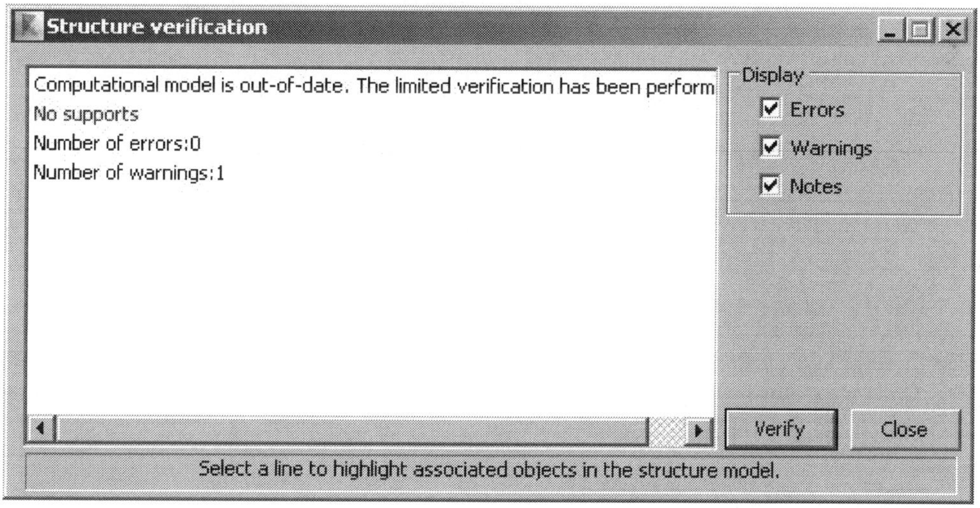

Note the message which indicates that the computational model is out of date. The full verification of the model cannot be run without generating the computational model. Remember Robot's physical model is used as a basis for generating the computational model (recall adding a calculation node under point loads? That results in an additional node in the computational model. Unless the computational model is generated, we can't fully verify the model). To generate the computational model, open the analysis types **dialog Analysis>Analysis Types...** and switch to the "Structure Model" tab. On that tab near the bottom you will find "Generation of computational model". Press that button to generate the computational model:

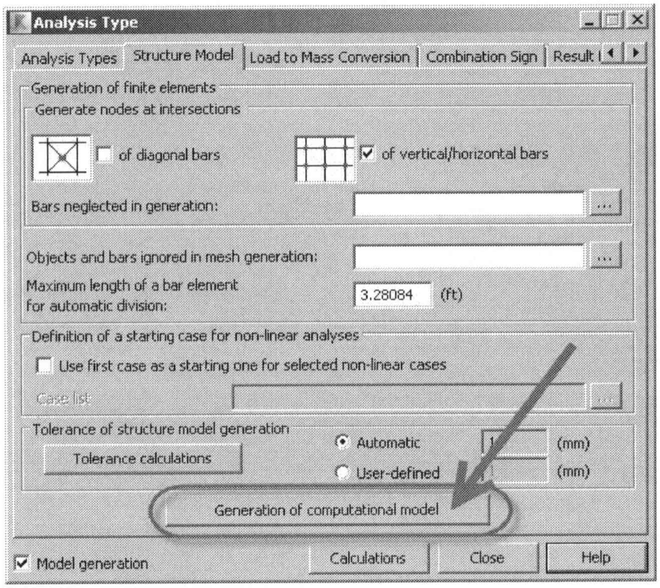

Now your verification dialog will contain all errors and/or warnings which can be determined from the model prior to running calculations and the dialog now looks like this:

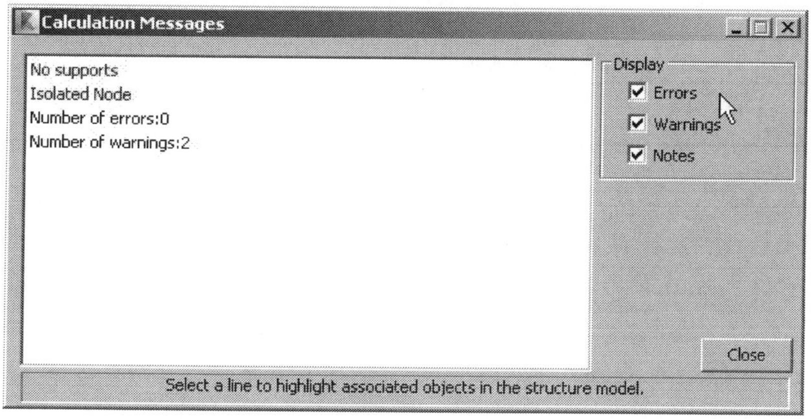

Here we have 2 warnings: 1) No Supports and 2) Isolated node. These errors are fairly easy to figure out (I deleted all the supports before I ran the verification!). "No Supports" can be resolved by adding some reasonable supports for the model and the Isolated node warning means that we somehow either left a node hanging in space or we may have a duplicate node somewhere. Notice that the dialog has a message "Select a line to highlight associated objects in the structure model" meaning if you select either "No supports" or "Isolated Node", the selection bar will populate with a selection of the elements in question. These are both warnings. If there had been errors the same applies. Warnings are typically things that Robot can ignore and finish the calculations by fixing some degrees of freedom in the model though the automatic actions will typically result in strange or unreliable results. It is best to understand each issue and resolve them so that you understand how your model is performing and that you have increased confidence that what you have modeled and configured is what is analyzed. You may use the checkboxes on the right to filter the list.

I resolved the "Isolated Node" by selecting the line in the verification dialog (which selects the elements in question) and pressing delete. You could also use a special feature of Robot called "Edit in New Window" 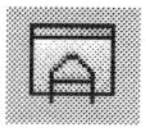 from the standard toolbar to open a new view with only the selected elements shown so that you can start to understand what might be the issues and start work to resolve them.

Running Calculations

Now that we have looked at adding geometry, supports, loads, load cases and load combinations we can start running calculations on our models. In the absence of the user selecting a non-linear analysis or adding a model element which would require non-linear analysis, all calculations will be linear-static. To run calculations, you can either use the calculations button  or the calculations option from the Analysis menu (**Analysis>Calculations**). Calculations may also be run from the Analysis Types dialog shown below.

Calculation notes

You can access calculation notes from the Analysis menu (**Analysis>Calculation Report**>). You may choose simplified or full note. This report will list information about the calculation date, the name of the model, the size of the model (number of nodes, bars, etc.) along with other information about the performance of the model, the solver selected (if you have job preferences set to automatic this is where you can see which solver was chosen by Robot. Check **Tools>Job Preferences** for more information). Some information is only available in the full report. The simplified note is a quick view of model information and analysis cases. Select Full Report for the most detailed information. This report can also be included in documentation via the screen capture functionality.

Sample Report (Simplified Note):

Project properties: **Structure8**

Structure type: Space frame

Structure gravity center coordinates:
X = 59.295 *(ft)*
Y = 27.490 *(ft)*
Z = 20.333 *(ft)*
Central moments of inertia of a structure:
Ix = 33850193.885 *(lb*ft2)*
Iy = 112088363.305 *(lb*ft2)*
Iz = 135093713.157 *(lb*ft2)*
Mass = 65806.347 *(lb)*

Structure description

Number of nodes:	44
Number of bars:	69
Bar finite elements:	69
Planar finite elements:	0
Volumetric finite elements:	0
No of static degr. of freedom:	168
Cases:	1
Combinations:	0

Table of load cases / analysis types

Case 1 : DL1
Analysis type: Static - Linear

Analysis Types

The Analysis types dialog allows us to view current settings for cases and combinations. Launch the Analysis Types dialog from the main toolbar or from the menu **Analysis>Analysis Types...**

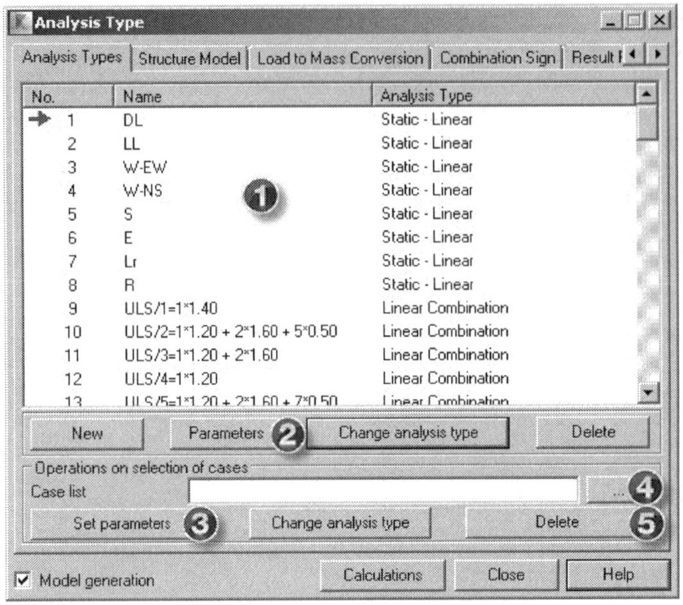

1- List of cases and manual combinations: All load cases and manually defined combinations will be listed here along with their current Analysis Type.

2- Parameters button: allows access to configure the parameters associated with an analysis type. For static cases and combinations, you can access the settings for P-Delta and Large Displacement analyses here. Select a case or combination and press Parameters to launch the Parameters dialog for P-Delta and Large Displacement settings:

a. 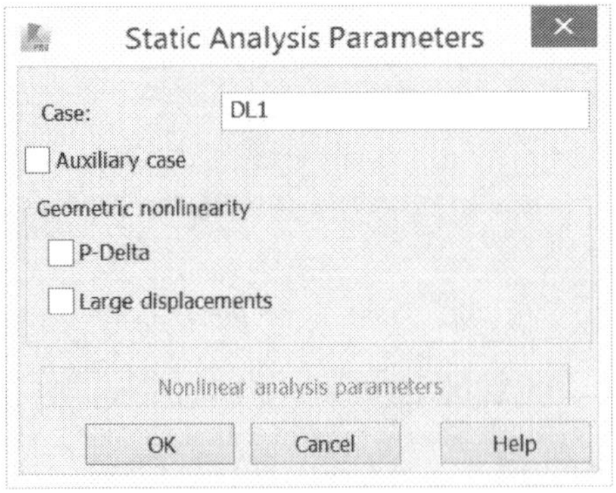

b. This dialog will be discussed in more detail below

3- Set parameters for a list of cases: enter the list of cases/combinations by number in the edit control above and then use Set Parameters button to configure multiple cases at once.

4- Case selector: in addition to manually entering cases in the case list you may also access the case selection dialog using the ellipsis button. Cases configured in the case selection dialog will be transferred back into this list.

5- Delete: In this section of the dialog we are operating on multiple cases. Enter a case list in the edit control or use the case selector (#4) to configure a list of cases which you would like to delete. Use this delete button as opposed to the one above to operate on multiple cases at once.

Non-Linear Analyses

In order to account for p-delta effects in structural analysis we will need to use the P-Delta analysis options. In the Analysis Types dialog, using the parameters button will launch the Non-Linear Analysis Parameters dialog:

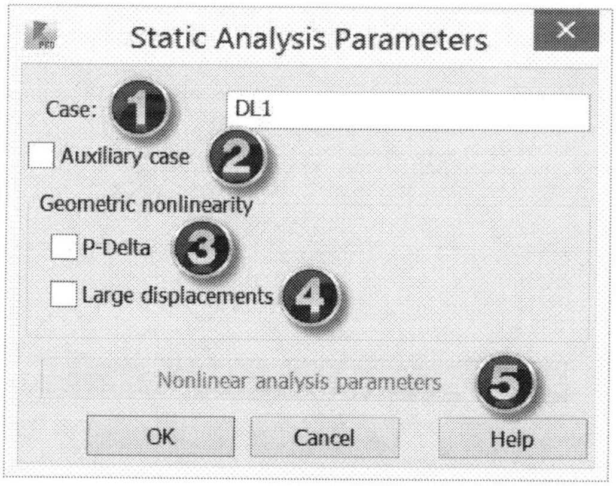

1- Case Name: the case name can be modified here if desired. Initially displays the current name of the case. In general, there is no need to change the case name here.

2- Auxiliary case: Making a case auxiliary means that it will be excluded from the calculations. This case and combinations including this case will not have any results available after running calculations. This is a tool which can help you reduce the time of the overall calculations if there are particular areas you need to focus on and iterate several times.

3- P-Delta option: Turns on the iterative solution of the system of equations. This will cause Robot to use the Newton-Raphson iterative solver to evaluate the system of equations. Once this option is enabled the "Parameters" (#5) button will be available with options for controlling the behavior of the solver.

4- Large Displacement analysis: This turns on the option to not only reconfigure the overall stiffness matrix at each Newton-Raphson iteration but to also include the effects of displaced geometry. Please continue reading below in "Understanding P-Delta and Large Displacement Analysis". If you select this option you will notice that the "P-Delta" checkbox also becomes checked. P-Delta analysis is required in order to perform a more advanced Large Displacement analysis. Robot's Large Displacement analysis is effectively a more advanced non-linear analysis.

5- Parameters: Here you can access the parameters which control the operation of the Newton-Raphson solver in Robot. These options can be used to fine-tune the algorithm in order to increase calculation efficiency or to tweak the convergence of the solver. Please see more detailed information in "Understanding P-Delta and Large Displacement Analysis" below.

AUTOMATIC NON-LINEAR ANALYSIS:

IT IS IMPORTANT TO NOTE THAT IN THE PRESENCE OF NON-LINEAR ELEMENTS IN A MODEL (E.G., RELEASED UPLIFT IN SUPPORTS, GAP SUPPORTS, TENSION AND/OR COMPRESSION ONLY MEMBERS, CABLES, NON-LINEAR RELEASES, NON-LINEAR COMPATIBLE NODES, OR PLASTIC HINGES) ROBOT WILL AUTOMATICALLY SPECIFY A NON-LINEAR ANALYSIS REGARDLESS OF WHETHER YOU HAVE IT CHECKED OR NOT:

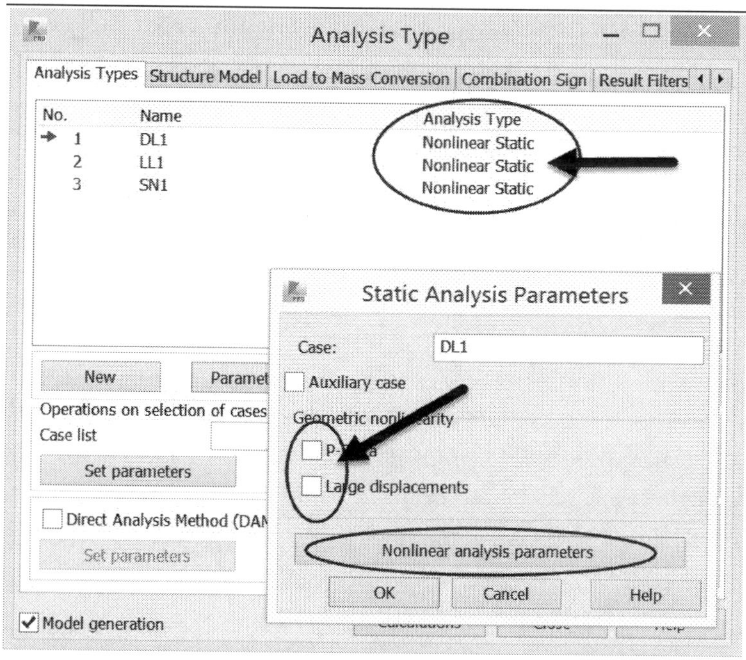

IN THE PRESENCE OF NON-LINEAR ELEMENTS THE CASE AUTOMATICALLY TURNS TO STATIC-NONLINEAR AND EVEN THOUGH THE CASE PARAMETERS FOR NON-LINEAR AND P-DELTA ARE UNCHECKED, THE NON-LINEAR ANALYSIS PARAMETERS BUTTON IS STILL ACTIVE AND WILL CONTROL THE NEWTON-RAPHSON SOLVER FOR THIS CASE.

Understanding P-Delta and Large Displacement Analysis

P-Delta and Large Displacement analyses are methods to account for the behavior of a structure in situations where there is either material non-linearity and/or large displacements which give rise to additional moments or corrective forces.

P-Delta analysis may, in fact, be required by Robot's calculation engines or can be additionally specified by the user. Any object with non-linear behavior in the model (e.g., released uplift in supports, gap supports, tension and/or compression only members, cables, non-linear releases, non-linear compatible nodes, plastic hinges) will result in non-linear analysis being required.

There are three main types of analysis in Robot:

1. Linear: All materials are treated as linearly elastic, no geometric affects are considered

2. P-Delta: Material non-linearity (in the form of non-linear hinges) is considered and calculations are preformed according to the Newton-Raphson method meaning that load is incrementally applied, the structure is allowed to stabilize and then additional load is applied until the total load is applied and stabilized (converged). This accounts for additional moments arising from vertical loads acting at a lateral displacement of the structure.

3. P-Delta + Large Displacement: Material non-linearity (in the form of non-linear hinges) is considered as well as the effects of locally displaced geometry meaning that the stiffness matrix, including geometry changes, is reassembled after every load increment (iteration).

Generally the P-Delta analysis option with linearly elastic materials corresponds to our traditional definition of P-big delta analysis (P-Δ). However, while the P-Delta option only is appropriate for a structural engineer's notion of P-Δ, it does not technically account for p-little delta (p-δ) or the effects of local member deflections. It is furthermore, only appropriate for relatively small displacements. (small displacements small strains)

Robot Large Displacement analysis accounts for additional stiffness and loads which result from the actual change of the local geometry of the structure caused by the applied loads. This is the so-called "stress-stiffening effect". Meaning that it accounts for an increase in bending stiffness in the presence of tension forces and decreased bending stiffness in the presence of compressive forces. This analysis is appropriate for large displacements and is the most accurate non-linear analysis provided by Robot for typical structures.

NEWTON-RAPHSON TECHNIQUE:

There are many papers available which detail this technique for the solution of non-linear problems. I want to give you the basic idea as some of us may not have studied advanced finite element analysis and even if so, may not be familiar with the solution of non-linear problem types. To boil this down, Newton-Raphson technique applies mainly the principle of virtual work to determine convergence of the model. What this says is that the total external virtual work must, at any converged state, be equal to the total internal virtual work. External virtual work is the work of the applied loads through distances that they have moved and the internal virtual work is the work of stresses over their associated strains. The solver begins with an assumption of the overall structure stiffness matrix and also portions the load to be applied into several divisions. For each division, an approximation of the displacement is calculated and a determination of the unbalanced loads is also determined. Knowing both of these quantities, we can make another approximation of the stiffness matrix and solve again (another iteration). When the vector of unbalanced loads is less than the specified tolerance, then we consider the division to have converged and the next load division is applied. This technique effectively turns a non-linear problem into a series of linear analyses which are run until convergence is achieved.

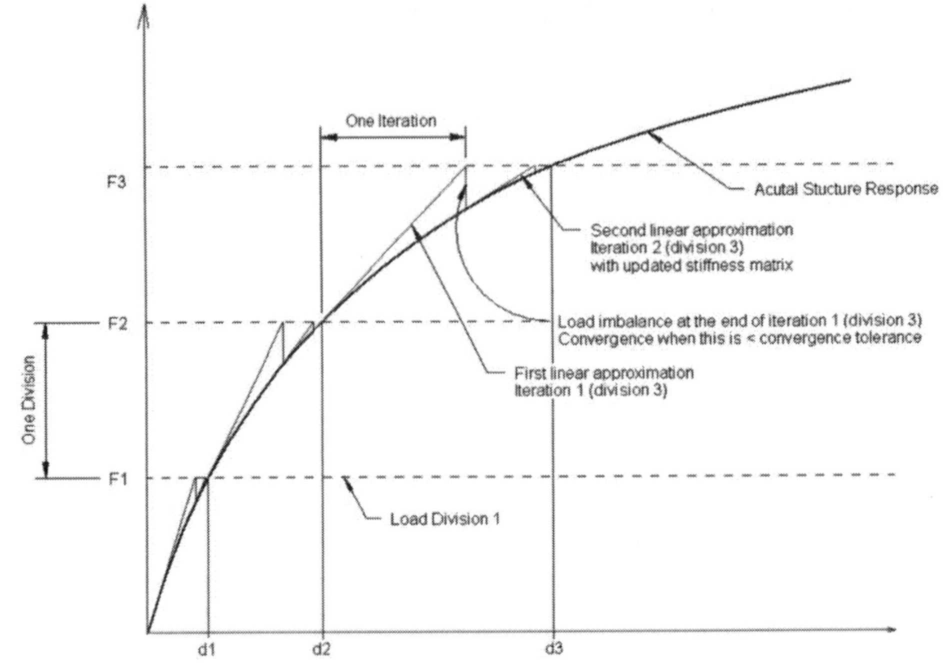

Full Newton-Raphson

As you can see, the actual structure response is approximated by a series of piece-wise linear solutions. First the load is divided into an equal number of divisions and after application of the first load division the structure is solved linearly with the original stiffness matrix. The imbalanced loads are calculated and found to be greater than the convergence tolerance. The stiffness matrix is updated and solved again. This process repeats until either convergence is reached or the maximum number of iterations has been reached. If the division converges, the next load division is applied and the solution is iterated until convergence is achieved again. This process continues until either no convergence is reached or the solution converges at the final value of load.

NEWTON-RAPHSON SOLUTION CONTROL OPTIONS:

Noting that reassembling the stiffness matrix after every iteration can be computationally intensive, a couple of computation efficiency measures have been implemented. The modified Newton-Raphson algorithm says: Only update the stiffness matrix at the end of each division. Additionally the stiffness matrix can be updated only at the beginning of the entire solution, known as the initial stress method. If you open the non-linear parameters dialog, you will find these controls at the bottom of the dialog:

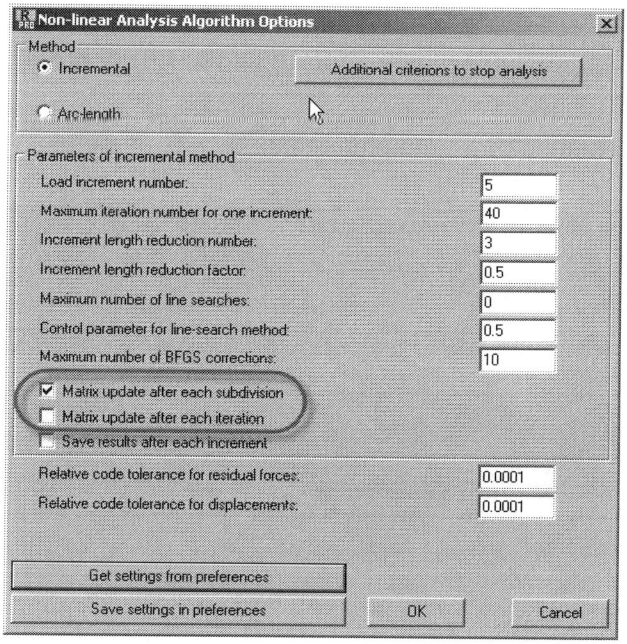

Initial Stiffness (stress) Method:

Takes the largest number of iterations to solve though each iteration is relatively less computationally intensive because the stiffness matrix is only calculated at the beginning. It is the least likely to converge easily. Set the options as follows:

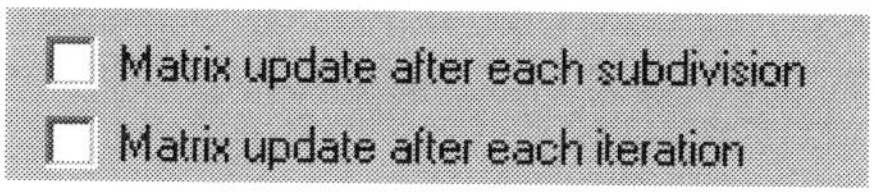

Modified Newton-Raphson

The stiffness matrix is only updated at the end of each load division which means that there will be fewer iterations than in the initial stress method and the solution will be more likely to converge though overall will be computationally more intensive due to the fact that the stiffness matrix will be recomputed at the end of each load division. Set the option as follows:

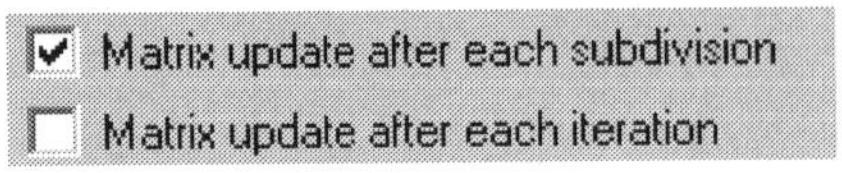

Full Newton-Raphson

The full Newton-Raphson solver will update the stiffness matrix after each iteration. This is by far the most computationally intensive of the options though will have the best chance of reaching convergence due to the enhanced approximation of the structure response at each iteration. Set the options as follows:

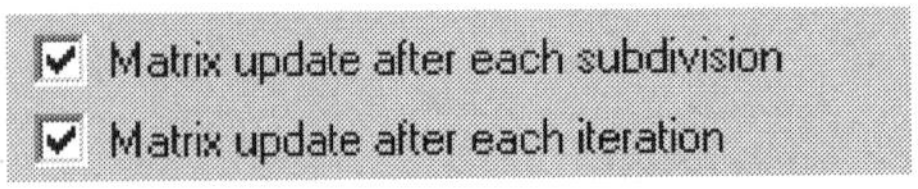

LOAD INCREMENT CONTROL:

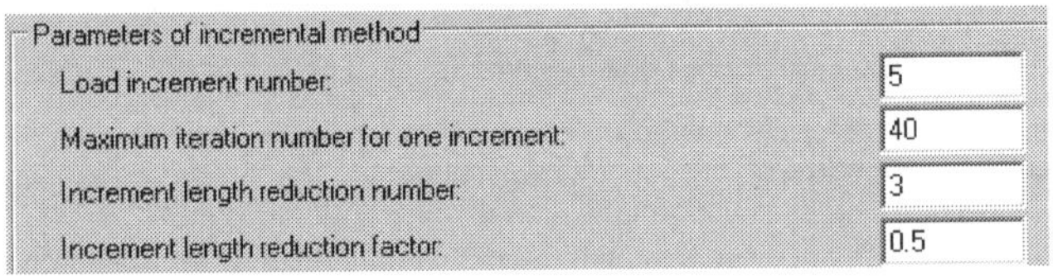

Load Increment Number: Specify the number of load divisions that will be applied to the structure. Often a problem which will not converge at a smaller number of divisions may have a better probability of converging at a larger number of divisions (smaller load increments).

Maximum Iteration number for one increment: For each loading division applied, several or many iterations may be required to achieve convergence of the model. This setting controls how many iterations may be performed before the analysis is failed (determined to not converge).

Increment Length reduction number: When the first number of load divisions fails to converge, Robot can automatically increase the number of load divisions and try again. This setting controls how many times Robot will increase the number of divisions to try again. It is initially set to 3. For a model which is not converging you can try increasing this number to allow Robot to successively increase the number of load divisions and try again.

Increment Length Reduction Factor: after the first failed attempt at convergence, Robot will automatically increase the number of load divisions and try again. Each time Robot does this it will reduce the load divisions by this factor. It is initially set to 0.5 so if Robot fails to find convergence, it will half the magnitude of each load divisions or effectively double the number of divisions. Using smaller divisions can sometimes increase the probability of achieving convergence.

ADVANCED PARAMETERS:

Line searches and BFGS corrections are both advanced methods of attempting to solve problems where the structure response might have characteristics which make standard Newton-Raphson techniques fail to find convergence. These types of problems frequently involve situations where the response involves rapid changes in the stiffness matrix, limit points, turning points or bifurcation points.

Maximum number of line searches:	0
Control parameter for line-search method:	0.5
Maximum number of BFGS corrections:	10

ARC-LENGTH OPTION:

Additionally, sometimes the use of Newton-Raphson by load increment method can result in non-convergence of the solution. Robot also provides another solution finding methodology called the Arc-Length method. This technique helps to deal with challenging response curves which might include bifurcation points or turning points. The solution proceeds to force convergence along an arc and can even handle situations where the slope of the load/deflection curve is zero or even negative because it controls both load and deflection.

This option, while powerful, is also very sensitive to settings. It is recommended to not use arc-length method without a thorough understanding of the settings which will not be covered in this text.

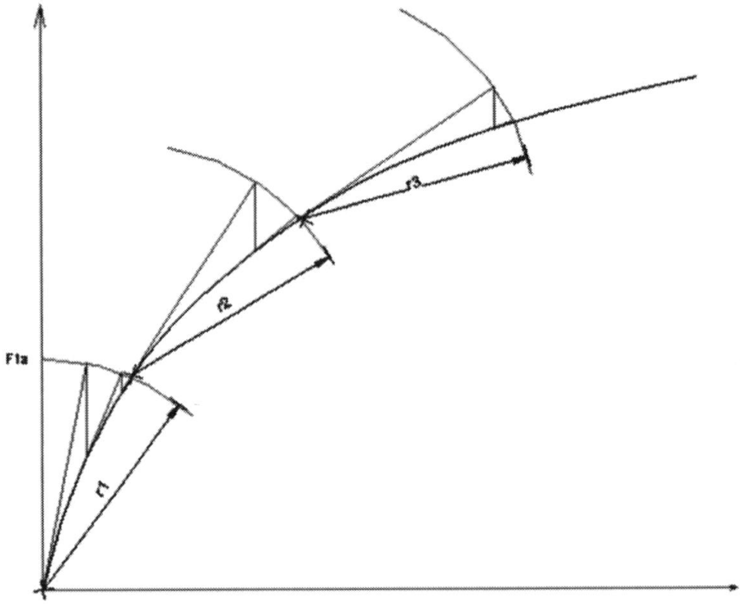

Full Newton-Raphson - Arc Length Method

P-Delta and Large Displacement examples:

Example 1: Simply supported member pinned at both ends

In the case of a beam pinned at both ends, notice that no horizontal forces arise due to the fixity of the supports in either Linear-static or non-linear analyses. Only once the Large Displacement option is enabled can we start to see the effects of the support fixity (developed horizontal reactions at the supports). This has no effect on the vertical reactions but begins to account for the fact that the deflection of the beam, caused by the load, actually results in horizontal reactions counterbalancing the sagging of the member. Also notice that for "Case 7" (Large Displacement) the deflection of the member is less than static-linear or non-linear. This is the effect of the supports resisting the deflection of the member.

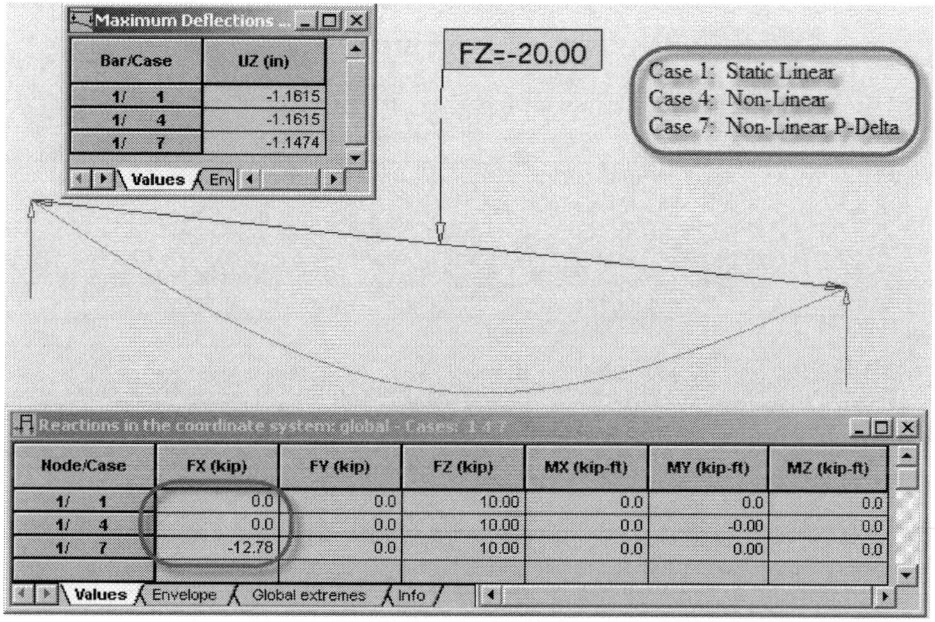

Example 2: Vertical Column with axial load and load at midpoint

In this example, note that the axial force in the member (FX) is the same for both static and P-Delta analysis though with the Large Displacement option added, part of that axial force becomes shear force (FY) due to the fact that Robot has considered the actual member internal deflections in the re-formulations of the stiffness matrix throughout the solution process. It is also interesting to note that in the presence of large deflections, simple non-linear analysis can drastically overestimate the member deflections and moments. For most engineering structures where structure drift must be limited to code specified levels, P-Delta analysis will usually be sufficient. You can always create an additional load combination with the P-Delta case result and a 1.0 factor and assign the Large Displacement option to compare results as I have done in these examples.

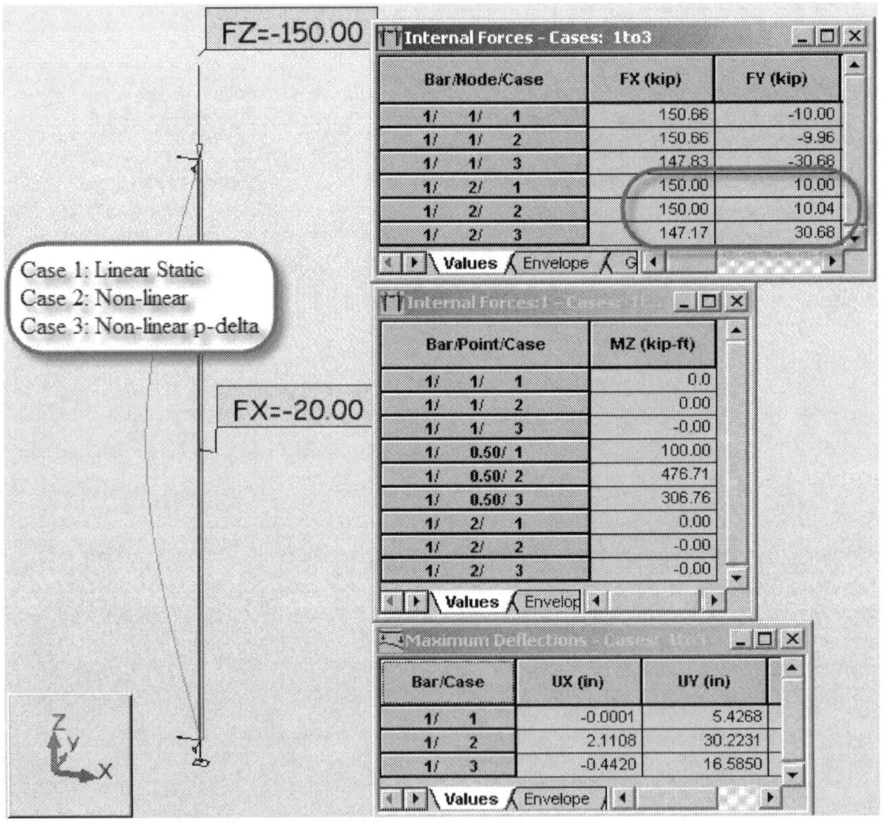

Bar/Node/Case	FX (kip)	FY (kip)
1/ 1/ 1	150.66	-10.00
1/ 1/ 2	150.66	-9.96
1/ 1/ 3	147.83	-30.68
1/ 2/ 1	150.00	10.00
1/ 2/ 2	150.00	10.04
1/ 2/ 3	147.17	30.68

Bar/Point/Case	MZ (kip-ft)
1/ 1/ 1	0.0
1/ 1/ 2	0.00
1/ 1/ 3	-0.00
1/ 0.50/ 1	100.00
1/ 0.50/ 2	476.71
1/ 0.50/ 3	306.76
1/ 2/ 1	0.00
1/ 2/ 2	-0.00
1/ 2/ 3	-0.00

Bar/Case	UX (in)	UY (in)
1/ 1	-0.0001	5.4268
1/ 2	2.1108	30.2231
1/ 3	-0.4420	16.5850

Example 4: Using Arc Length to solve a mechanism

This is a classic example of a non-linear problem which can be solved with the Arc Length control for the Newton-Raphson solver. This is admittedly not a terribly practical problem but it's a very clear example of how the arc-length method can solve a problem that load control simply cannot. Take a look at the model below, you'll notice that there are two mechanisms. Bars one and two form a mechanism and are supported on the right side by bars three and four which also form a mechanism. As this model is loaded, bars three and four will have a point of no resistance to the lateral load applied by bar two, this is normally an unstable model and cannot be solved

Direct Analysis Method

Another new feature for Robot Structural Analysis 2016 is the facility for producing a full Direct Analysis Model for use in steel design per AISC 360-10. A full direct analysis consists of the following components:

1. Application of notional loads to the structure to account for initial imperfections

2. Full P-Delta analysis (including both P-Δ and P-δ) NOTE: Robot does not offer the provision to consider only P-Δ in analysis as provided in AISC 360-10 Part C2.1-2. The DAM analysis will be full second order including both effects.

3. Consideration of flexure, shear, and axial member deformations

4. Stiffness reduction to account for material stressed into the plastic range.

Robot's implementation of the Direct Analysis Method includes all of the requirements for Direct Analysis as well as a provision for automatically maintaining the Direct Analysis model as a separate model. Having the separable model allows you to fluidly leverage the model for DAM to design steel but also to look at the base (original) model for serviceability design or other material design in a combined model. Let's look at how to configure the direct analysis method in Robot.

Enabling Member Shear Deflections

In order to perform a complete Direct Analysis Method analysis you will need to enable consideration of shear forces in the members. The shear deformations are enabled from the "Advanced Properties" dialog. You can access advanced properties from **Geometry>Additional Attributes...>Advanced Bar Properties**. This will bring up the Advanced Properties Dialog:

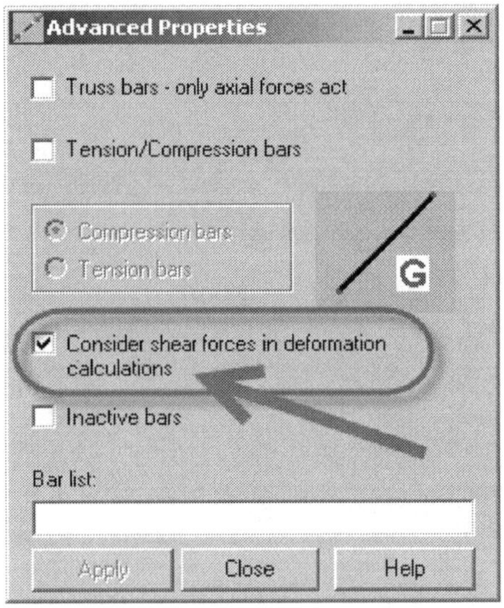

Check "Consider shear forces in deformation calculations" and then enter a list of bars to which this property should be applied. Once you have completed this step, your p-delta analysis will include the effects of member shear deformations.

Configuring Direct Analysis Method Parameters

Once you have a model constructed, from the Analysis menu select Analysis>Analysis Types... to access the Analysis Types dialog. At the bottom portion of the dialog, check the "Direct Analysis Method (DAM)" checkbox and then press the "Set Parameters" button as shown below:

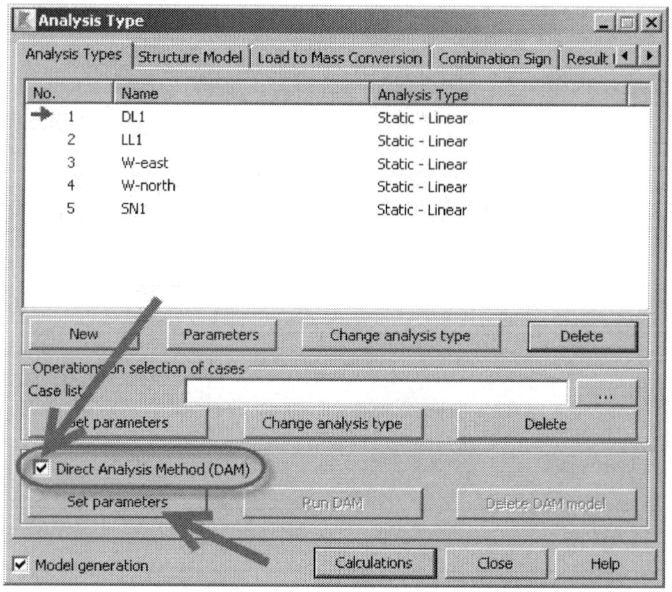

Once you press "Set Parameters" you should see the Direct Analysis Method Parameters dialog. This is where we will configure settings for both notional loads as well as reduced stiffness and second order analysis parameters.

NOTE: IF YOUR MODEL IS IN "VIEW DIRECT ANALYSIS MODEL" MODE THE FOLLOWING DIALOG WILL NOT ALLOW YOU TO MAKE AND SAVE CHANGES! STRANGE BUT TRUE. JUST SWITCH YOUR MODEL BACK TO MAIN MODEL VIEW AND THEN YOU CAN MAKE CHANGES HERE.

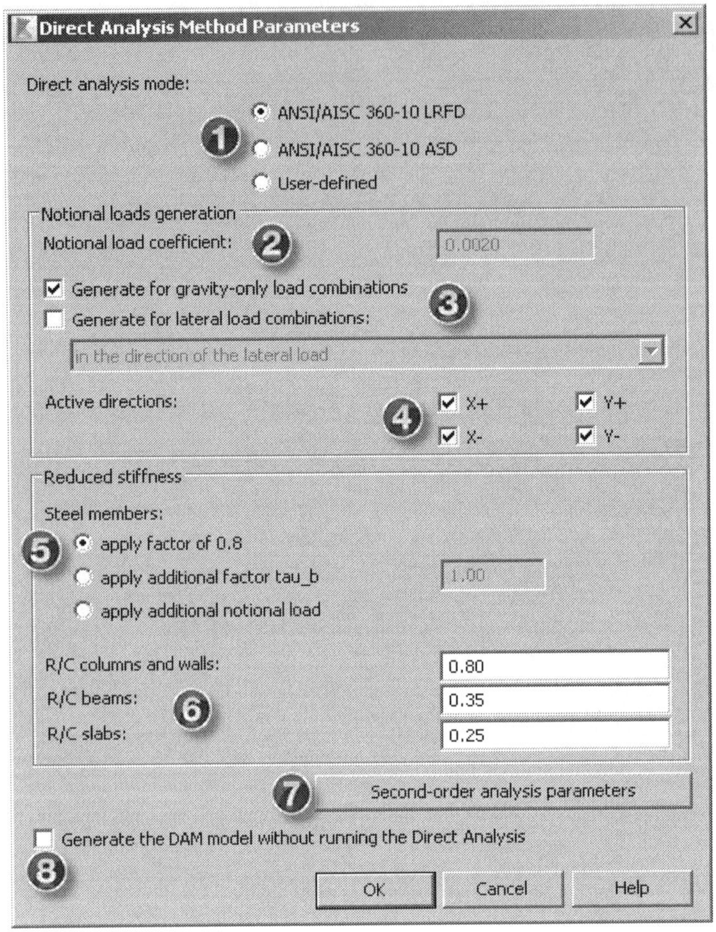

1. Direct Analysis Mode: Robot's direct analysis method is preconfigured for AISC 360-10 LRFD or ASD depending on your analysis/design preference. You may also select "User-defined" to configure your own notional load coefficient and steel member stiffness reduction. If you select one of the preconfigured options Robot will use 0.002 for notional load coefficient and offer you the options detailed in item 5 below for stiffness reduction.

2. Notional Load Coefficient: This option can only be modified if you opt for "User defined" in the "Direct Analysis Mode". If you select one of the pre-configured options this text box will be greyed-out.

3. Load Combination options: Robot gives you control over which load combinations have notional loads applied for analysis. In the DAM model, load cases matching your main model load cases are generated and analyzed. These option control which of those load combinations have notional loads applied for the analysis. The choice is yours. You can see this reflected when you switch the model to DAM View (click the button in the upper left hand corner of your view). Now the load combinations drop down looks like this:

    a.
    1 : DL1
    2 : LL1
    3 : WIND1
    4 : ULS/1=1*1.40
    5 : ULS/2=1*1.20 + 2*1.60
    6 : ULS/3=1*1.20
    7 : ULS/4=1*1.20 + 2*1.00 + 3*1.00
    8 : ULS/5=1*1.20 + 3*1.00
    9 : ULS/7=1*0.90 + 3*1.00
    10 : ULS/8=1*0.90
    11 : SLS:STD/1=1*1.00 + 2*1.00
    12 : SLS:STD/2=1*1.00
    13 : SLS:STD/3=1*1.00 + 3*1.00
    14 : SLS:STD/4=1*1.00 + 2*1.00 + 3*1.00
    15 : NL X (DL1)
    16 : NL X (LL1)
    21 : DAM - ULS/1=1*1.40
    22 : DAM - ULS/2=1*1.20 + 2*1.60
    23 : DAM - ULS/3=1*1.20
    24 : DAM - ULS/4=1*1.20 + 2*1.00 + 3*1.00 + NL X+
    25 : DAM - ULS/5=1*1.20 + 3*1.00 + NL X+
    26 : DAM - ULS/7=1*0.90 + 3*1.00 + NL X+
    27 : DAM - ULS/8=1*0.90

    b. Notice that NLX has been generated for both gravity-only load cases but only applied to the combinations involving lateral loads because I unselected the first option.

4. Active Directions: These are the directions active for application of notional loads on the structure. For each combination type selected in item 3 above, the set of notional loads will be applied in each of the "active directions" specified here. This means that, for instance, a lateral load combination involving loads to the positive X direction, will have two combinations in the DAM one with notional loads applied in the positive X and one with the notional loads applied

in the negative X direction. This will increase computation time so if your model is large enough that this becomes an issue, consider running the DAM with certain cases and directions active at a time.

5. Reduced Stiffness for Steel Members: If you select one of the pre-configured AISC options you will have the option to decide which of the allowable AISC methods you will apply:

    a. Apply 0.8 factor: The standard option for member stiffness reduction

    b. Apply additional factor tau_b: This method applies the 0.8 factor with an additional τ_b factor calculated according to AISC.

    c. Apply additional notional load: This will invoke the provisions of AISC 360-10 C2.3(3) where the notional load applied will have a coefficient of 0.003 and load combination options will be fixed to apply to all combinations.

6. Reduced Stiffness for Concrete Members: Here you can configure stiffness reduction for concrete members in the DAM. AISC requires stiffness reductions for all members in the model including concrete members.

    a. **NOTE**: PLEASE BE AWARE: Stiffness reductions applied to concrete members in the section properties dialog will be applied **in addition** to any reductions applied here in this dialog! Avoid double application of stiffness reduction by checking your concrete section properties as we discuss in Chapter 13: Beginning Concrete Design: On Concrete Elastic Displacements.

7. Second-order analysis parameters: The Direct Analysis Method requires a full p-delta analysis (Robot options non-linear + p-delta). This button will bring up the parameters you can use to configure loading steps and convergence criteria for the non-linear p-delta analysis. Please see more information earlier in this chapter on "Understanding Non-Linear Analysis".

8. Generate the DAM model without running the Direct Analysis: Check this option if you simply want to generate the model and investigate the loadings, load combination modifications. You can also do some initial investigation of the main model to determine if the model is performing more or less as you would expect and do so without the overhead of calculating several additional non-linear load combinations which can take significant time in a larger model. Once you are ready to run the DAM for use in steel design, just uncheck this option and re-run the main analysis.

Direct Analysis Method Report

Once calculations are run including a Direct Analysis Method Model, you will first be presented with the Direct Analysis Method Report. When the Direct Analysis Method Model view is active, you can access this report from the Analysis Menu: **Analysis>Calculation Notes....>Full Note** , or **Simplified Note**. Some key information is presented in the Direct Analysis Method Report including the notional load coefficient used, the maximum 2nd order to 1st order drift ratio, the maximum Pr to Py ratio, as well as the calculated τ_b required. In the case that you selected only 0.8 stiffness reduction and not the additional τ_b factor, this report value will let you know if you would be required to either revise your initial column sizes or further reduce your member stiffness by the τ_b factor:

Direct Analysis Report

ANALYSIS PARAMETERS :

Mode :	ANSI/AISC 360-10 LRFD
Notional loads generation :	
For combinations which contain gravity loads only	YES
Notional load coefficient :	0.0020
For combinations which contain lateral loads	YES
In the direction of the lateral load	
Notional load coefficient :	0.0020
Active directions	X-X+Y-Y+
Reduced stiffness :	
Steel members	
reduction method	apply factor of 0.8
Reduced stiffness :	0.80
Reduced stiffness of non-steel members	
R/C columns and walls :	0.80
R/C beams :	0.35
R/C slabs	0.25
others :	0.80

DAM MODEL :

Number of combinations generated from combinations which contain gravity loads only	16
Number of combinations generated from combinations which contain lateral loads	3
Nonlinear P-Delta analysis	COMPLETED
Maximum 2nd order drift to first order drift ratio	2.91
for the combination :	33/2
Maximum Pr/Py ratio	0.04
For elemenet No / Load case No	11 / 25
Tau_b value required	1.00

Navigating the Direct Analysis Method View

In a model which includes a Direct Analysis Method model, your views will include a button in the upper left hand corner which will toggle between the Direct Analysis Method model view and the normal model view as shown here:

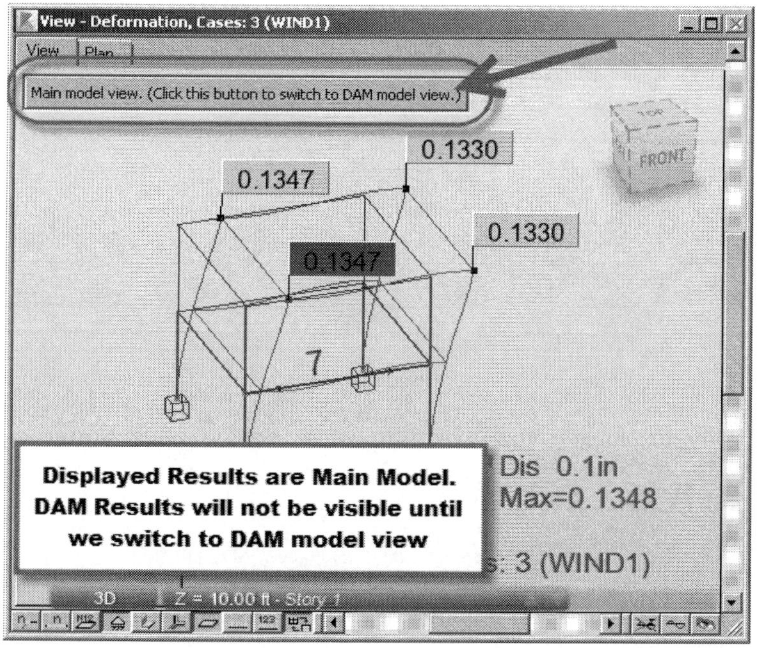

After we click the button to switch to DAM model view, our DAM displacements will be shown which include notional loads applied to the combinations, member stiffness reduction, shear deformations, and p-delta effects:

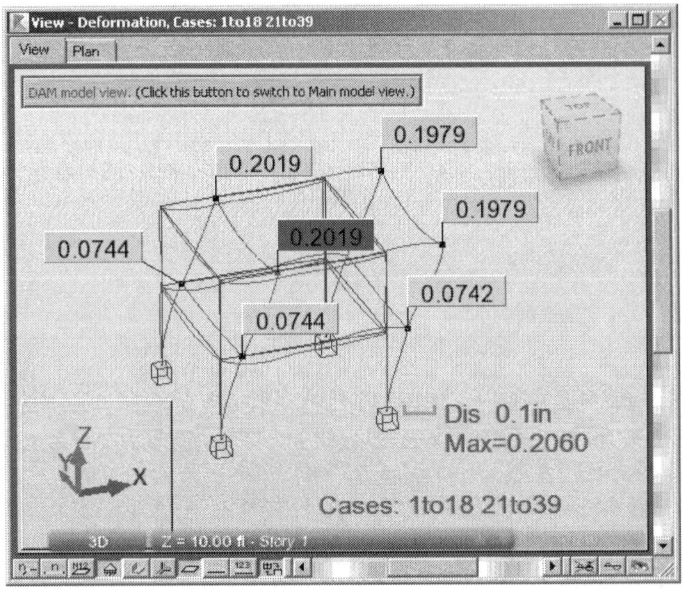

Note the increase in displacement as a result of the notional loads and p-delta effects, etc. Each view or table will have a red border around it when viewing the DAM model so that you can distinguish between the two different model results. See Chapter 8 for changes in the steel design workflow around the Direct Analysis Method.

Chapter 6 - Exploring Preliminary Results

Once calculations have been successfully run, you can begin digging into the analysis results. You can look at member displacements, forces, reactions, stresses, etc. To get started with results exploration, you can either select from the results menu or use one of the results layouts from the layout selector.

Diagrams for Bars

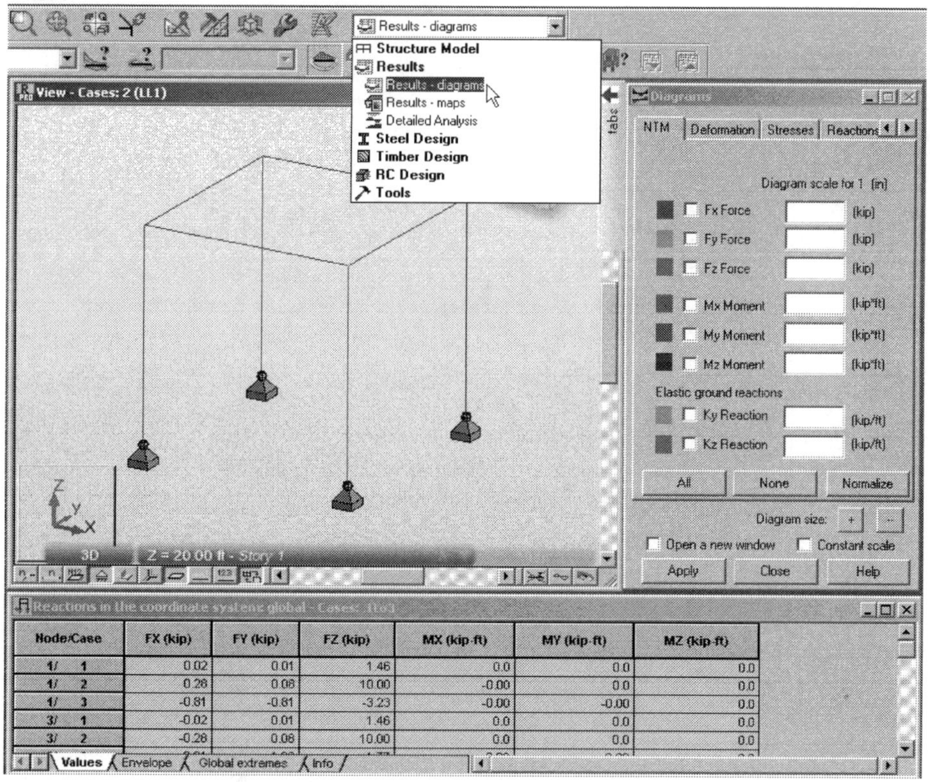

Using the results diagrams layout offers both the diagrams display control dialog as well as the reactions table to begin your investigation.

The results dialog for bar diagrams allows you to view member force diagrams, structure displacement, reactions as well as stresses and reinforcement requirements. Here is the general operation of the diagrams results dialog:

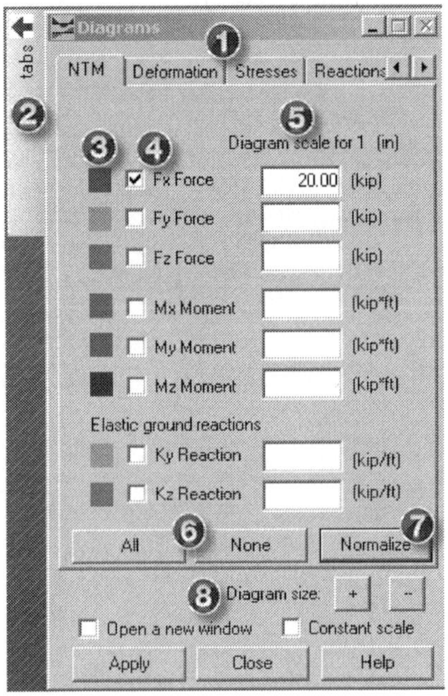

1- Result Type Tabs: Results are divided into logical groupings of member forces (NTM), Deformations, Stresses, Reactions, etc. Select the tab for the type of results you wish to display.

2- Quick access tab fly out: Hovering over this will offer a fly out menu of all the tabs available in the dialog. This saves you from needing to use the left/right buttons ◄ ► to access other tabs.

a. 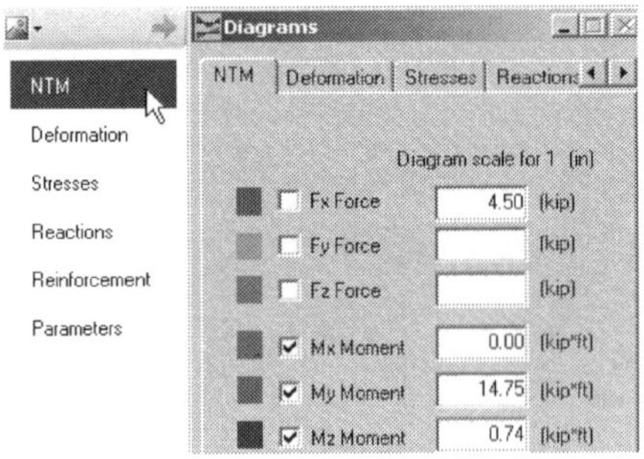

3- Color for specified result: Clicking on the color swatch next to any calculated quantity will allow you to change the displayed color for this quantity.

4- Calculated Quantity: Select the checkbox next to the quantity you wish to display. It will not be visible in the view until you press either "Normalize" or "Apply"

5- Diagram Scale: Using the "Normalize" button below will automatically determine a reasonable scale for diagrams. However, you can also enter a custom scale to increase or decrease the size of the diagrams. Enter a smaller number for larger diagrams and a larger number for smaller diagrams.

6- All/None: Use these buttons to select or unselect all calculated quantities on this tab at once.

7- Normalize: Pressing this button will apply your current settings for this tab to the view (unless "Open a new window" is selected) and the scale will be automatically adjusted to fit the diagram nicely on the view.

8- Additional Settings:

   a. Diagram Size: Press the + or – buttons to increase or decrease the scale. Changes will not take effect until you press "Apply". Normalize will reset the scale

   b. Open a New Window: Opens a new view window containing the structure and the results settings you have configured. You may open as many view windows as you like to be able to view multiple quantities at once in separate view windows.

NTM Tab

The NTM tab is used to display diagrams on bars which correspond to the major force components of the cross-section. All forces are oriented according to the member local axes as shown below.

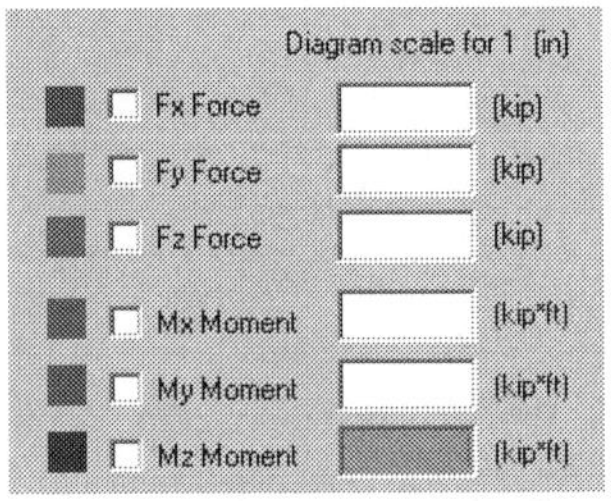

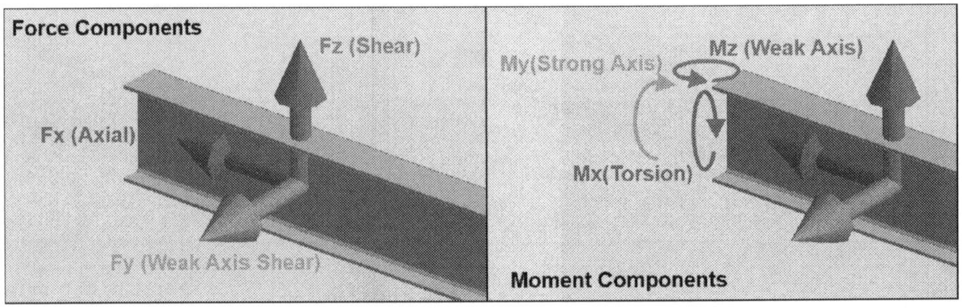

Or as shown on the cross section:

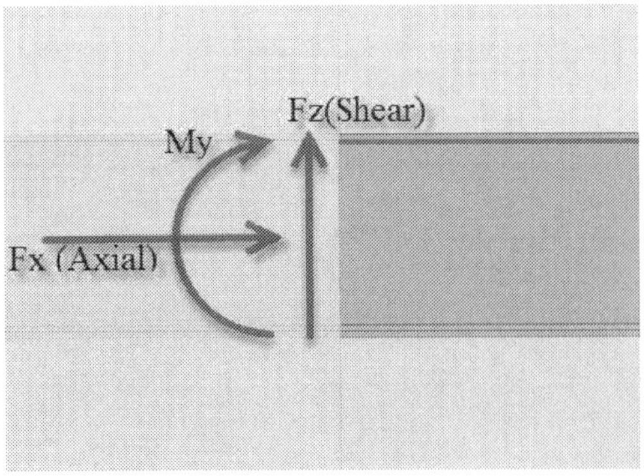

If Elastic Ground Supports have been defined for members (**Geometry>Additional Attributes>Bar Elastic Ground...**) then quantities Ky and Kz can be used to show these reactions as diagrams.

Normalize: It is probably worth noting that the normalize button is particular to this tab and applies only to diagrams configured on this tab

Deformation Tab

The deformations tab displays not only structure deflection, but will also display mode shapes of the structure if a modal analysis has been performed. You can also use this tab to animate the deflected shape to better visualize the performance of the model.

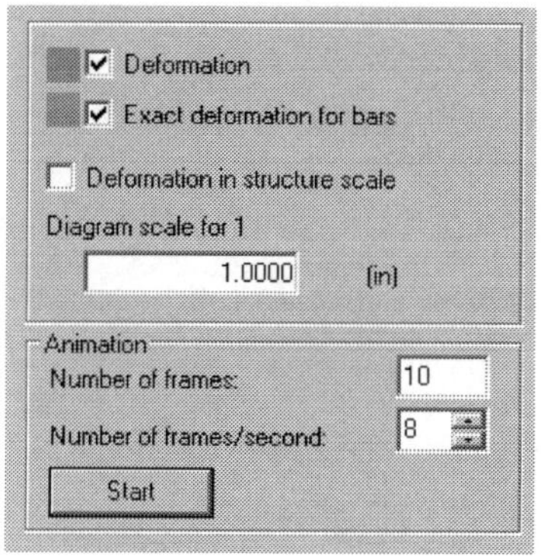

There are two options for displaying bar deformation: "Deformation" and "Exact Deformation for bars". The main difference between these two is the inclusion of the distributed bar forces in calculating the deformation. If "Deformation" is selected, only the nodal displacements and rotations will be used in determining the deformed shape of the bar. If "Exact Deformation for bars" is selected, the deformed shape is not approximated, but is determined from the nodal displacements and rotations as well as the internal forces, which is somewhat more computationally intensive. To help clarify the difference, look at this beam with mid-span alternating loads, specifically chosen to illustrate this difference:

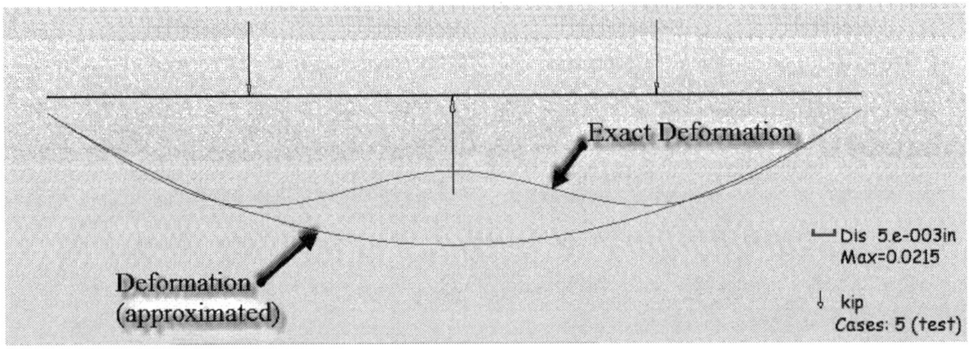

Please note that these loads were not added with a calculation node and we have already discussed issues which can arise if a calculation node is not generated under concentrated loads. If they are added with the option to "Generate a calculation node" at the load point, these two curves will almost perfectly coincide. In most cases the general deformation display will be adequate, but to visualize the exact deformation considering mid-span loading, choose "Exact Deformation".

The Normalize button will select an appropriate scale for the diagram. Specifying a larger number in the scale edit control will reduce the size of the diagrams and a smaller number will increase the size

Animations

To view an animation of the displaced shape, specify the number of frames and the number of frames to be displayed per second in the animation.

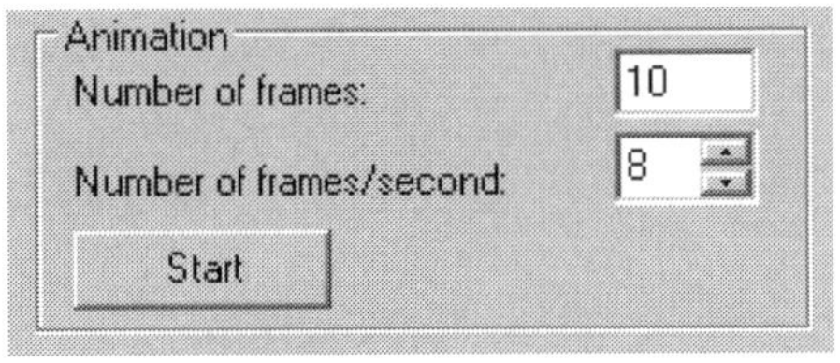

A larger number of frames will create a smoother animation and number of frames per second will adjust the speed of the visualization: specify a smaller number for a slower animation and a larger number for a faster animation.

Once you press the "Start" button, Robot will assemble the animation for you and once completed, will start to play the animation. The animation is an AVI file which Robot has created and layers on top of the view window. (This means that all view controls will be unavailable during playback though it will seem like they are available.) During animation playback you will see this toolbar in the Robot environment:

The Animation view controls from left to right:

1- Open a saved AVI animation file
2- Save current animation to an AVI file
3- Play
4- Pause
5- Stop
6- Rewind (limited functionality)
7- Fast-forward (limited functionality)
8- Previous frame
9- Next frame

As long as this toolbar is open, the animation will continue to be layered on top of the view window and you will not have access to the view. Simply close this toolbar to work with the view in a normal manner again.

Reactions Tab

The main purpose of this tab is to display reactions at supports, but it can also display inter-element forces (Residual forces) or forces which have been applied to the structure that arise from total modal contributions of mass participation in seismic loading (Pseudostatic Forces). Additionally, reactions along linear supports can be displayed as diagrams rather than individual nodal reactions.

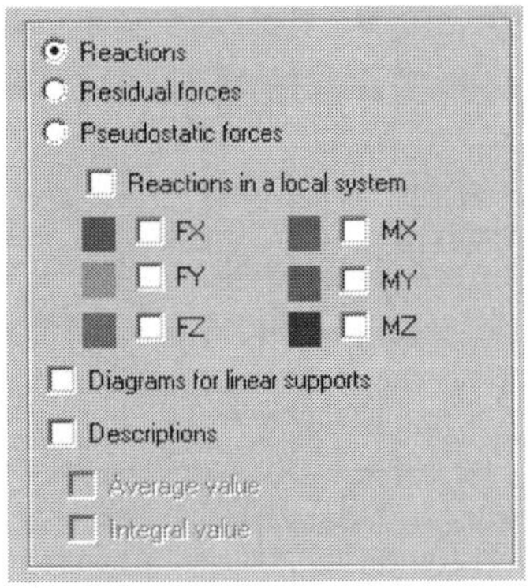

Reactions:

Selecting reactions and force or moment components (FX, FY, FZ, MX, MY, or MZ) will display arrows and/or moment vectors associated with the reactions at supports. Choose "Descriptions" for tags which call out the reaction values at the supported node.

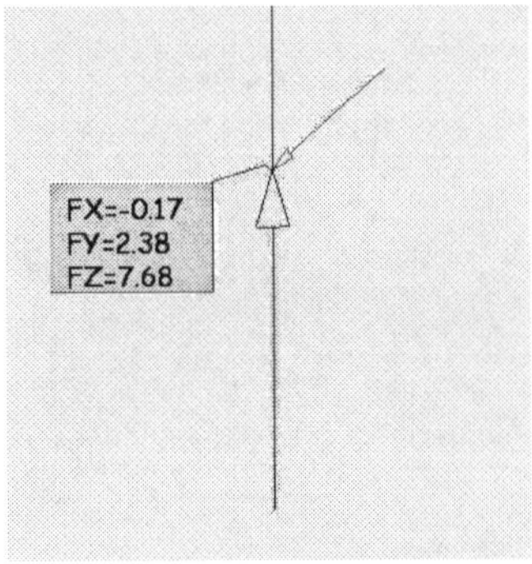

Reactions in a local system

Select Reactions in a local system to display the components relative to a support which has been rotated relative to the global coordinate system. Otherwise, Robot will display all reactions in the global coordinate system.

Residual Forces:

These are the sum of forces at each node in the structure. They are used as a visual check of the equilibrium of the structure. Expect these to be zero unless there is a problem with the model. Sometimes if a mesh is not fine enough, the discretization can give rise to unbalanced force sums at nodes and these residual forces can help you identify areas where the model might not be performing as you expected. For surface elements, it can be useful to plot pressure bands to look for discontinuities which would suggest similar issues with a mesh not fine enough for the analysis. Descriptions will also work with Residual Forces.

Pseudostatic Forces:

"Pseudostatic forces" refers to nodal loads which arise from loads indirectly applied to the structure. For example, the mass of the structure subjected to a spectral loading or other seismic type analysis or loads applied to mid-span of elements (not at nodes). Selecting "Pseudostatic forces" will show you loads which have been applied to nodes of the structure resulting from seismic or other loading. Below is an example of a seismic load generated from modal analysis and the pseudostatic forces applied to the structure nodes as a result.

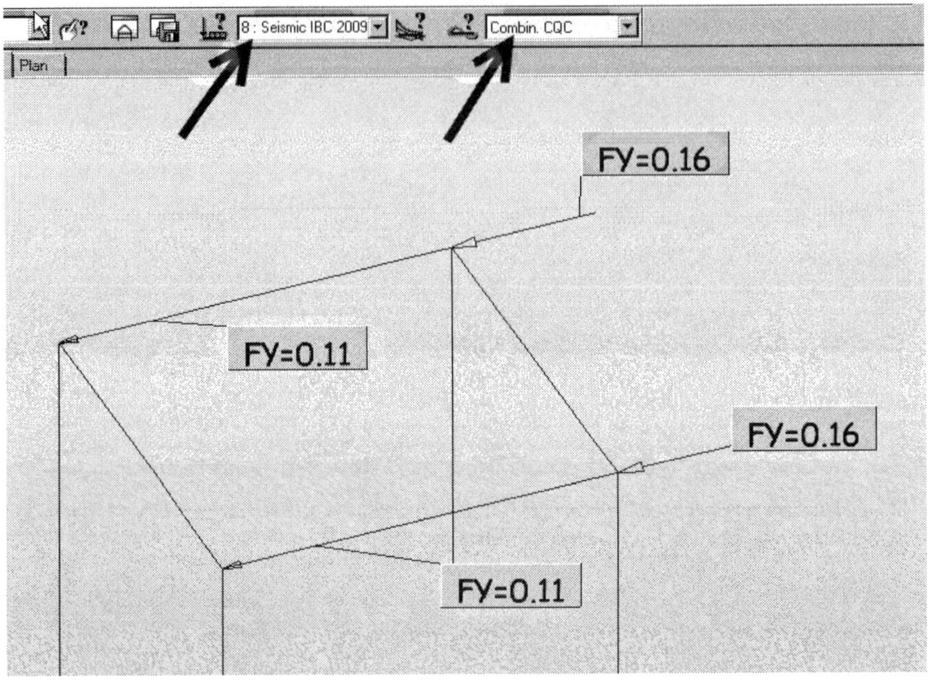

Pseudostatic "reactions" will also show you where Robot has generated nodal loads resulting from applied loads.

In the example below, a 20k load has been placed at mid-point of a beam element (again, without generating a node! A little contrived, but a great example of what Robot is showing you). I've isolated just this one element for illustrative purposes:

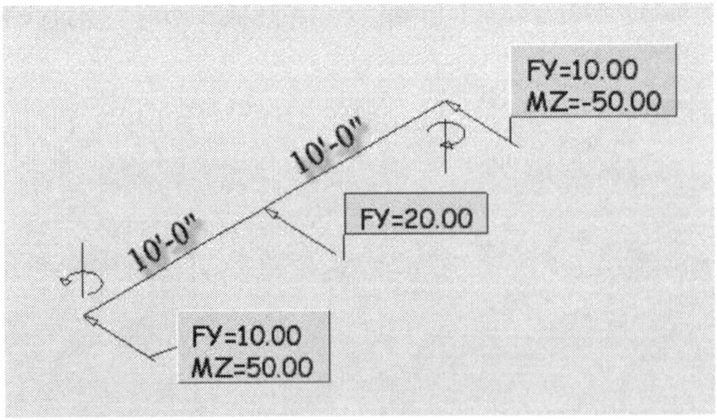

Notice the “Pseudostatic” forces shown by Robot for this case: FY=10 and MZ=50 at each end. Recall the fixed end moments and forces for a concentrated load at mid-span:

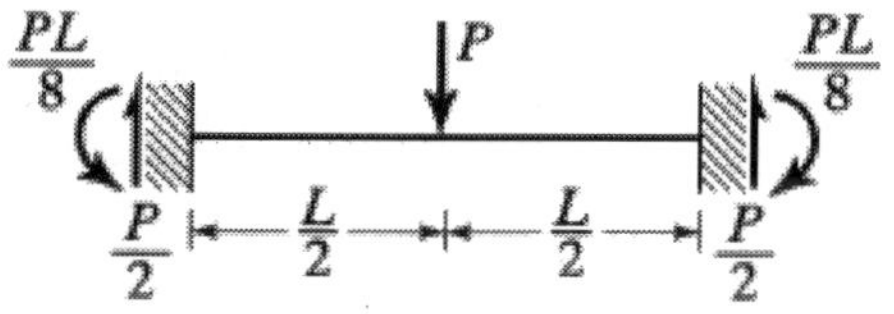

MZ is P*L/8 = 50 ft-k combined with P/2 = 10k in the FY direction.

I hope that gives you a fairly clear picture of what pseudostatic forces are. The terminology is somewhat unusual, but the functionality is actually very useful for investigating exactly what is going on with your model and how it is behaving.

Parameters Tab

The parameters tab offers options for configuring the display of diagrams on the structure. Most are self-explanatory or are well covered in the help file.

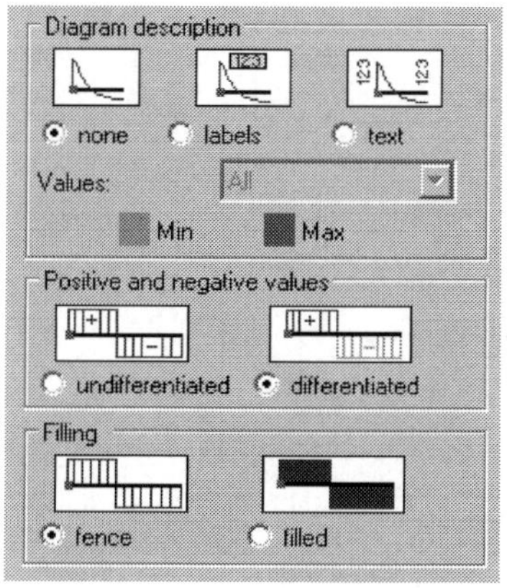

Detailed Analysis

Detailed Analysis can be used to investigate just one bar for several diagrams at once and also to simultaneously view the bar deflections.

Select a bar of interest and start the detailed analysis from the results menu (**Results>Detailed Analysis...**). This will launch the Detailed Analysis dialog:

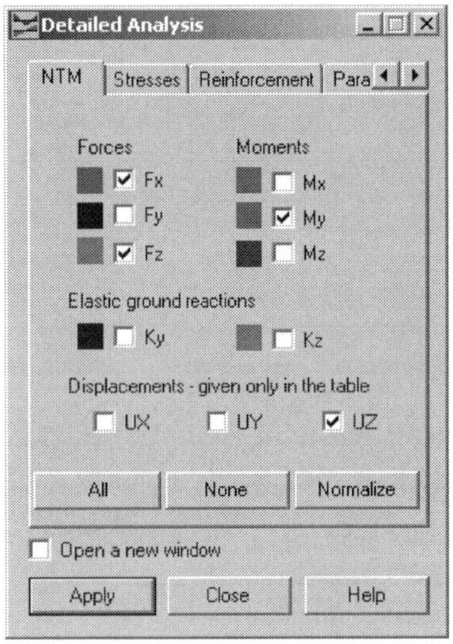

Notice that this is very similar to the diagrams dialog we have just discussed. This dialog works in conjunction with one selected element. If you selected an element before launching this dialog, then you will see your element in the detailed analysis window though no diagrams will be shown until you select some quantities of interest and press “Apply”. Once in this dialog, you can return to the view and select another element. Using “Open in a new window” and “Apply” you can have multiple windows of detailed diagrams for different elements. Shown below is an example of one element with axial, shear and moment diagrams shown. Robot by default shows the bar element, and the load associated with the current load case. Bar deflections, if desired are only shown in the table list below. Control the number of deflection points shown on the “Division points” tab of the Detailed Analysis dialog.

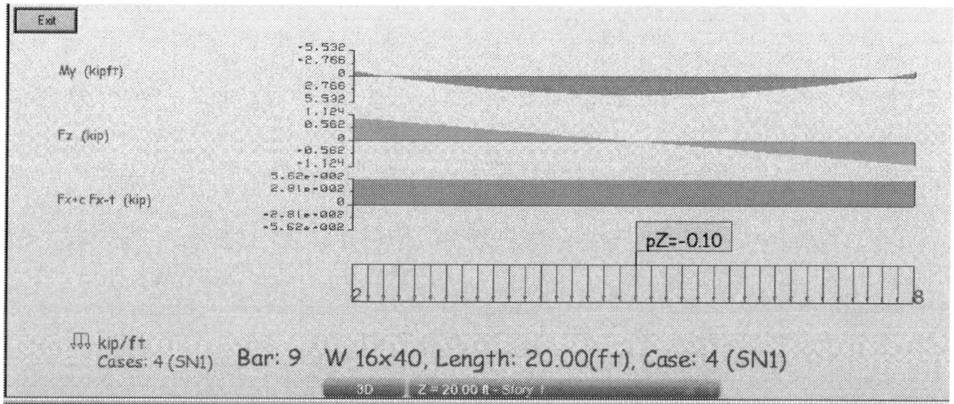

Bar / Point (ft)	FX (kip)	MY (kip-ft)	UZ (in)	FZ (kip)
Current value	0.06	-1.10	-0.0017	1.00
for bar:	9			
in point:	x=0.0 (ft)			
9 / origin	0.06	-1.10	-0.0017	1.00
9 / point x=5.00	0.06	2.65	-0.0140	0.50
9 / point x=10.00	0.06	3.90	-0.0193	-0.00
9 / point x=15.00	0.06	2.65	-0.0140	-0.50
9 / end	0.06	-1.10	-0.0017	-1.00

The Secret Gem: Object Properties

An awesome way to review member properties as well as results and design checks for a single member is to use the “Object Properties...” tool. Once you have an element selected, you can access the object properties either from the right-click menu choosing “Object Properties...” or from two other somewhat obscure locations: **Edit>Substructure Modification>Object Properties...** OR **Results>Properties>Bar Properties...** These will launch the object properties dialog for a single bar element offering you several different ways to view the information about a single element:

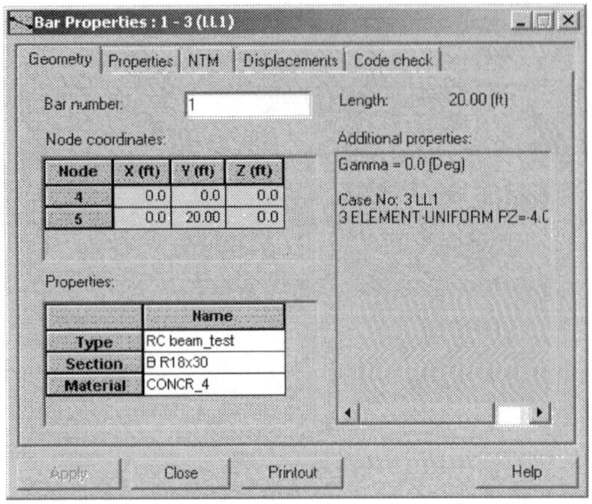

As you can see, along the top of the dialog you will find several tabs to organize the information about the member. "Geometry" and "Properties" provide information about the physical characteristics of the member (anything not greyed-out can be modified via this dialog):

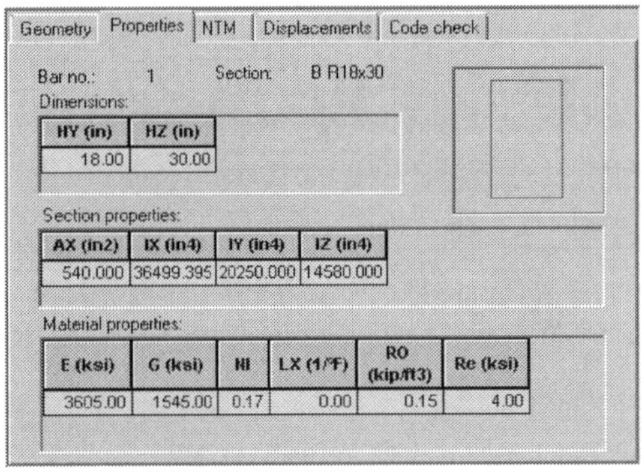

And, if calculation results are available, you will also have access to view individual force and displacement diagrams for the member:

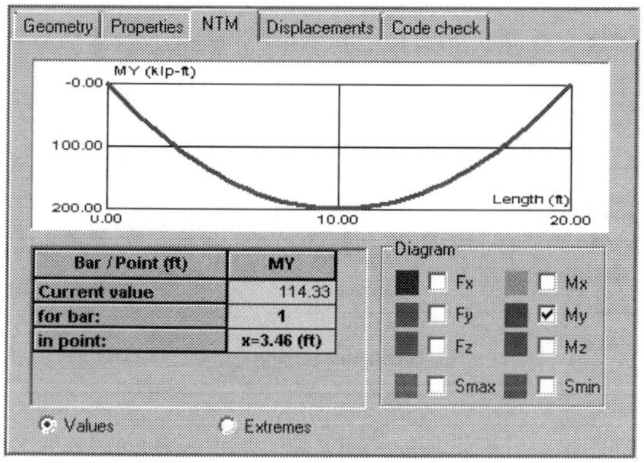

Finally the "Code check" tab will show current results of any code check calculations which have been run for the element

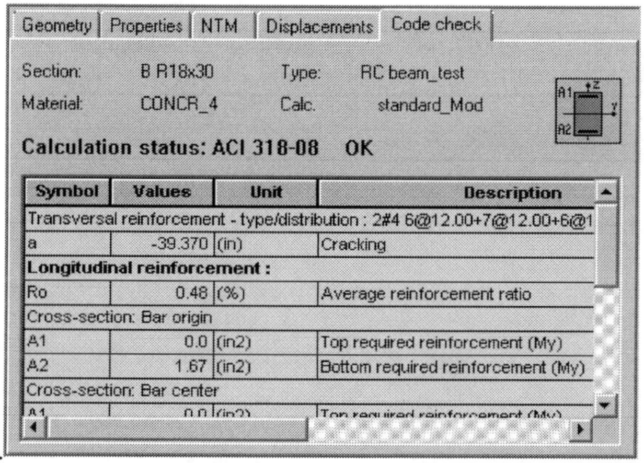

Quite possibly the most interesting part of this functionality is the report which can be generated for the element which includes member info, diagrams for results and the code check information. As with other reports you can include them in the printout composition. See section on printout composition below.

Reactions and Displacements Tables

The Reactions table (**Results>Reactions**) and the Displacements Table (**Results>Displacements**) are simply the nodal results table pre-configured for displaying reactions or displacements. In other words this table is all about calculated quantities at nodes: Reactions, displacements, pseudostatic forces, node coordinates, support information, etc.

NOTE: THE REACTIONS TABLE IS PRE-FILTERED TO NODES WITH REACTION VALUES! THE DISPLACEMENTS TABLE IS ALSO FILTERED TO NODES WITH DISPLACEMENTS. KEEP THIS IN MIND AS YOU WORK WITH THE REACTION AND DISPLACEMENT TABLES.

Along the bottom of the nodal results table are 4 tabs:

Values | Envelope | Global extremes | Info

The Values Tab:

When you first open the Reactions Table it will be on the Values tab which is configured to show force and moment reactions at each supported node:

Node/Case	FX (kip)	FY (kip)	FZ (kip)	MX (kip-ft)	MY (kip-ft)	MZ (kip-ft)
1/ 1	-0.17	2.38	7.68	-0.00	-0.00	0.0
1/ 2	0.03	0.11	4.00	0.0	0.0	0.0
1/ 3	0.0	0.0	0.0	0.0	0.0	0.0
1/ 4	0.02	0.06	2.00	0.0	0.0	0.0

Reading the left-most column is as shown at the top: Node / Case. So, in this model, the results 1/3 are read Node 1, Case 3. Then for each row simply read across for information pertaining to the particular node and case. Manual combinations and Automatic combinations will also be listed here if they have been generated. For example:

Node/Case	FX (kip)	FY (kip)	FZ (kip)	MX (kip-ft)	MY (kip-ft)	MZ (kip-ft)
1/ 1	-0.17	2.38	7.68	-0.00	-0.00	0.0
1/ 2	0.03	0.11	4.00	0.0	0.0	0.0
1/ 3	0.0	0.0	0.0	0.0	0.0	0.0
1/ 4	0.02	0.06	2.00	0.0	0.0	0.0
1/ ULS+	-0.15	3.33	16.62	-0.00	-0.00	0.0
1/ ULS-	-0.24	2.14	6.91	-0.00	-0.00	0.0
1/ 11 (C)	-0.21	2.86	9.22	-0.00	-0.00	0.0

Where ULS+ and ULS- are the maximum and minimum values of the evaluated automatic combination components (See section on automatic load combinations) and 11(C) is a manually created load combination (automatically generated manual combinations would appear in a similar manner to case 11 here).

The Envelope Tab:

The envelope tab gives you the max and min for all currently selected cases (see filtering below).

Node/Case	FX (kip)	FY (kip)	FZ (kip)	MX (kip-ft)	MY (kip-ft)	MZ (kip-ft)
1/ 2	**0.03>>**	0.11	4.00	0.0	0.0	0.0
1/ ULS-	**-0.24<<**	3.33	10.75	-0.00	-0.00	0.0
1/ ULS+	-0.24	**3.33>>**	10.75	-0.00	-0.00	0.0
1/ 3	0.0	**0.0<<**	0.0	0.0	0.0	0.0
1/ ULS+	-0.15	3.06	**16.62>>**	-0.00	-0.00	0.0
1/ 3	0.0	0.0	**0.0<<**	0.0	0.0	0.0
1/ 2	0.03	0.11	4.00	**0.0>>**	0.0	0.0
1/ ULS-	-0.24	3.33	10.75	**-0.00<<**	-0.00	0.0
1/ 2	0.03	0.11	4.00	0.0	**0.0>>**	0.0
1/ ULS-	-0.24	3.33	10.75	-0.00	**-0.00<<**	0.0
1/ 1	-0.17	2.38	7.68	-0.00	-0.00	**0.0>>**
1/ 1	-0.17	2.38	7.68	-0.00	-0.00	**0.0<<**

The rows in this table are read in the same way as the Values tab: Node/Case, but the table is arranged to show max and min values for each force/moment at each node. The max and min are highlighted and have >> or << added to indicate max or min respectively. In the example above, for Node 1 the Maximum FY is 3.33 as indicated by the shading and the >> after 3.33. It occurs in the ULS+ case (meaning somewhere in the composed combinations although the particular component is not specified. To see exactly which case component generated the maximum/minimum, use Table Columns, Extremes tab to turn on detailed information to see which exact component generated the maximum. See Automatic Combinations for more information about composed case combinations). Correspondingly, the minimum value for FY is 0.00 suffixed by << to indicate that it is the lowest value (or all other values are greater than 0.00). Max and min are presented for all forces, moments or displacements currently shown.

The Global Extremes Tab:

The global extremes tab will show the worst case maximum and minimum for either reactions or displacements of all nodes and the associated case (and case component if detailed information is selected in the Extremes tab of table columns). This tab is effectively the boiled down version of the Envelope tab allowing you to quickly identify the node with the worst deflection and/or reaction and which load cases/components are controlling.

	FX (kip)	FY (kip)	FZ (kip)	MX (kip-ft)	MY (kip-ft)	MZ (kip-ft)
MAX	0.32	3.33	16.62	0.00	0.00	0.00
Node	7	1	1	3	5	5
Case	ULS+	ULS+	ULS+	ULS+	ULS+	2
MIN	-0.32	-0.11	-1.06	-0.00	-0.00	-0.00
Node	3	7	7	7	1	5
Case	ULS-	2	ULS-	ULS-	ULS-	ULS-

Here we can see that for FY that we were previously investigating, 3.33 is, in fact, the global maximum for the entire model, not only for node 1.

The Info Tab:

In the nodal results table, the Info tab will give you a quick view of any filtering that has been applied to the table. The Envelope and Global Extremes tabs will only show envelope and extremes for the nodes/cases selected so it is very important to check and verify that you have included all nodes and cases you wish to consider before using this information for design or other considerations. Here is an example of the Info tab where I have applied filtering to only show nodes 1, 3, and 5 and also filtered down to cases 2 and 4 only.

Filtering	Node	Case
Full list	1to8	1to7 11
Selection	1 3 5	2 4
Total number	8	8
Selected number	3	2

Look at the difference in the Global Extremes tab from previous:

	FX (kip)	FY (kip)	FZ (kip)	MX (kip-ft)	MY (kip-ft)	MZ (kip-ft)
MAX	0.03	0.11	4.00	0.0	0.00	0.00
Node	1	3	3	1	3	5
Case	2	2	2	2	2	2
MIN	-0.03	-0.11	2.00	[illegible]	[illegible]	0.0
Node	3	5	1	[illegible]	[illegible]	1
Case	2	2	4	[illegible]	[illegible]	2

Was 3.33!!

Be aware that Robot table filtering applies to all tabs and all information in the table. This can lead to very undesirable conditions. Be sure to check that any filtering applied to your table is exactly what you desire and that Robot is evaluating all of the cases and nodes you wish to consider. It is a good idea to keep a full table and a filtered table as a check to make sure that no unexpected maxima or minima are missed by application of filtering to the table.

Filtering:

You can filter the view of results tables by using the node/bar/case selectors on the selections bar when the table is active. If I select node 3 and case 2 in the selections bar while in this table, here is the filtered view:

Node/Case	FX (kip)	FY (kip)	FZ (kip)	MX (kip-ft)	MY (kip-ft)	MZ (kip-ft)
3/ 4	-0.02	0.06	2.00	-0.00	0.00	0.00

This filtering is not limited to only one node and/or case:

Node/Case	FX (kip)	FY (kip)	FZ (kip)	MX (kip-ft)	MY (kip-ft)	MZ (kip-ft)
1/ 2	0.03	0.11	4.00	0.0	0.0	0.0
1/ 4	0.02	0.06	2.00	0.0	0.0	0.0
3/ 2	-0.03	0.11	4.00	-0.00	0.00	0.00
3/ 4	-0.02	0.06	2.00	-0.00	0.00	0.00

You may have multiple versions of this table open at once by selecting **Results>Reactions** (or displacements) again from the menu. Filtering only applies to one table so each table can be filtered differently. You can further access filtering from the right-click context menu in the table view which will bring you to the selection dialog where you can use more advanced creation of selection criteria for both nodes and cases:

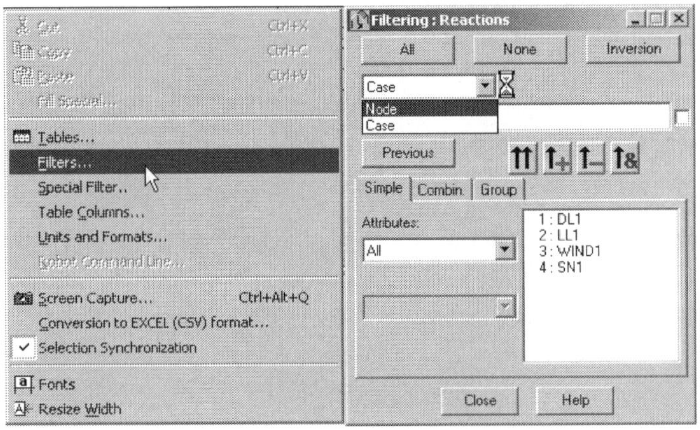

Note that "Special filter..." is different. This type of filter is also available from "Table Columns" and offers you a way to filter on particular values of interest. More on that subject in table columns below.

Table Columns:

We briefly covered table columns in our review of table functionality, but here in the Reaction and force result tables are where "table columns" really starts to have a ton of functionality.

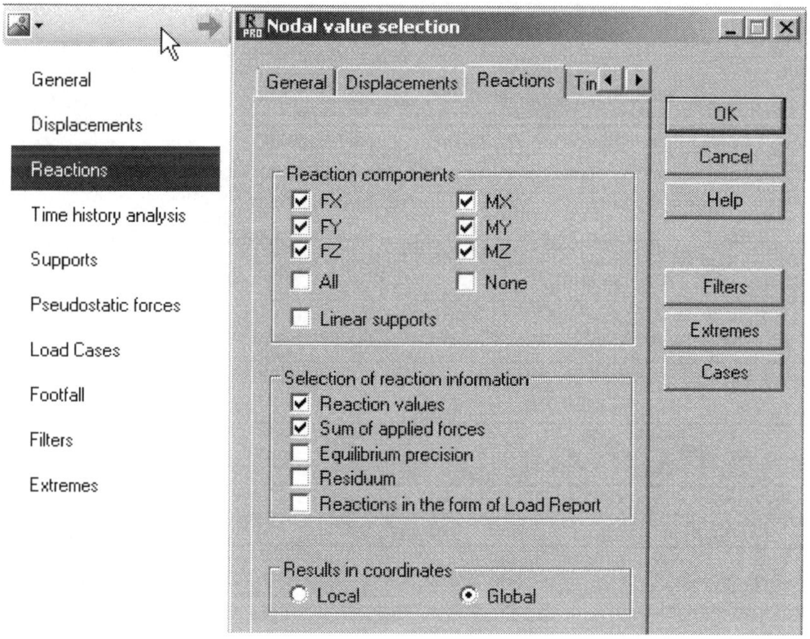

In the default configuration, all reaction components are checked on as well as "Reaction values" and "Sum of applied forces". Notice also that there are many tabs along the top allowing you to add additional columns or information to be displayed about each node and some additional tabs to provide filtering and control over the display of Extreme value information. By default the table appears this way: (I've filtered to only nodes 1 and 3, and cases 2 and 4)

Reactions in the coordinate system: global - Cases: 2 4

Node/Case	FX (kip)	FY (kip)	FZ (kip)	MX (kip-ft)	MY (kip-ft)	MZ (kip-ft)
1/ 2	0.03	0.11	4.00	0.0	0.0	0.0
1/ 4	0.02	0.06	2.00	0.0	0.0	0.0
3/ 2	-0.03	0.11	4.00	-0.00	0.00	0.00
3/ 4	-0.02	0.06	2.00	-0.00	0.00	0.00
Case 2	LL1					
Sum of val.	-0.00	0.22	8.00	-0.00	0.00	0.00
Sum of reac.	-0.00	0.00	16.00	160.00	-160.00	-0.00
Sum of forc.	0.0	0.0	-16.00	-160.00	160.00	0.0
Check val.	-0.00	0.00	0.00	-0.00	-0.00	-0.00
Case 4	SN1					
Sum of val.	-0.00	0.11	4.00	-0.00	0.00	0.00
Sum of reac.	0.0	0.00	8.00	80.00	-80.00	-0.00
Sum of forc.	0.0	0.0	-8.00	-80.00	80.00	0.0
Check val.	0.0	0.00	0.0	-0.00	-0.00	-0.00

The upper part of the table corresponds to the "Reaction values" setting and the lower part corresponds to the "Sum of applied forces" setting on the Reactions tab.

The General Tab:

All of the information about a node can be included in the table of results for nodes including things like the coordinates of the nodes (in any coordinate system you choose below), the story associated with the node, bars that join the node etc.

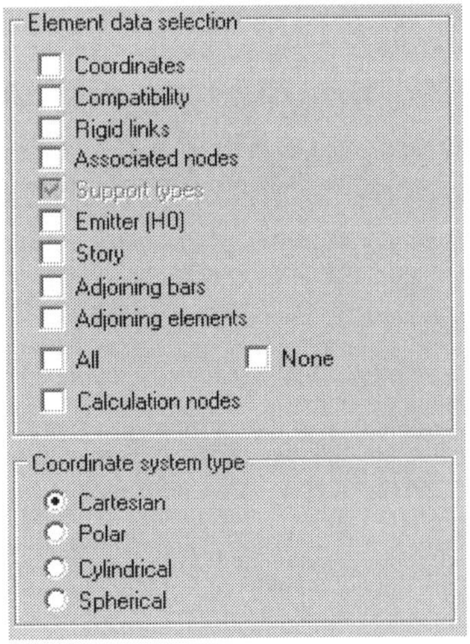

The Displacements Tab:

Here you can additionally show any displacements associated with nodes in the model. You are not limited to showing either reactions or displacements, you can actually show them all in one huge table if desired. The Envelope and Global extremes tabs will also locate the maxima and minima for displacement quantities as well. However, be warned that the "Reactions" version of the table will filter the table to only those nodes which contain reaction forces, even if you have asked for displacements too. Take a look at the Info tab to see what filtering Robot has automatically applied. Use the Displacements table instead to view all displacements of all nodes in the model.

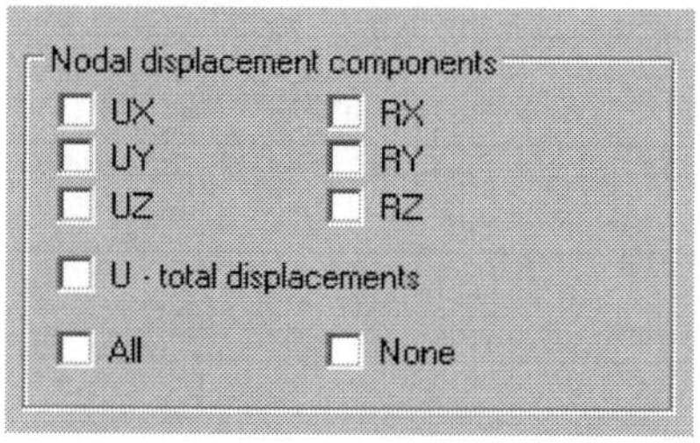

Displacements are given in global coordinates.

The Reactions Tab:

Select which components of reaction you wish to view in the table along with the style of information in the Selection of reaction information section. In order to view Pseudostatic forces for the model nodes, choose Residuum and configure the pseudostatic forces on the Pseudostatic tab.

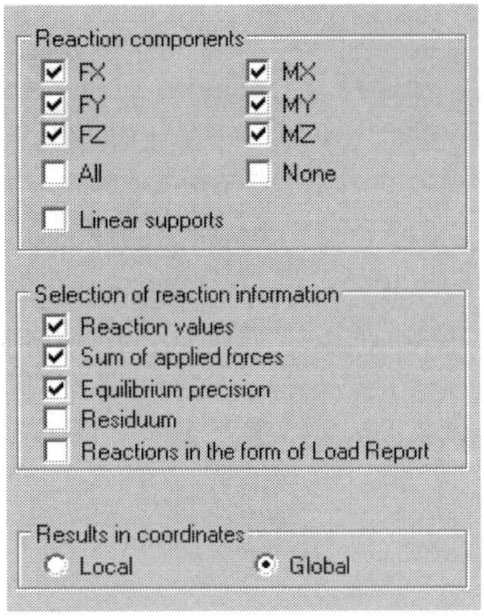

Results in a local system setting on the diagrams for bars results view is equivalent to selecting "Results in coordinates" setting of "local" where reaction can be viewed in relation to any rotation or inclination of supports.

The Supports Tab:

You can optionally include information in the table about the support conditions at supported nodes. Select Label or Support Code in the Displayed information tab to include this information as additional columns in the table.

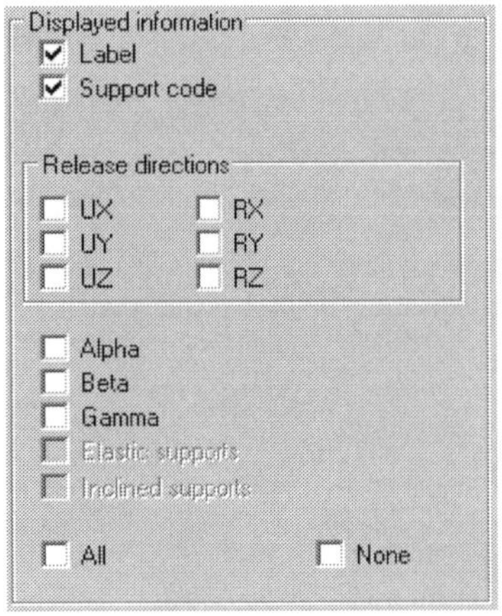

The Load Cases Tab:

You can also include information about the case or combination as additional columns in the table. Choose options here to add columns of information to the table for case or combination label and/or definition:

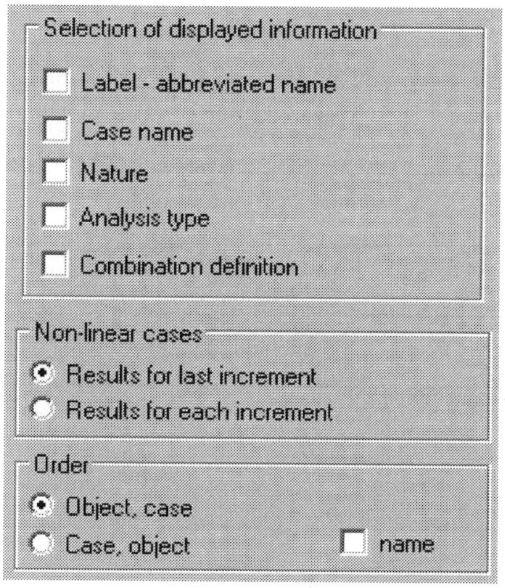

For non-linear cases, you can view results either only for the last increment (the increment at which the case finally converged at full load factor/displacement factor or the increment at which the analysis was stopped for some reason) or for each increment if you have selected "save results of each increment" in the non-linear analysis parameters. This is a method of investigating the non-linear behavior/response of a model during the analysis, not just at the final iteration.

The Filtering Tab:

You can use this tab to set up filtering of the table based on result values. For instance, you want to filter the reactions table to only FX reactions greater than 20kips. In the filtering tab, you can select FX, configure the type of filtering and supply values for filtering as shown here:

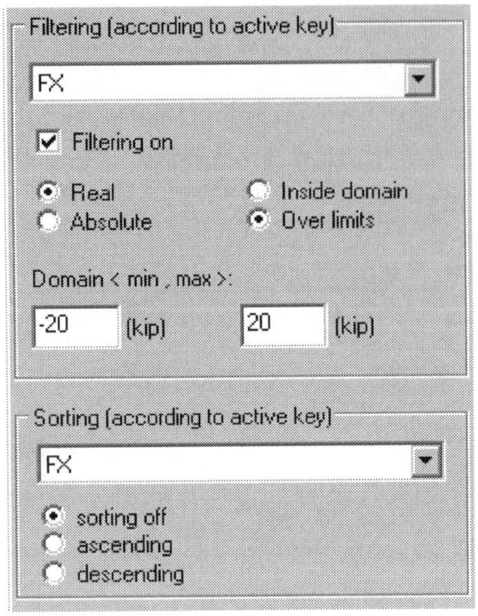

You can only filter on one quantity at a time. Choose the quantity on which you want to filter from the drop down list, select "Filtering On" and configure the filtering options:

1. Real vs. Absolute: Consider sign or do not consider sign of each value

2. Inside Domain vs Over limits: Show either values which fall within the values specified in "Domain" below or show only values which fall outside of the domain specified below.

3. Domain: The range of values for the selected quantity above.

You can further apply sorting to the table based on this quantity if desired. Select either ascending or descending to have the table sorted based on this quantity. You can also double-click on a column header to sort according to that column.

The Extremes Tab:

This tab allows you to configure the behavior of the table on the Envelope and Global Extremes tabs:

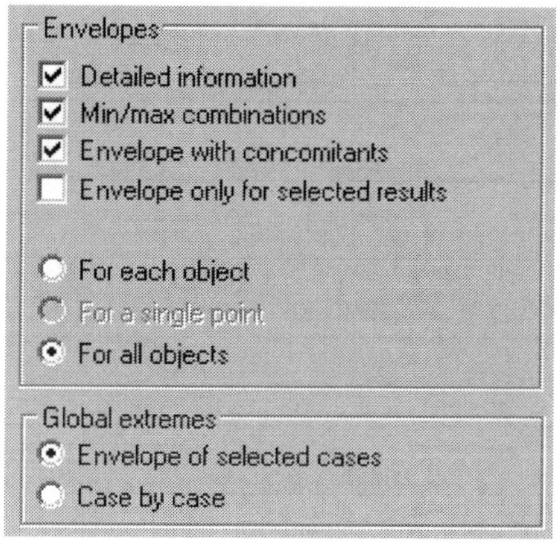

1- Detailed information: Shows which case component is responsible for the max or min. If unchecked, Robot will simply display ULS+ or ULS- if you are using composed cases.

2- Min/Max Combinations: Add a column to the envelope tab which shows the definition of the combination which resulted in the max or min.

3- Envelope with Concomitants: On the envelope tab, show the associated forces/moments or displacements for the node which also occur with the max or min for the current quantity. This effectively collapses the table down to only the envelope values. Here is the difference:

a. With:

Node/Case	FX (kip)	FY (kip)	FZ (kip)	MX (kip-ft)	MY (kip-ft)	MZ (kip-ft)
1/ 2	**0.03>>**	0.11	4.00	0.0	0.0	0.0
1/ ULS/1	**-2.16<<**	25.91	52.09	-0.00	0.00	0.0
1/ ULS/1	-2.16	**25.91>>**	52.09	-0.00	0.00	0.0
1/ 3	0.0	**0.0<<**	0.0	0.0	0.0	0.0
1/ ULS/1	-2.16	25.91	**52.09>>**	-0.00	0.00	0.0
1/ 3	0.0	0.0	**0.0<<**	0.0	0.0	0.0

b.

c. Without:

d.

Node	FX (kip)	FY (kip)	FZ (kip)	MX (kip-ft)	MY (kip-ft)	MZ (kip-ft)
1 / MAX	0.03	25.91	52.09	0.00	0.00	0.0
Case	2	ULS/1	ULS/1	4	ULS/1	1
1 / MIN	-2.16	0.0	0.0	-0.00	-0.00	0.0
Case	ULS/1	3	3	ULS/1	4	1

4- Envelope only for selected results: If you select this option, a new button is available: Envelope:

a.

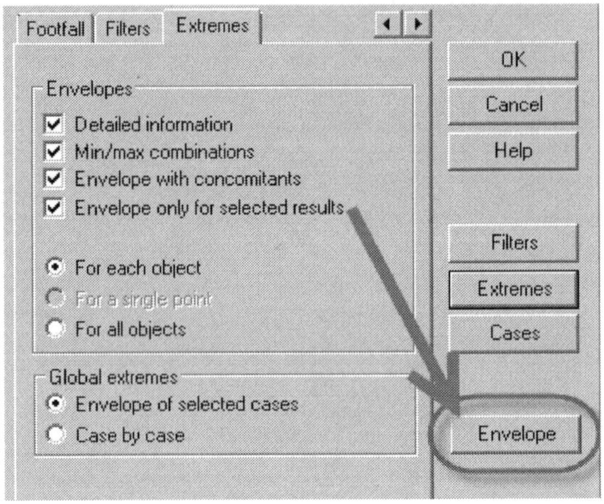

b. Select this envelope button to choose for which calculated quantities you wish to have Robot calculate an envelope.

c.

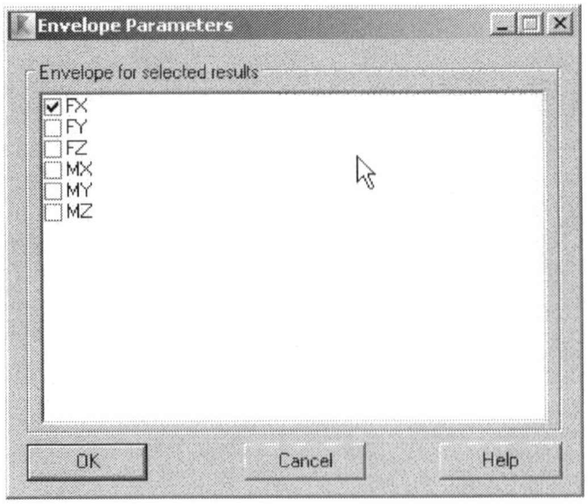

d. By default, all are selected, but you may narrow down the envelope list by unselecting calculated quantities. Here is the result of only selecting FX:

e.

Node/Case	FX (kip)
1/ 2	0.03>>
1/ ULS/1	-2.16<<
3/ 3	0.0>>
3/ ULS/1	-2.22<<
5/ ULS/1	2.16>>
5/ 2	-0.03<<
7/ ULS/1	2.22>>
7/ 3	0.0<<

5- For Each Object: Lists envelope results for each force or displacement component at each node.

6- For All Objects: Lists envelope for each force and/or displacement component for the global maxima and minima: Only lists the node with the maximum or minimum, does not show values for each node.

In the Global section, these options affect the Global Extremes tab of the nodal results table. Envelope will show the global envelope whereas case by case will show the global envelope for each case/combination/or composed case in the model.

Forces and Deflections Tables

The functionality of the Forces and Deflections tables is similar to the nodal results tables except that these tables present results for bar elements. The major difference here is that you additionally need to indicate where you would like to know the values of forces or deflections in bars. By default, you are given either maximum deflections or all six force components (FX/Y/Z, MX/Y/Z). Note that the forces and deflections tables also have 4 tabs at the bottom which function exactly like the nodal results tables. Please see the Reactions and Displacements tables section for more information on how these tabs work. We will only cover differences in functionality between these two result table styles here as opposed to repeating the same information.

The deflections table:

Open the deflections table from the Results menu (**Results>Deflections**). By default the maximum deflection values for each bar are shown:

Maximum Deflections - Cases: 1to4 6 7 11

Bar/Case	UX (in)	UY (in)	UZ (in)
1/ 1	0.0001	-1.2850	3.3021
1/ 2	-0.0000	0.0266	0.0197
1/ 3	0.0	0.0	0.0
1/ 4	-0.0000	0.0133	0.0098
1/ ULS+	0.0001	-1.1565	4.6229
1/ ULS-	0.0001	-1.7990	2.9719
1/ 11 (C)	0.0001	-1.5420	3.9625
2/ 1	0.0001	-1.3223	0.2928

In Table columns for the bar results tables, you will find the following options for bar deflections/displacements:

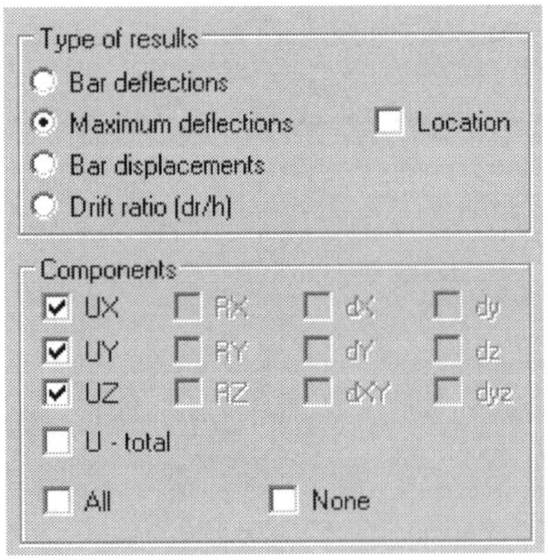

To better understand the first three options take a look at this diagram:

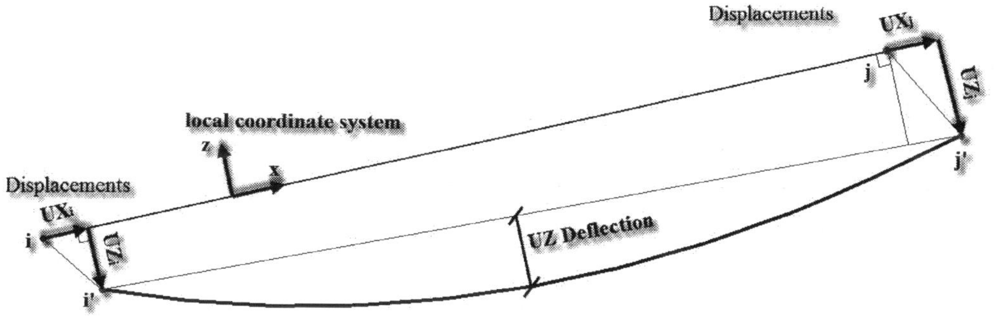

In this diagram, we ignore the y-component and look only at x/z components. The bar is initially located between points (i,j) with the local coordinate system oriented as shown. After analysis, the bar undergoes a translation as well as a deformation. This gives rise to both a displaced location/shape as well as a deflection from centerline. Even though the displacement may be large, the deflection may be small. Deflection is measured in the local coordinate system of the bar relative to its new position (i', j') whereas displacements are measured with respect to the bar's original position (i,j). The x-component of deflection is measured relative to the new position of the bar (i',j') and will always be 0 at the ends. Relative displacement of the bar within its length only, will be reflected by the x-component of deflection.

The quantity U is the vector sum of Ux,Uy and Uz.

The last option "Drift ratios" are a great way to investigate the lateral performance of columns in particular. Robot will take the relative displacements of the member ends and divide by the member length to determine the ratio. The results can be presented in the global system (best for vertical columns) or local system. For vertical columns, these will be the same, but for other elements (e.g., sloped columns, compression struts) you might want to look at the local calculation which will give you the drift of the member ends relative to itself.

The Forces Table:

This version of the bar results table is somewhat simpler than the deflections version. Again, the tables are effectively the same though they have been streamlined for the particular results. In a similar way to the nodal results table, you can display deflections along with forces in the same table, but clicking on table columns will bring you to the forces tab by default as opposed to the deflections tab. Unlike the nodal results tab which does do stealth filtering, there is no stealth filtering of bar elements from the table between the forces version and the deflections version.

The forces tab:

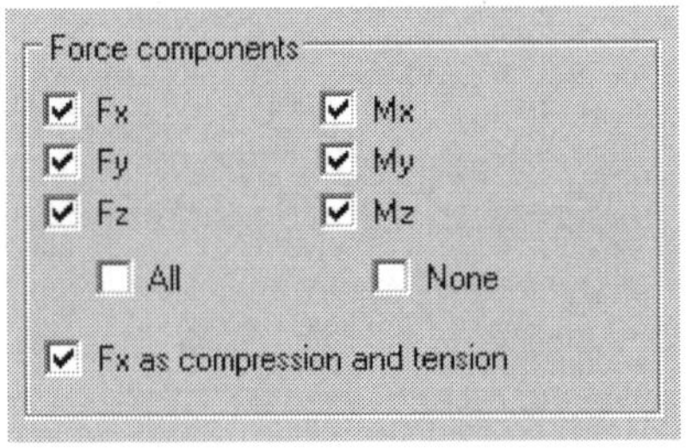

The information here is very straightforward with the additional option of having Fx or axial forces displayed in separate columns based on whether they are compressive forces or tensile forces:

Internal Forces - Case: 1 (DL1)

Bar/Node/Case	Fx compression	Fx tension (kip)	FY (kip)	FZ (kip)
1/ 1/ 1		-53.72	-0.01	-0.00
1/ 2/ 1		-53.72	-0.01	-0.00
2/ 2/ 1	57.34		0.03	0.00
2/ 3/ 1	57.34		0.03	0.00

Values | Envelope | Global extremes

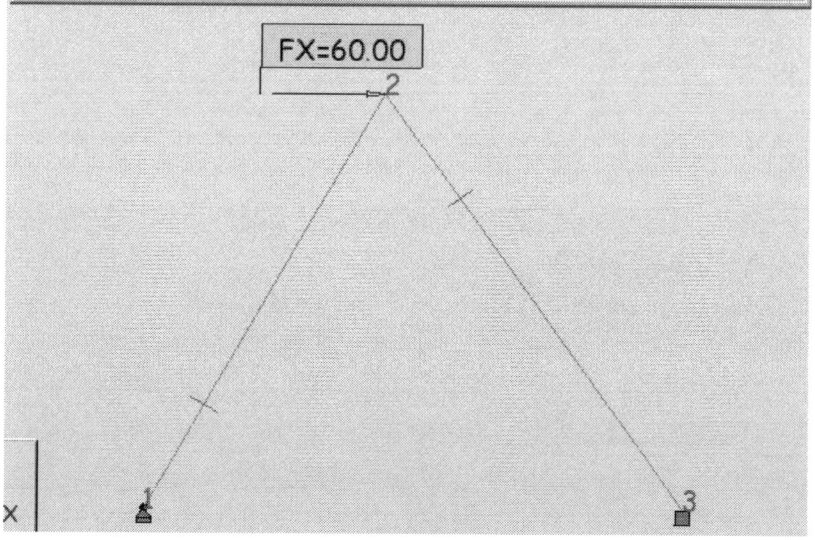

The Points Tab:

This tab has a major difference from the nodal results tab and the forces/deflections tab. Here you can configure exactly which points along the member for which Robot will list the result values.

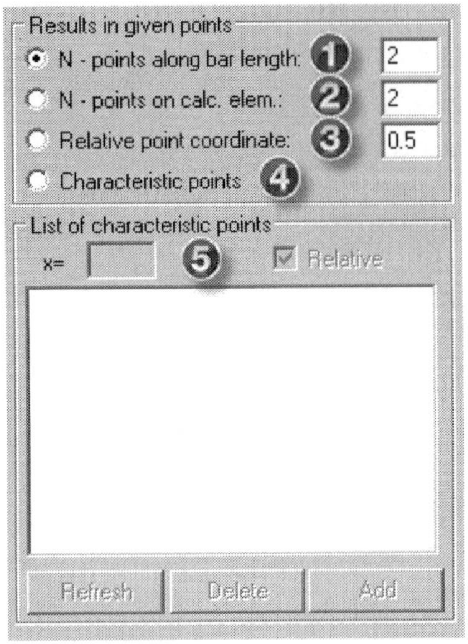

1- N-points along bar length: The entire bar is considered and results are presented at points evenly distributed along the bar length. Provide a number of points in the edit control to the right, the default number is 2 (the bar ends)

2- N-points along calculation element: If a calculation node has been added to a bar element (either by adding a point load and creating a calculation node at the same time or by dividing a bar without dividing the bar element), then you can have Robot display results at an even number of points along the calculation element. Take a look at the model below. There are only 3 bars elements in this model. Bar 1 has a calculation node, added manually by me, (See **Edit>Divide...**) along its length whereas the right side has two distinct bars (bars 3 and 4). The results are presented at 3 equal divisions "along the calculation element". So bar one has 6 results (because it contains 2 calculation elements) and bars 3 and 4 have 3 results each because bars 3 and 4 only have one calculation element each. Bars can be divided into as many calculation elements as needed while retaining their sense of being one bar in the end.

a. 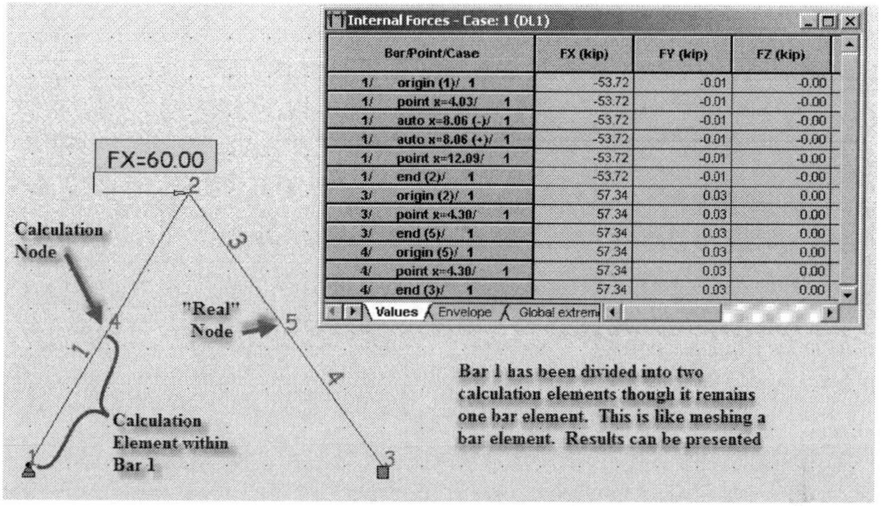

3- Relative Point Coordinate: Specify one single coordinate location along the length of the bar element relative to its length. Results will only be presented for this one point. Relative coordinates are: desired position/length. (0<x<1).

4- Characteristic points: Specify your own custom list of points where results will be presented. These points can either be real distances or relative distances as measured along the bar in the local bar coordinate system. Each time you specify a coordinate and "add" it to the list, it will be added for each bar in the model. You can further refine this list be unchecking individual points or deleting them entirely from the list.

a. 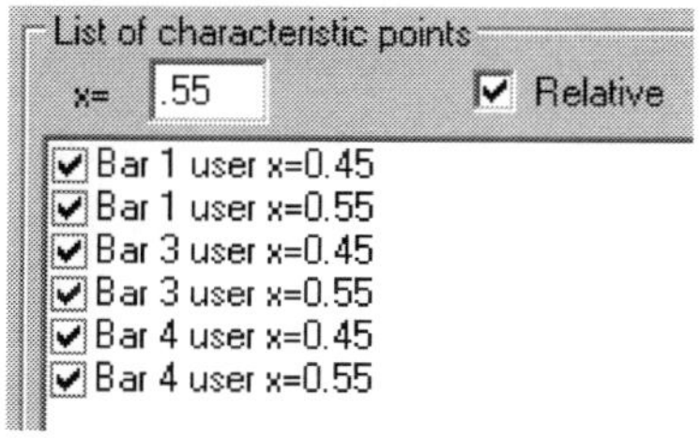

Other tabs in this table columns dialog will offer you the opportunity to add additional columns of information to the table, e.g. load case information, bar properties, material information, and section information. This additional information can be used to further investigate the results, sort and filter the information to help digest it. Check out the other tabs in the table columns dialog and see what kind of information is available.

Chapter 7 - Printout Composition

Robot has a very powerful report composition tool. Probably the most powerful part of it is the screen capture functionality. We have only alluded to this feature till now, but I hope you gain a thorough appreciation for it over the next few pages. Reports for all basic model data, table information, views as well as calculation results and member design reports can all be included in the printouts providing a very comprehensive reporting tool for documenting your design.

Screen Captures (more than just pretty pictures)

As opposed to starting with the report composer, "screen captures" are so important in Robot documentation that I thought it best to explain them first. As you're thinking about preparing your documentation, you will probably want to include images of the model, selected results visualization and any number of other things. Robot screen capture tool is great for not only views of the model, but it can also capture table views too. You can capture views with results presented and/or with model information documented. You may have as many screen captures as you desire to adequately document and communicate your analysis and code-check/design.

Model View Captures:

In any view of the model, notice the screen capture button on the standard toolbar:

 Pressing this button will launch the Screen Capture dialog where you can

configure the parameters of the screen capture. The screen capture will include only the visible portion of the current view for model view screen captures.

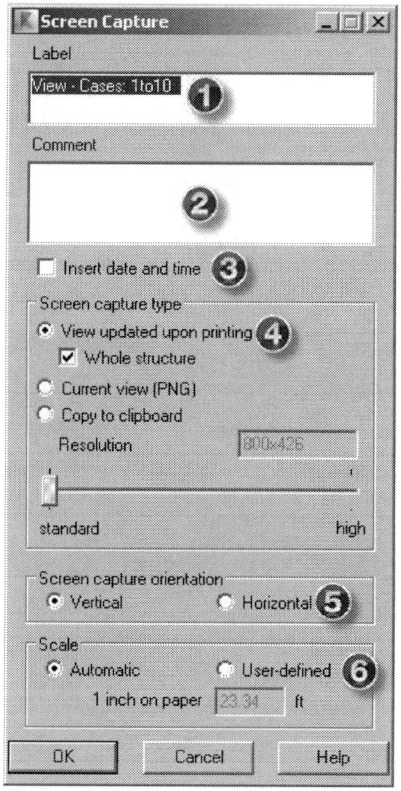

1- Label: A label is suggested for you which includes information about the load case currently displayed or other suggested information. Customize this description to match your concept of this screen capture. This label is how you will identify the screencapture when you are assembling the report. (This field is not available if copy to clipboard is selected, see #4 below)

2- Comment: An additional piece of information which will be displayed in the report when it is printed or previewed. Use this field to add descriptive text to the image. (This field is not available if copy to clipboard is selected, see #4 below)

3- Insert Date and Time: Self-explanatory. It is recommended to check this box so that you can have an additional visual verification that the screen-captures are produced at the same time as your other documentation. If View Updated on Printing is selected, this will always be the time of report generation, if Current View (PNG) is selected, the date/time will be the date and time of the capture and this option is not available at all if copy to clipboard is selected, see #4 below.

4- Screen Capture Type: This is the magic location for screen captures.

a. View Updated on Printing: This option causes Robot to effectively regenerate the screen capture with the most current information whenever it is printed or previewed. If you had captured without this option, you would have a static screen capture equivalent to a JPG file; just a still image. With this option, at the time of printing, no matter how many iterations you've been through, the screen-capture will have the latest and most current information. This is the option to use most often as old screen captures with out of date information are not typically worth much in an analysis documentation.

i. Whole Structure: With this option, new elements added to the model after the screen capture was taken will also be included automatically in the screen capture. Without selecting this option, *only* the elements in the model/view at the time of capture will be updated. Note that all geometry will be accurate, only new elements will be excluded from the screen capture when it is updated.

b. Current View (PNG): A static PNG file will be added to the list of available views for the printout composition. You can select a screen resolution for this image with the slider at the bottom of this section. There is no relationship between the current state of the model and the screen capture using this option.

c. Copy to Clipboard: If you want to use the capture in another document or report outside of Robot, you can use the copy to clipboard option to transfer the view to another program. The resolution slider will also be available for this option. There is no continuing relationship between the current state of the model and the screen capture using this option. (I.e., It is a static image)

5- Screen capture orientation: These options refer to the orientation of the image in the printout set. So vertical and horizontal refer to the paper orientation while viewing the image. Vertical means Portrait, Horizontal means Landscape.

6- Scale: In general, unless you need an image at a particular scale, it is best to leave this at Automatic. If you select a specific scale Robot will apply that scale and anything which does not fit on the page simply will not be shown. Remember that only the visible portion of the screen will be captured so the scale can be appropriate for the visible portion of the structure. Automatic scale works pretty well, but will not be scale-able easily on paper.

Table View Captures:

In a similar manner, table views when on the "Values" tab, can also be "Screen captured". As we have seen, tables can be configured in many different ways to show information in different ways. Any time you have spent effort to configure a table, that exact configuration can be screen captured including automatic updating to include in report documentation. There are fewer options with table captures as you can see from the table screen capture dialog:

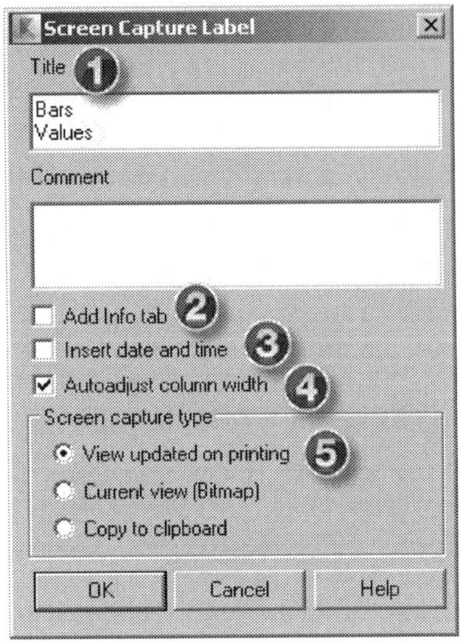

1- Title and Comment: This is the same as for model view screen captures. It is not available for copy to clipboard.

2- Add Info Tab: Include a screen capture of the info tab of the table. This is recommended so that you have a verification of which elements were included in the table for documentation. This option is not available for copy to clipboard.

3- Insert Date and Time: If capture type is set to "View updated on printing", then the date and time of printing will be inserted with the table. If Current view (Bitmap) is selected, then the date and time of capture will be in the image. If copy to clipboard is selected, this option is still available but the date and time do not actually get copied to the clipboard.

4- Auto Adjust Column Width: Automatically shrink column widths to the most suitable size for efficient use of the report space. Columns which will not fit on the page will be sent to the next page similar to how popular spreadsheet programs work.

5- Screen Capture Type: This is similar to model view.

a. View updated on printing: This option causes Robot to effectively regenerate the table capture with the most current information whenever it is printed or previewed.

b. Current View (Bitmap): Produces a still image of the table which can be inserted in a report but will not be updated with current information. Only the state of the table at the time of capture will be shown.

c. Copy to clipboard: Copies a paste-able version of the table into the clipboard. This can easily be inserted into word processing or spreadsheet programs for use elsewhere. With this option the capture will not be available for use in the Robot report.

NOTE: TABLE VIEWS CAN ONLY BE CAPTURED ON THE "VALUES" TAB OF THE TABLE OR THE "INFO" TAB OF THE TABLE. THE "EDIT" TAB CANNOT BE SCREEN CAPTURED AND THE SCREEN CAPTURE BUTTON AS WELL AS THE RIGHT-CLICK MENU OPTION WILL BE GREYED OUT.

Stealthy Screen Captures:

There are many other areas in Robot with information which you might wish to include in a report. Although there are some interesting options in the composition dialog in the simplified printout tab, I want to introduce to you the concept that "screen captures" can come from many different areas in Robot. Here are a few which you can include in reports and also report templates which you can apply to future projects.

1. Calculation notes:

While viewing calculation notes (**Analysis>Calculation Report>Simplified** or **Full Note**), take a look in the file menu of the text editor of the calculation notes report:

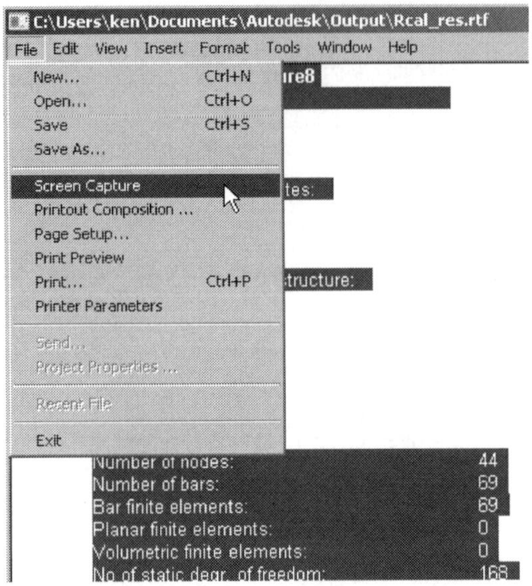

Notice the "Screen Capture" option in the file menu. Selecting this will open a dialog for capturing text reports such as this one and offer options for the "screen capture":

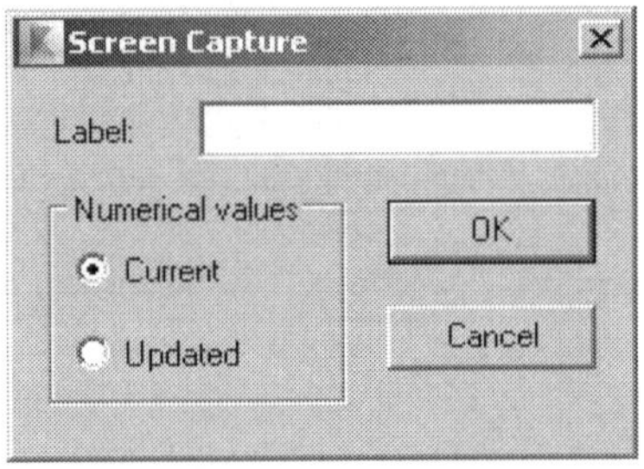

You provide a label which will be used to identify this "screen capture" in the printout composition dialog and allow you to include it with your documentation. Although the "Numerical Values" section should allow you to have this report updated at the time of printing similar to table and view screen captures, this functionality does not, as of this printing, work. Regardless of this setting, as of this printing, you will only have a static copy of this report. There is no option to copy to clipboard, if you wish, you may simply select the text and copy to clipboard with Ctrl+C or just save this report as a text file from the file menu of the report editor.

2. Steel Member Design

In the steel member verification and design dialogs you can also screen capture the design reports as well as the output table. In the Member Verification table, select "Calc. Note" to access report capture options:

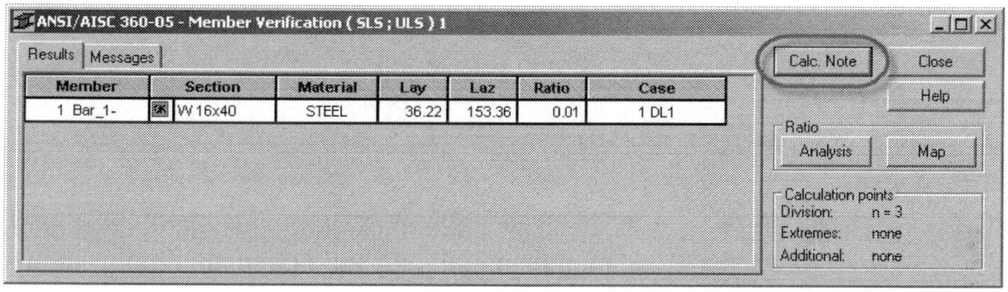

This will launch the Calculation Note dialog:

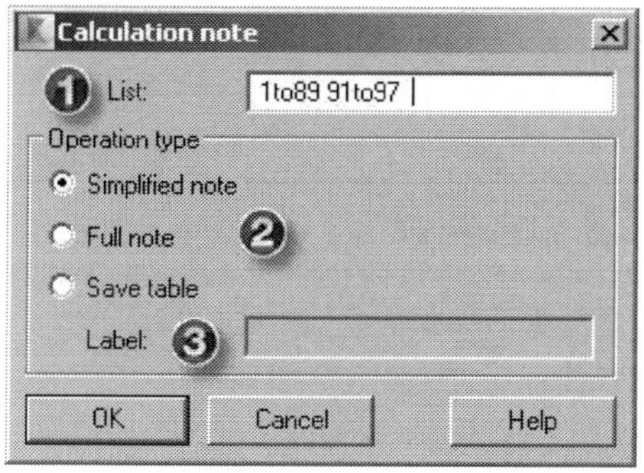

1- List: Enter a list of members for which you wish to prepare either the calculation notes or a screen capture of the table of design or verification results. Only the members you select will be included in this table or calculation note. (You can use the bar selection dialog to populate this list by pressing the bars selection icon on the selection bar. Refer to "Making Selections" earlier in the text.)

2- Choose the type of output you would like to capture

a. Simplified Note: Prepares a print preview of the list of steel members design. This is similar to saving the table as a screen capture, but is intended for more immediate printing

i. .

Autodesk Robot Structural Analysis Professional 2013
Author:
Address:

File:
Project: Structure

Member	Section	Material	Lay	Laz	Ratio	Case	Ratio(uy)	Case (uy)	Ratio(uz)	Case (uz)
1 Bar_1-	W 16x40	STEEL	36.22	153.36	0.02	1 DL1	0.00	1 DL1	0.00	1 DL1
2 Bar_2-	W 16x40	STEEL	18.81	79.62	0.00	1 DL1	0.00	1 DL1	0.00	1 DL1
3 Bar_3-	W 16x40	STEEL	36.22	153.36	0.02	1 DL1	0.00	1 DL1	0.00	1 DL1
4 Bar_4-	W 16x40	STEEL	18.81	79.62	0.07	1 DL1	0.00	1 DL1	0.01	1 DL1
5 Bar_5-	W 16x40	STEEL	36.22	153.36	0.09	1 DL1	0.00	1 DL1	0.02	1 DL1
6 Bar_6-	W 16x40	STEEL	18.81	79.62	0.07	1 DL1	0.00	1 DL1	0.01	1 DL1

b. Full Note: Extended detailed information regarding each member and parameters of design as well as the results of each code check performed for the member. This is the most comprehensive steel design report. Here is an example of a single member design report:

STEEL DESIGN

CODE: *ANSI/AISC 360-05 An American National Standard, March 9,2005*
ANALYSIS TYPE: Member Verification

CODE GROUP:
MEMBER: 1 Bar_1- **POINT:** 1 **COORDINATE:** x = 0.00 L = 0.00 ft

LOADS:
Governing Load Case: 1 DL1

MATERIAL:
STEEL Fy = 36.00 ksi Fu = 58.00 ksi E = 29000.00 ksi

SECTION PARAMETERS: W 16x40

d=16.01 in	Ay=7.065 in2	Az=4.883 in2	Ax=11.800 in2
b=7.00 in	Iy=518.000 in4	Iz=28.900 in4	J=0.790 in4
tw=0.30 in	Sy=64.710 in3	Sz=8.263 in3	
tf=0.51 in	Zy=73.000 in3	Zz=13.000 in3	

MEMBER PARAMETERS:

Ly = 20.00 ft	Lz = 20.00 ft	
Ky = 1.00	Kz = 1.00	Lb = 20.00 ft
KLy/ry = 36.22	KLz/rz = 153.36	Cb = 1.00

INTERNAL FORCES: / **NOMINAL STRENGTHS:**

Tr = 0.00 kip*ft	frvy,mx = 0.00 ksi		
	frvz,mx = 0.00 ksi		
Pr = 0.00 kip		Fic*Pn = 113.35 kip	
Mry = -1.94 kip*ft	Vry = -0.01 kip	Fib*Mny = 118.72 kip*ft	Fiv*Vny = 137.34 kip
Mrz = -0.03 kip*ft	Vrz = 0.52 kip	Fib*Mnz = 35.10 kip*ft	Fiv*Vnz = 105.47 kip

SAFETY FACTORS
Fib = 0.90 Fic = 0.90 Fiv = 0.90

SECTION ELEMENTS:
UNS = Compact STI = Slender

VERIFICATION FORMULAS:
Pr/(2*Fic*Pn) + Mry/(Fib*Mny) + Mrz/(Fib*Mnz) = 0.02 < 1.00 LRFD (H1-1b) Verified
frvy,mx/(0.6*Fiv*Fy) = 0.00 < 1.00 LRFD (G2-1) Verified
frvz,mx/(0.6*Fiv*Fy) = 0.00 < 1.00 LRFD (G2-1) Verified
Ky*Ly/ry = 36.22 < (K*L/r),max = 200.00 Kz*Lz/rz = 153.36 < (K*L/r),max = 200.00 STABLE

i.

c. Save Table: This option functions similarly to the table screen capture. You are effectively saving a static screen capture of this table for use in a report. It will not be updated so it is best to attach a date and time to the name (label) for the screen capture as a memory aid.

3- Label: If you have selected "Save Table" enter a name (and optionally a date/time) here which will be used to identify the screen capture in the printout composition dialog.

3. Required Reinforcing Code Parameters Definition

In the Required Reinforcing section the calculation parameters for concrete reinforcing can have a note created and screen captured from an html report window. To create this screen capture, in the concrete reinforcing parameters dialog, use the Note button in the bottom left hand corner. Launch from the menu: **Design>Required Reinforcement of Beams/Columns - Options>Code Parameters...** Then in the R/C Member Type dialog, open one of the definitions:

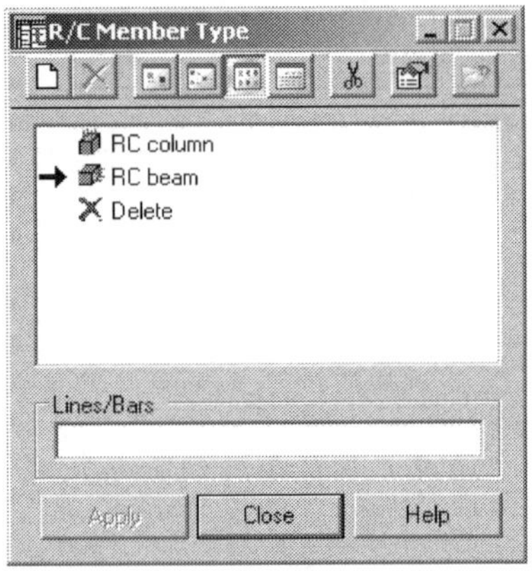

To open the member type definition dialog and notice the "Note" button at the bottom left hand corner:

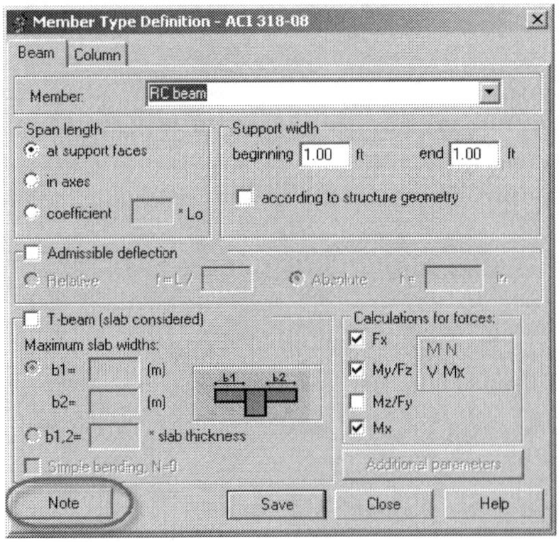

Press "Note" to generate a report of member code parameters definitions such as this:

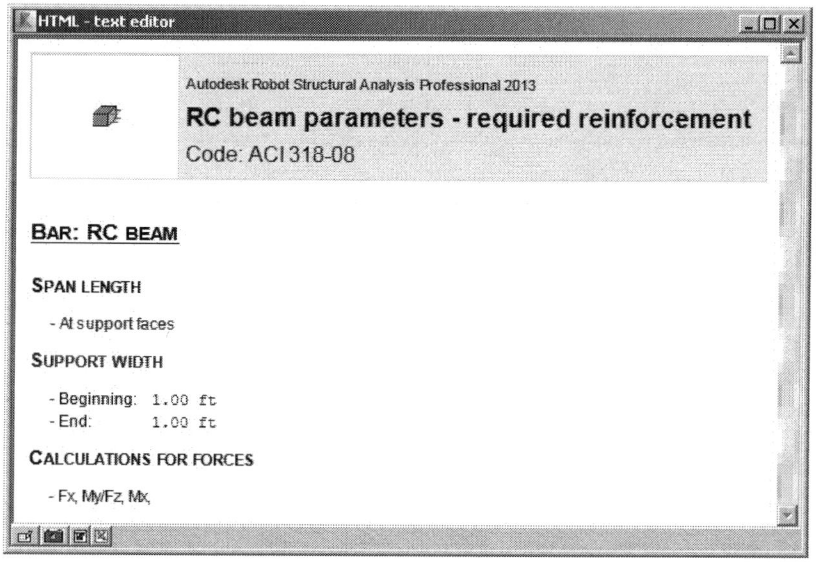

In the lower left hand corner are 4 buttons:

From left to right: Edit, Screen Capture, Export to word processor, Export to spreadsheet

Use the screen capture button  to save a screen capture of the HTML report. This report will not be updated so remember to add date/time/analysis run information into the screen capture label. In the HTML Screen capture dialog, enter the label which will be used to identify this screen capture in the printout composition dialog.

4. Required Reinforcing - Calculation Parameters Definition

This is almost identical to the above workflow so I will only mention briefly that in the Required Reinforcing design options menu: Launch from the menu: **Design>Required Reinforcement of Beams/Columns – Options>Calculation Parameters...** launches the Calculation Parameters dialog and if you open one of the parameter definitions or create a new one, in the Calculation Parameter Definition Dialog there will also be a "Note" button in the lower left hand corner allowing you to add a report of configured calculation parameters to your report. It is the same HTML style report and screen capture as above and will not be updated so use care in naming the screen capture to be sure you have the most current version.

The Calculation Parameters Dialog:

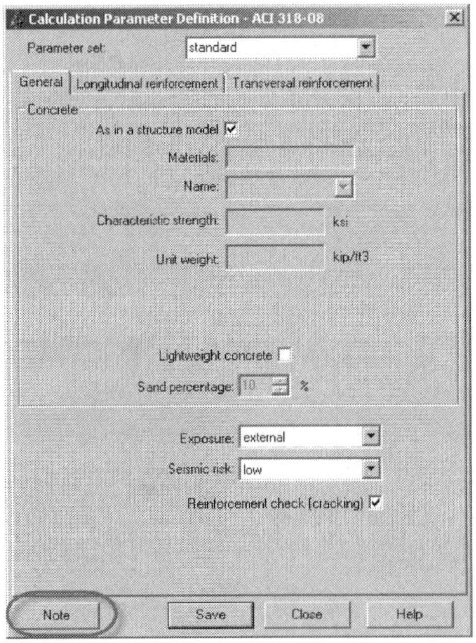

"Note" will create an HTML report of the calculation parameters for inclusion in your report. The same four buttons as above are available to facilitate screen captures, editing or export to word processing or spreadsheet.

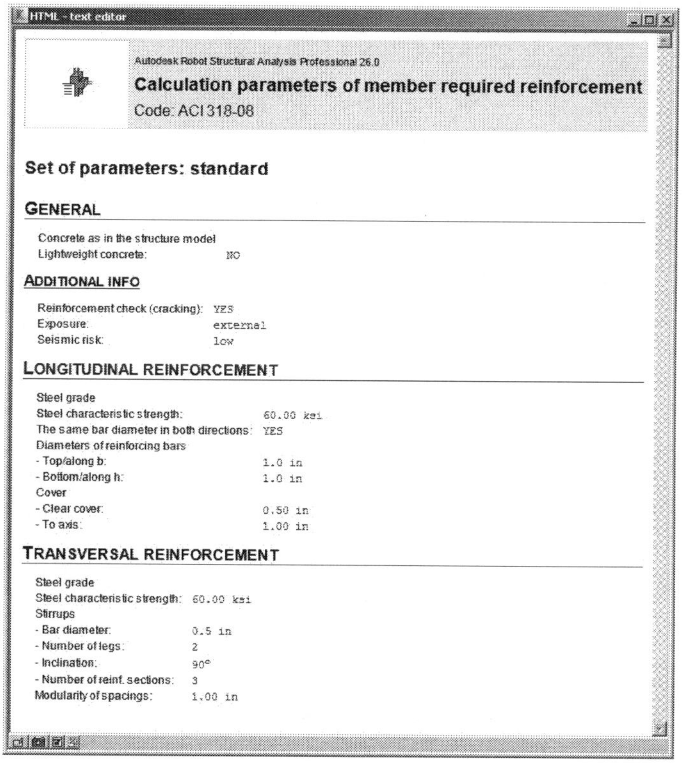

NOTE: BE VIGILANT WITH REGARD TO SCREEN CAPTURES. IN TERMS OF PREPARING YOUR CALCULATIONS AND REVIEWING YOUR DESIGNS, IT WILL BE SAFEST TO NOT EXPECT ANY SCREEN CAPTURES OTHER THAN TABLES AND VIEWS TO UPDATE ON PRINTING, REGARDLESS OF WHAT THE OPTIONS OFFER YOU IN TERMS OF UPDATING. STANDARD PRINTOUT COMPONENTS IN THE PRINTOUT COMPOSITION DIALOG WILL UPDATE ON PRINTING, BUT EVERYTHING ELSE BESIDES TABLE VIEWS AND MODEL VIEWS IS SUSPECT AND MUST BE CHECKED. IN THE CASE OF CALCULATION NOTES NOT UPDATING, THIS IS A BUG. HOWEVER, CARE MUST BE TAKEN IN CHECKING YOUR PRINTOUT RESULTS TO SEE THAT CURRENT INFORMATION IS ACCURATELY REFLECTED.

Printout Composition Dialog

In this dialog, we will select the elements of the report, order them, and decide on formats and spacing. There are several ways to access this dialog. From the File menu: **File>Printout Composition...**, or from the standard toolbar with the printout composition button: or even from a calculation or other note text editor window file menu (also **File>Printout Composition...**). Note that the dialog will open to the Screen Captures tab if you have screen captures. It will open to the Templates tab by default if you do not have any screen captures.

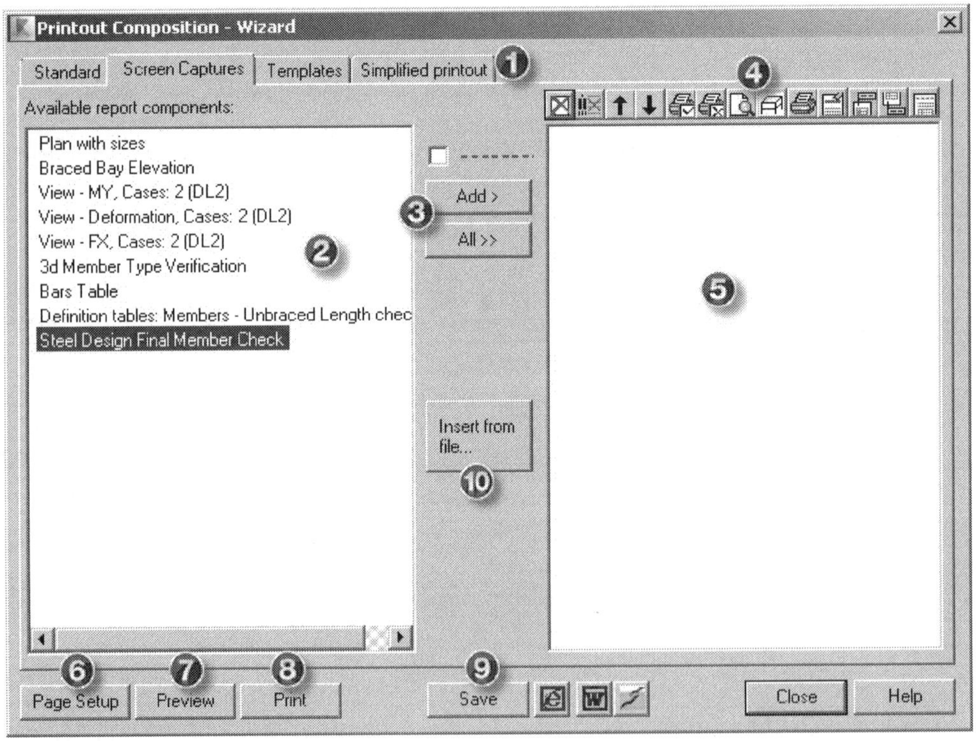

1- The Printout composition wizard contains 4 main tabs which offer different options for composing your report. They simply allow you access to different types of components which can be added to the printout component list (#5 below)

a. Standard: Choose from predefined typical components. E.g., node table, bar table.

b. Screen Captures: List of screen captures you have recorded.

c. Templates: Predefined and saved templates for inclusion in the report.

d. Simplified Printout: A report generator allowing quick access to standard components with added filtering capabilities.

2- Selectable Components window: This list contains components or templates which may be added to the printout component list (#5 below). The list changes based on which tab is selected at the top.

3- Add and Add All: After selecting a component in the component list, move it into the printout component list by using either Add or Add All if you want all components to be added to the report. The dashed bar with a check box allows you to specify that you want a page break added before the component you are adding to the report.

4- Report Component Configuration: These buttons operate on the report components in the right hand list. The button functions are basically self-explanatory via the tool-tips. They can be used to delete individual components, delete all components, move components, turn on or off printing of particular components, adjust titles, add headers and footer notes and add page breaks.

5- Report Composition List: All components currently added to the report are in this list. Page breaks and page templates will also be shown in this list. Use the buttons at the top to adjust components. Right-clicking on one of the components will offer access to apply a page template to the component. (use a page break before the template or the template won't work properly)

6- Page Setup: Access the page template dialog:

a. 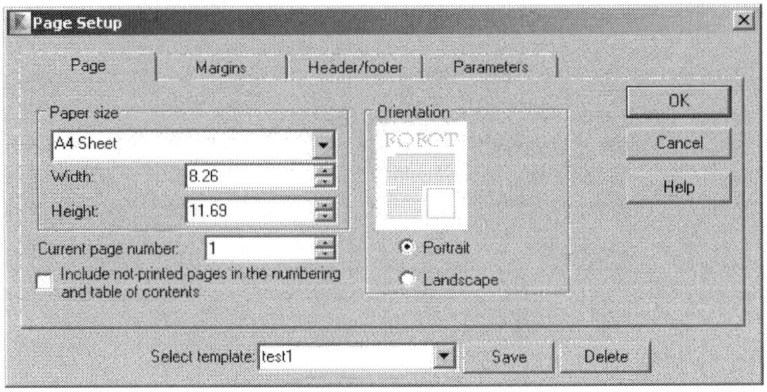

b. Page templates can be saved (as "test1" in this capture) and can be used to control the look of components in the printout.

7- Preview: Opens a report previewer where you may preview all selected components as configured and also edit components, adjust page setup and also print the report.

8- Print: Sends the report to the printer.

9- Save/export buttons: The save button will allow you to save a copy of this report in RTF format. The following buttons:

a.  creates an HTM file and open your browser from which you can view and/or save the HTM file. If you do not have an internet browser, this option will not work.

b.  creates a Microsoft Word file (.docx) and if you have Microsoft Word installed, will launch Microsoft Word where you can review the file, make any changes and save it. If you do not have Microsoft Word installed, this option will not work.

c.  creates an OpenOffice.org file. If you do not have OpenOffice installed, this option will not work.

10- Insert from file: any .RTF file can be inserted in the report. Other file types may work but the feature works best with .RTF files.

The Printout Composition Tabs

Your printout or calculation package will be composed of components you select from these tabs. As you select components and move them to the right hand side, the right-hand side of the dialog becomes your printout outline. Your printout will most likely be composed partially of standard options, screen captures and perhaps some elements from the simplified printout tab.

The Standard Tab offers a set of predefined table views and reaction tables which may be included in your report. These are effectively standard screen captures of table views and will be updated on printing. Select any of the available components and move them into the outline with the "Add" button (or the "All" button if you want all components added to the outline).

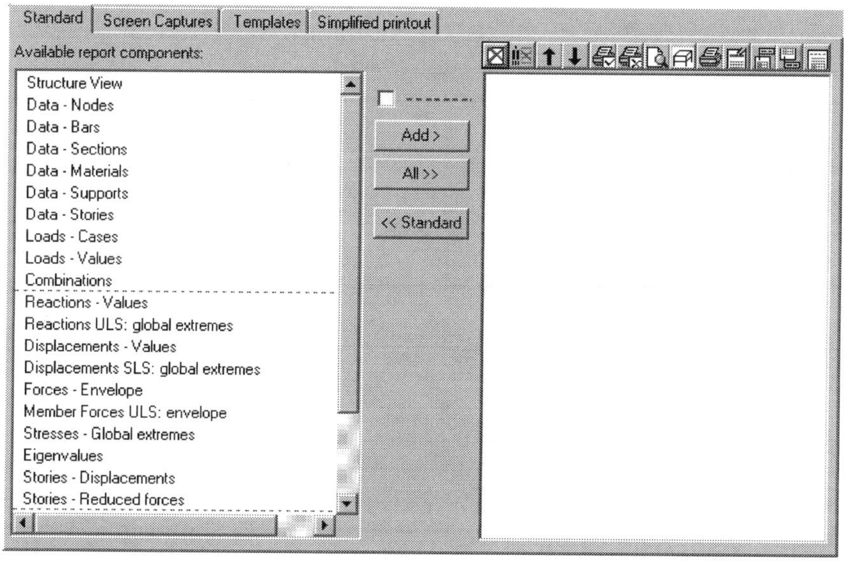

The "Standard" button can be a bit unwieldy: This takes the current outline from the right hand side and replaces all components on the standard tab (left-hand side) with the current outline (the right-hand side). While this can be useful to create a standard set of components for your calculations, it cannot be undone so be sure you really want to erase all components from the default standard tab. If you wish to create your own standard set of components, don't hesitate to create a template of the default (add all components to the outline, switch to the Templates tab and use the <New Button). The only other way to recover the defaults would be to rename Cfguser in your Roaming AppData folder. (See Trouble-shooting topic for more information).

The Screen Captures Tab lists all screen captures you have saved from the current project. The list on the left is also the screen capture management area. Screen captures saved here can be deleted from this list on the left hand side. To add a screen capture to the report outline, simply select the screen capture and use the "Add>" button. (Or the "All>>" button if you want to add all screen captures to the report outline.) Only the screen captures you create will be listed here, if you have not yet created any, the list on the left hand side will be empty.

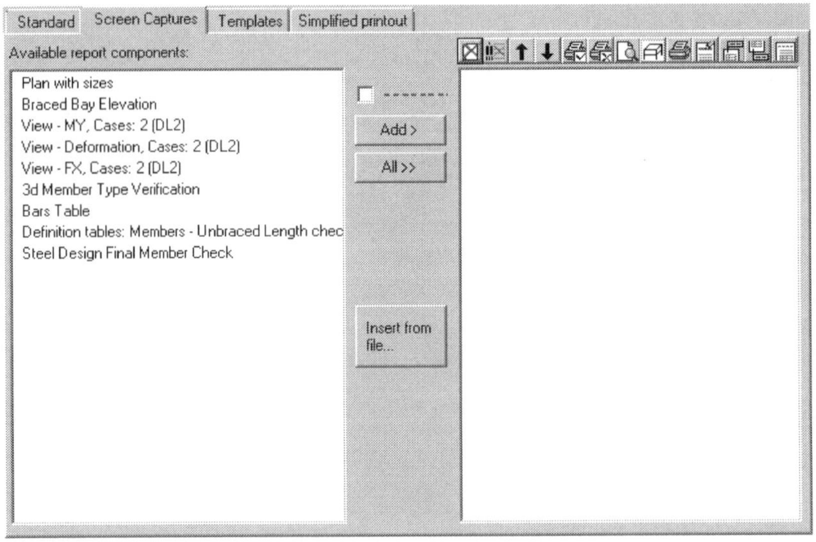

"Insert from file" can be used to add any .RTF (rich-text) file on disk to the report outline as well. As you create screen captures, it will help to provide illustrative names so that they can be more easily identified in this list.

The Templates Tab is a great tool for creating re-usable report outlines. Initially this tab will have several default templates and you can add as many of your own as you like.

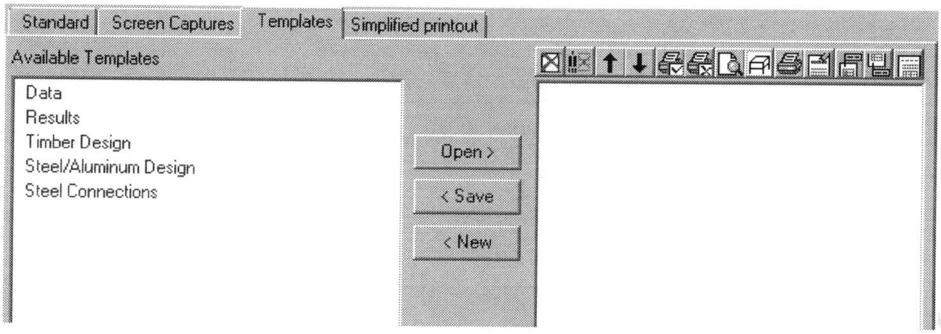

Templates in blue are default templates and cannot be deleted though other templates you create here can be deleted by selecting the template and pressing the delete key. Use the "Open>" button to add the components of the selected template to the report outline. Use the "<Save" button to save the current report outline to a selected template, replacing the contents of that template. Use the "<New" button to create a new template composed of the components currently defined in the report outline on the right hand side. When you press "<New" you will need to supply a template title by typing the name in the new line:

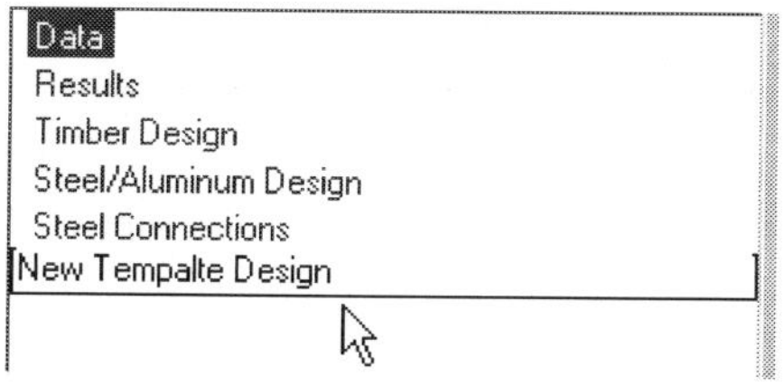

Spelling errors in the template can be quickly fixed by simply clicking once on the template title to select it, then clicking again on the title will allow you to rename it.

The Simplified Printout Tab is quite different from the other tabs and is basically a simplified report wizard which allows you to optionally select several different items and then "Save Template" which does not, contrary to the button name, create a new template but rather adds the components configured to the report outline. This tab is really only good for the basic inputs and outputs of analysis. Steel and concrete design information and results will need to be added to the report from screen captures or the standard tab. When you click over to the Simplified Printout Tab it looks like this:

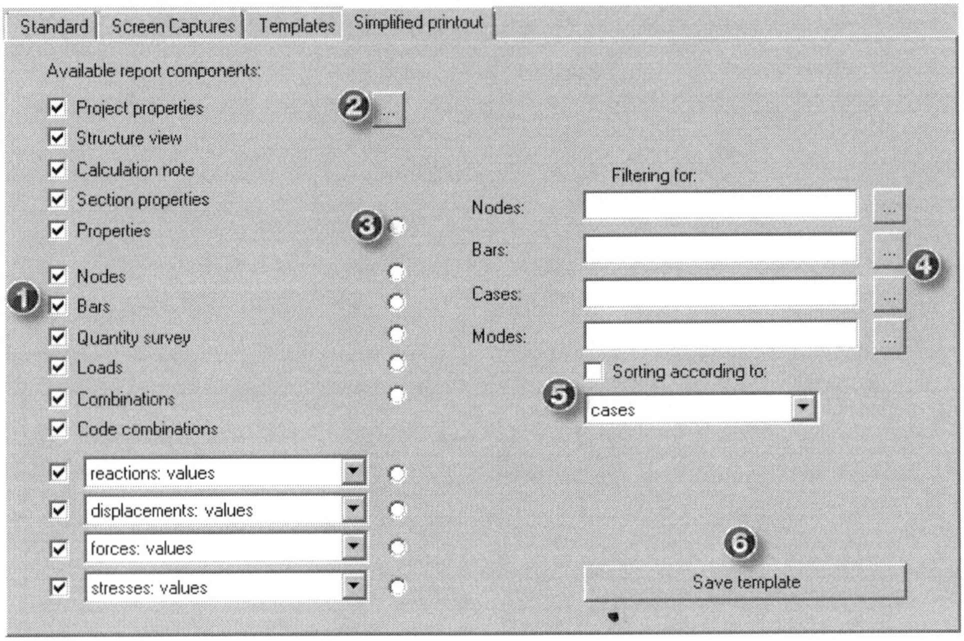

There are several standard options for quickly composing a report of basic inputs and basic outputs. There are also advanced options for filtering and sorting on the right hand side which can be applied to any element which has a radio button to the right of the component name.

1- Component checkboxes: Select which report components you would like generated. Some of these options are quite handy. For instance, Calculation notes can be added quickly here as opposed to opening the calculation notes and creating a screen capture. The same could be said of project properties and section properties etc. NOTE: If you want to apply filtering for any component which can be filtered, you *must* select the partially checked (grayed) state of the checkbox: 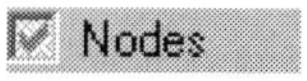

not the fully checked state: 

. Even if you have filtering options defined, the fully checked state will ignore filtering. In the results components, you have the option of generating all values, envelope values or global extremes:

a. E.g., 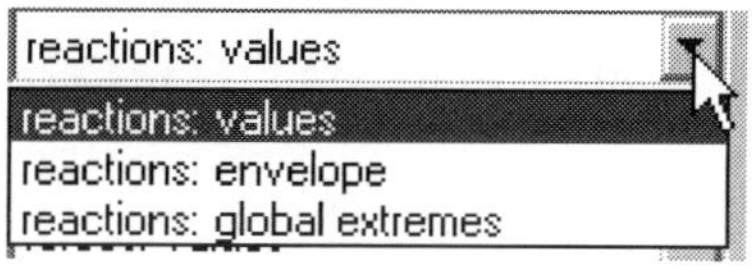

2- Project Properties access. Use this ellipsis button to take a shortcut to the project properties (otherwise **File>Project Properties...**) where you can make adjustments without leaving the printout composition dialog.

3- Filtering Radio buttons: 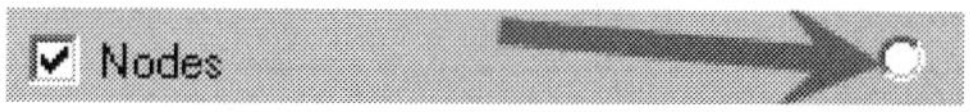

The filtering settings in the "Filtering for:" area of the dialog (#4) apply to individual components. These radio buttons control which filtering settings are currently shown in the "filtering for" area. Select a radio button to configure filtering options for the component to the left of the radio button.

4- Filtering for: This area allows you to configure filtering for the table components selected on the left. The radio button next to the component controls which settings are currently visible and allows you to edit them. Depending on which component radio button you select, only items which pertain to the table you are filtering will be available (not grayed out). Use the ellipsis buttons to the right to access the selection dialog to facilitate selection of elements to be included.

5- Sorting applies to all component tables, it is not particular to any one component selected to the left. You may sort based on cases or groups.

6- Save Template: This really means "Add these components to the report outline". It will not create a template in the templates tab. You are more than welcome to create a template from the output of this simplified printout wizard, but it is a manual operation.

Working with Printout Composition Outline

The right-hand side of the printout composition dialog is where you configure the selected components of the printout. You can move them around, add page breaks, apply page templates, delete components as well as rename titles and add headers and footers to each component. Below is a sample outline for a simple report combining screen captures, simplified and standard printout components.

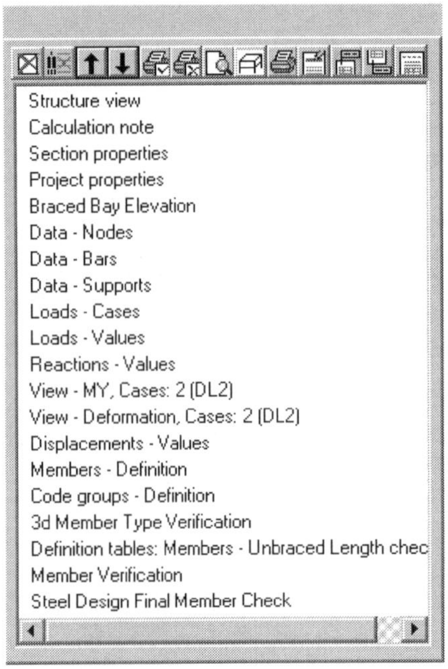

Managing Components: Individual components can be moved with the up/down arrows: . Components can be unhidden/hidden (respectively) from the printout by using the following buttons: . Edit the title of a component with this button: . Add header or footer notes to any selected component with these buttons: . Add a page-break before any selected section by using the add/remove page break button:

A page-break will appear as a red dotted line before the selected component as shown here:

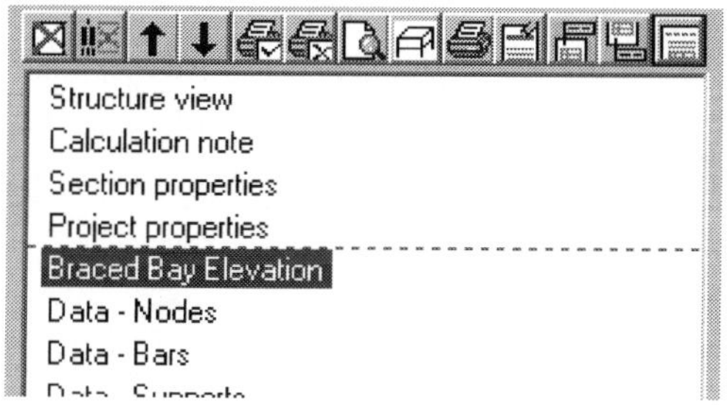

Page templates can also be assigned to components. Page templates will allow you to adjust the page orientation for a component or to control the page size of a component if desired. A word of note here: page templates can include title page configuration. Only the first template applied will govern the report title page. Subsequently applied templates will not affect the title page. Another note: If you wish to apply a new template it is recommended to use a page break before the new template. Robot will not have consistent results if a page break is not inserted before the new template.

Using page templates: When you have selected a component, right-click the component and notice that there is an option for page template in the context menu:

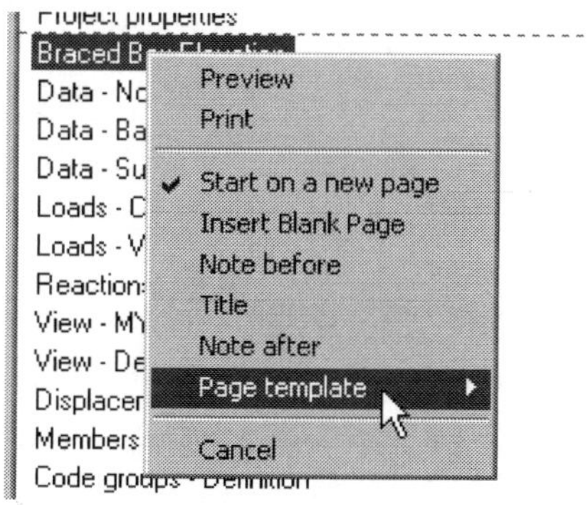

To configure page templates, from the main printout composition dialog, select the "Page Setup" button. This will bring you to the Page Setup dialog:

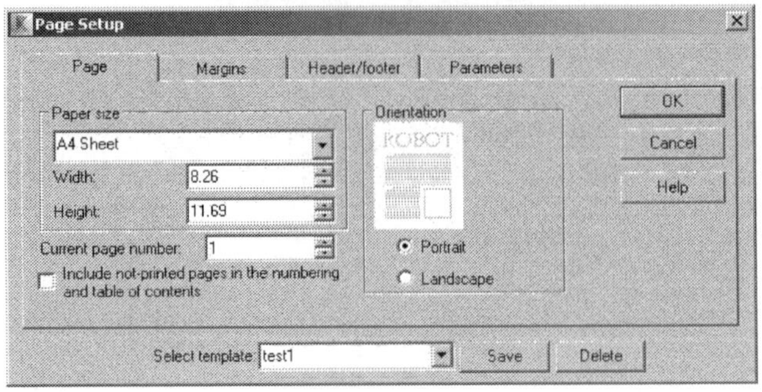

As far as the title page, headers and footers are concerned, the template files live in the following locations:

Standard Page Template: C:\Program Files\Autodesk\Autodesk Robot Structural Analysis Professional 2016\System\Template\

RP_H_001.RTF – Header file

RP_F_001.RTF- Footer file

RP_T_001.RTF – Title page file

Customized Page Templates:

C:\Users\<username>\AppData\Roaming\Autodesk\Autodesk Robot Structural Analysis 2016\Template\

The file names for customized page templates are similar to the standard ones. If you are having strange behavior in your title page or header/footers, check these files to see their contents. This can give you a way to track down what is happening with your printout.

After selecting a page template, your report outline will indicate the page template by enclosing the template name in angle brackets as shown here:

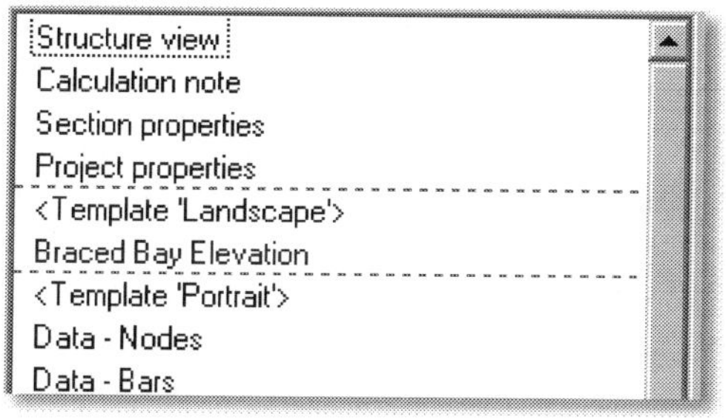

To return to a previous template (or to select another template) add a new page break and specify the original template. Once a template is applied, it is considered the active template for all following components in the outline until it encounters another template specification.

Chapter 8 - Basic Steel Design Workflow

Assuming that load combinations have previously been generated, in Robot the basic steel design workflow is as follows:

1- Assign code parameters to all elements you wish to check/design
2- Create initial code groups (group similar members at first and refine later)
3- Configure Calculation Configuration Options
4- Perform code group design for strength with optimization
5- Update members based on strength design
6- Perform Member Verification for Ultimate and Service, note any that fail for service
7- Perform serviceability design for code groups with failing members
8- Update members based on serviceability design.
9- Perform member verification to check all members for both strength and serviceability

a. Optionally review utilization ratio analysis and identify groups of members which can be re-designed more optimally, create new code groups for these members and iterate their design.

10- Iterate any individual members which still fail by creating a new code group specifically for each type of member which needs special attention.

11- Perform final member verification

NOTE: STRUCTURE WILL NEED TO BE RE-ANALYZED AFTER EACH MEMBER SIZE CHANGE OPERATION BECAUSE THE DEAD-LOAD AND THE OVERALL ELASTIC RESPONSE OF THE STRUCTURE WILL HAVE CHANGED AS A RESULT.

Code Parameters – AISC 360-10

As we have discussed previously, labels are groups of parameter settings which can be assigned to individual elements. In this case, we will be configuring parameters related to member steel code checking and assigning them to individual elements. Robot can use these parameters along with information about the member and the results of the analysis to perform code checking, design and verification of the elements in your model relative to the selected design code. For purposes of this book, we will be working with the AISC 360-10 code version as it is the latest supported US steel design code available in Robot. Other code configurations will be different and particular to the code of interest, but the overall functionality is similar across all steel design codes.

Configuring Code Parameters

To launch the AISC 360-10 code parameters, first make sure that you have the AISC 360-10 code selected for steel design in the job preferences (**Tools>Job Preferences...**>Design Codes) for more information on this refer to project setup. Then from the menu, select **Design>Steel Members Design - Options>Code Parameters...** to launch the AISC 360-10 Member Type label dialog:

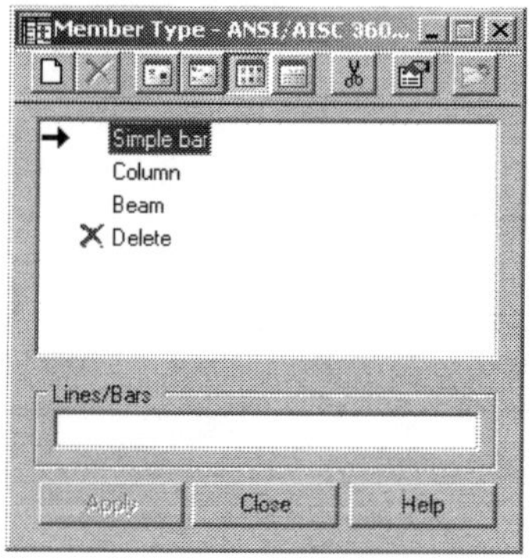

There are three default member types which cannot be deleted from the label list: Simple bar, Column, and Beam. You can use these as starting points for your own member types or create new ones from scratch.

Click the "New" button to start a new member type definition:  This will launch the AISC 360-10 Code Parameters Dialog:

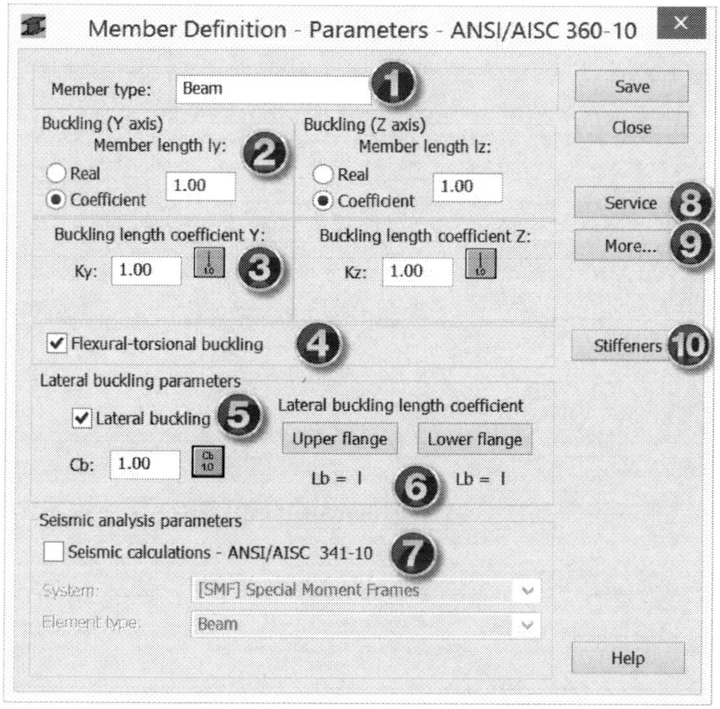

1- Member Type Name: Reserved names are Beam, Column and Simple Bar. Supply your own descriptive name for the set of parameters you want to assign to members. With exception of the reserved names, saving an existing name will overwrite the definition. (You will be prompted to overwrite the definition if you do. Note that any changes overwritten will affect all members to which that set of code parameters has been assigned.) If you supply a new member type name and then save, that label will be added to the list of member type labels though there is no feedback that this has been done. You'll have to check the Member Type labels dialog to see it and apply it to elements.

2- Base member length for compressive loading check (Y-Axis is Major Axis Buckling, Z-Axis is Minor Axis Buckling). Either specify an actual length of the member or by a coefficient applied to the member length. This is the unbraced member length for use in calculating buckling and will be modified by the "Buckling Length Coefficient" (K factor) in item 3 below.

   a. Specifying this length in a hard number can be disadvantageous unless you are certain the length will not change during the design. If the member grows longer and/or spacing between points of bracing increases you could potentially produce an unsafe design. Leaving it set to a coefficient of 1.0 is the best practice.

   b. Note: There is a setting in the "Buckling Length Coefficient" options where you can specify more options for unbraced length which may override settings here. See "Buckling Length Coefficient" Below.

3- Buckling Length Coefficient: This is the K-factor to modify the base member length for end conditions (pinned, fixed, free, etc.). Pressing the coefficient button  will launch the Buckling Diagrams dialog:

a. 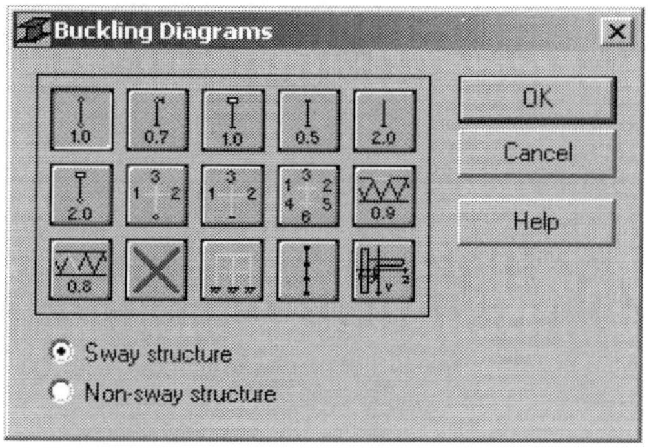

b. This dialog provides quick access to specify buckling coefficients for common end conditions.

c. 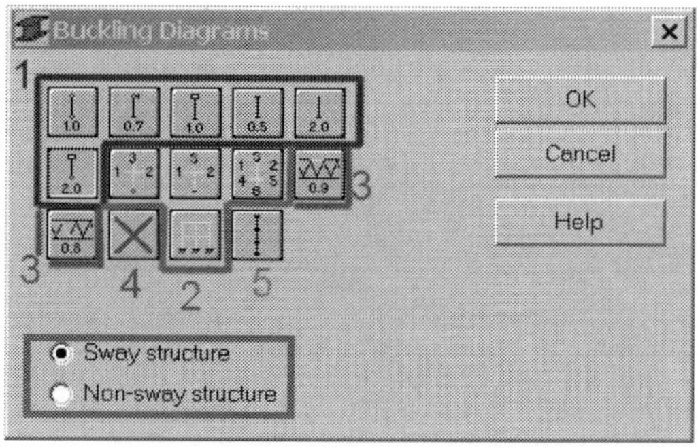

i. **Section 1** provides a quick select for standard buckling coefficients.

ii. **Section 2** provides manually configured or automatically calculated buckling length coefficients based on surrounding members.

1. Use 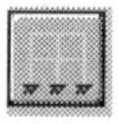 for Automatic calculation of member K based on surrounding members in the model.

2. Use 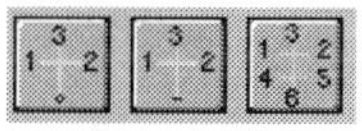 for Manual Configuration

3. Sway vs. Non-Sway works with these choices

iii. **Section 3** provides standard buckling coefficients for truss elements

iv. **Section 4:**  Eliminates buckling about this axis from the design and member checks! Literally Robot will ignore buckling about this axis in calculations.

v. **Section 5:** 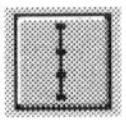 For specifying intermediate or manually configured bracing including automatically determining bracing length based on adjoining members. See Internal Bracings Dialog section below.

d. If you supply your own value for K factor, the Icon to the right will indicate a "hand" which lets you know that you have specified a user value. That icon looks like this:

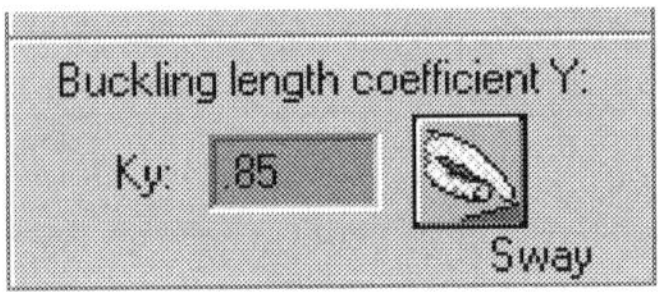

To specify a different value or use the Buckling Diagrams dialog simply press the hand icon.

4- Flexural Torsional Buckling: This causes Robot to apply special rules for performing code checks on mono-symmetric sections from chapter F specifically tees, channels and angles. If this is unchecked, Robot will not accurately check mono-symmetric sections and you will get an inaccurate design.

5- Lateral Buckling: This option must be checked in order for Robot to apply the code rules for members where the unbraced length is longer than maximum unbraced length for full plastic section capacity (Lp). If this is unchecked and your member's unbraced length is greater than Lp, you will have an un-conservative design. Make sure that you have this checked unless you are absolutely certain that the unbraced length is less than Lp. Checking this box when unbraced length is less than Lp will not adversely affect the design.

a. Cb button 

Press this button to access common settings for Cb or you may specify a value directly by typing it into the edit control Note that if you supply your own value, the icon will indicate the hand showing that a value has been manually entered 

b. Parameter Cb dialog:

c. 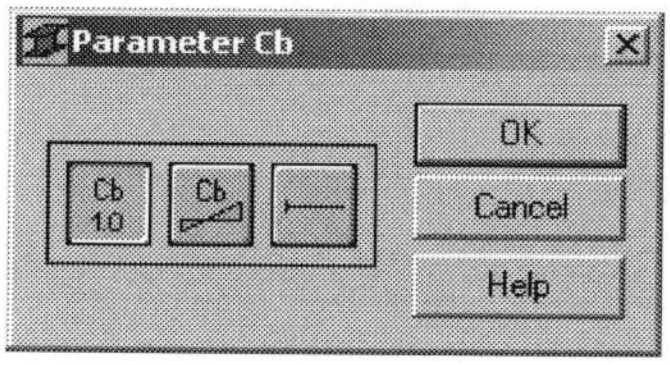

i. 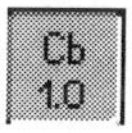

to specify Cb = 1.0 (Conservative)

ii. 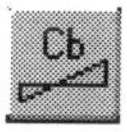

to have Robot automatically calculate and report Cb based on moments in the member

iii.  to specify a cantilever situation (also Cb = 1.0)

6- Lateral Buckling Length Factor (Top and Bottom flange bracing): Robot calculates lateral buckling for the compression flange whichever side of the element is in compression. Therefore, you have the ability to specify end conditions and bracing lengths for the top and bottom flanges individually. Press the "Upper Flange" or "Lower Flange" buttons to access the Lateral Buckling Length Coefficient settings:

a. 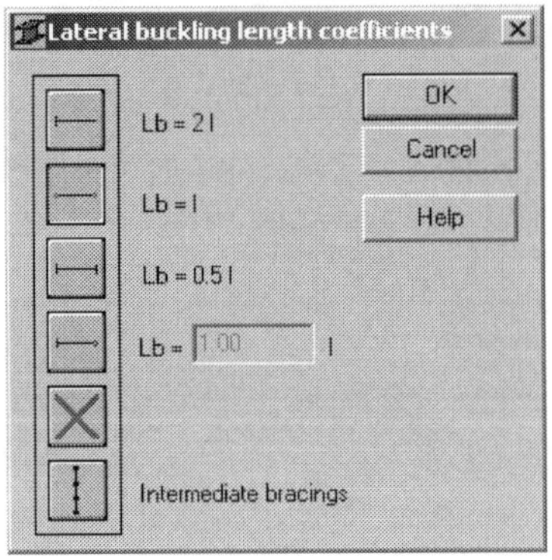

i. Use the first 3 buttons for standard configurations if the flange is unbraced along its length.

ii. Use the fourth button to enter a custom value.

iii. Use the fifth button  to ignore the lateral buckling calculations for this flange.

iv. Use the 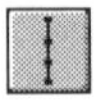 button to launch the Internal Bracing dialog (see below).

7- AISC Seismic calculation settings: When required select this option and specify the lateral system designation. Robot then uses the provisions of the AISC Seismic design provisions to check this member.

8- Service: This launches a dialog Settings Serviceability– Displacement Limiting Values dialog where you can specify which serviceability criteria you wish to have checked and their limiting values.

a.

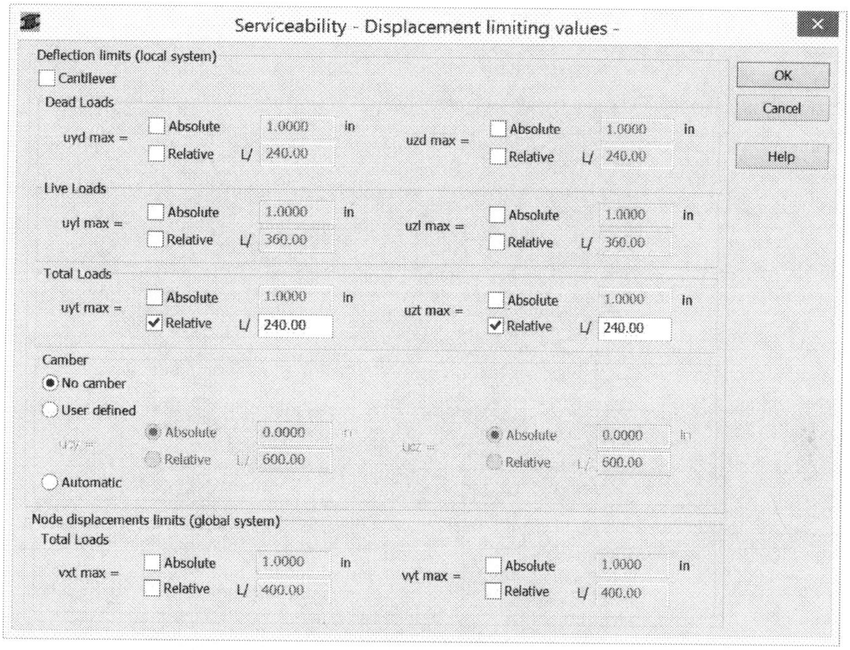

b. Select the deflection checks you wish to make and specify the limiting values. Recall that uy in member local systems is lateral (weak axis) deflection and uz is vertical or strong axis deflection. You may provide both relative and absolute limits to be checked for this member type definition. If the checkbox is not checked next to a deflection specification, no check will be made.

c. NOTE: While member deflection can be used in design routines, nodal (global) displacement will be only be checked and reported during member checking: it will have no effect on member designs. Unfortunately, it is frequently not obvious what changes to member sizes will impact global nodal displacements and while it is technically possible, Robot does not provide this functionality. Check structure drift manually and make adjustments manually to your lateral resisting system as necessary to control displacement.

d. The last option to check members considering camber allows you to specify a camber value manually for this member type. This is a quite manual process but if you intend to specify a precise value for camber for members this is a great place to indicate that and have it considered in the member deflection calculations. Selecting automatic camber will require that you specify which load case to use for camber values in the steel design calculations configuration dialog:

e. 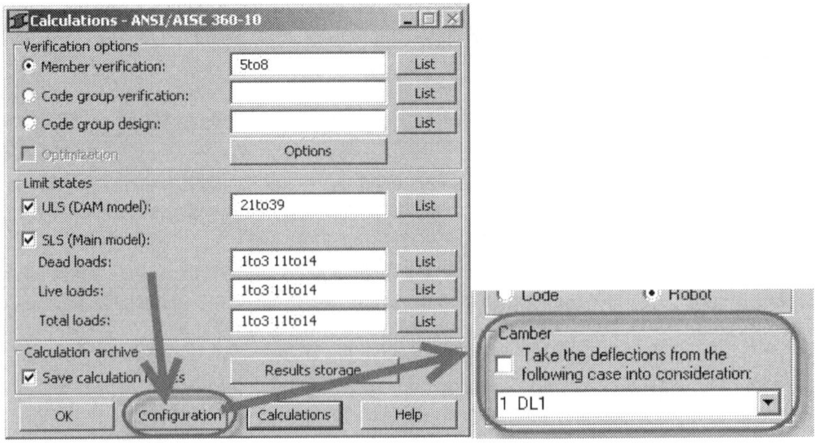

We will cover the calculations configuration dialog below.

9- More…: This is where we can specify reduced areas for bolt holes for tension design considerations as well as shear lag factors:

a. 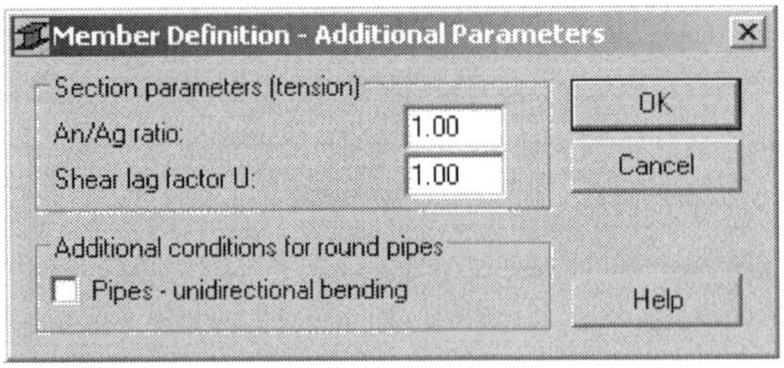

b. Supply a net/gross shear ratio to consider reduced area for tension design or a shear lag factor to account for welded tension connections. Refer to AISC 360-10 section D3.

c. NOTE: If you redesign the member to be larger or smaller, this ratio you provide here may not still be accurate! If you choose to use this option for tension design, make sure to verify that the ratio is still correct after any member size changes.

d. Pipes – unidirectional bending: Simply looks at the vector sum of My and Mz as the design moment for the pipe.

10- Stiffeners:

a. 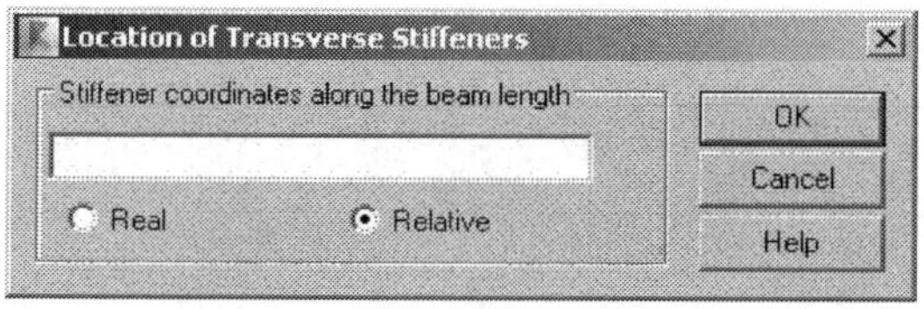

b. Note: The vast majority of AISC sections do not require transverse stiffeners for shear on the general section (They may still be required at concentrated loads or supports and still need to be checked by hand in those cases.).

c. Unless you specifically configure it in the AISC 360-10 calculation configuration options, Robot is not performing calculations considering tension field action per section G3 (Tension Field Action). It is only using spacing of transverse stiffeners to perform calculations per section G2.1 (Members with unstiffened or stiffened webs: shear strength) for calculation of (kv), but note that it is *not* performing any of the checks per section G2.2 (Members with unstiffened or stiffened webs: Transverse Stiffeners)

d. If you do choose to consider tension field action in the AISC 360-10 calculation configuration option, please note that you will need to manually conduct all checks of the stiffeners and ensure your design meets all criteria in section G3.1 (Tension Field Action: Limits on the

Use of Tension Field Action) and G3.3 (Tension Field Action: Transverse Stiffeners)

Internal Bracings Dialog:

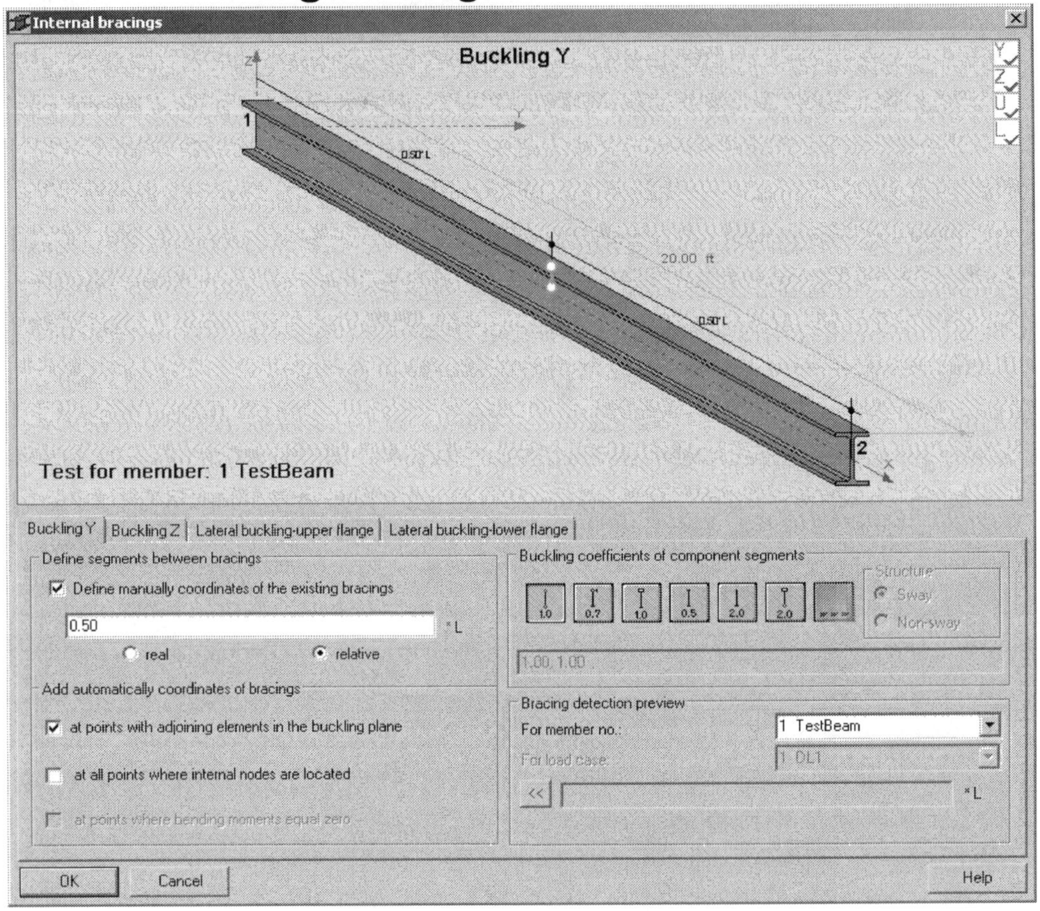

The four tabs in this dialog correspond to the four buckling length options in the member types dialog here:

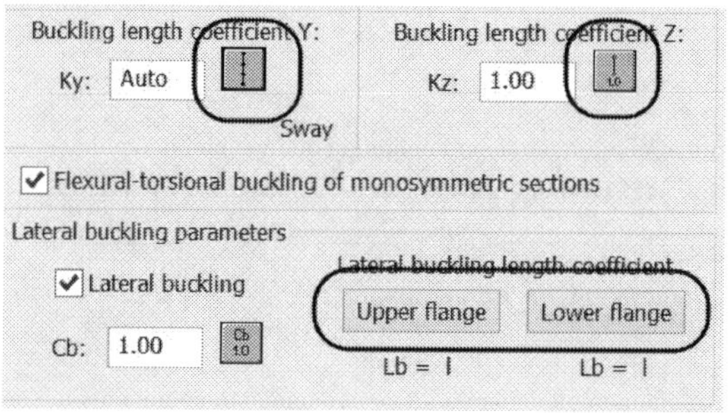

Changes made in the Internal Bracings Dialog will affect each of these if you adjust settings in the corresponding tab.

Each tab of the dialog is similar in function:

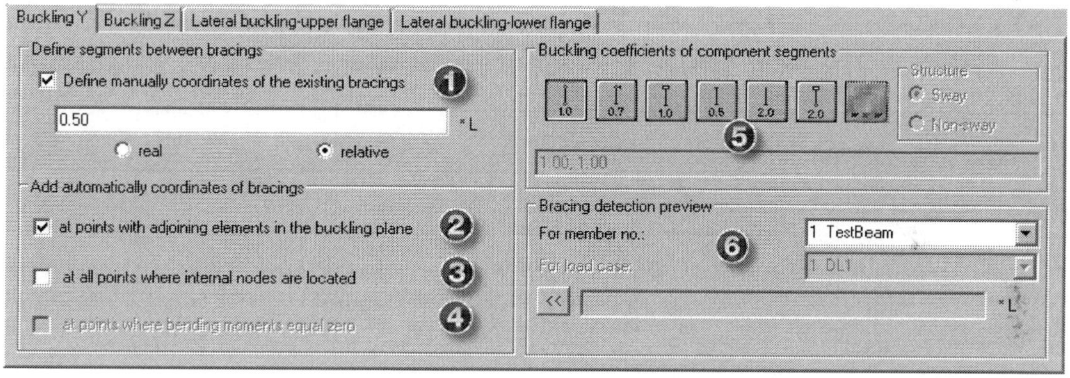

1- Select the first option to manually specify coordinates (relative or absolute) along the member which will be considered braced.

2- All points with adjoining members in the model will be considered braced for this direction if this option is selected. Adjoining member must lie in the plane of buckling or it will be ignored as a brace point.

3- Any calculation nodes will be considered bracing points. These would be any added nodes, any members divided without splitting the member, locations where point loads have been added with a calculation node, or any adjoining member regardless of whether it is in the buckling plane or not.

4- For top and bottom flange bracing, points of zero moment may be considered as braced with this option.

5- Configure option for K factors at the extreme ends of the members. Either supply them directly by editing the end segment K factors or use the buttons to automatically configure them based on standard member end conditions.

6- For options where bracings are based on adjoining members etc., you can review sample members to see that the bracing locations are as expected by selecting different members from the model and/or load cases to see a preview of the bracings in the beam view above.

Assigning Code Parameters:

Once you have configured enough member types to adequately represent your design considerations above, you will be back in the Member Type – AISC 360-10 label dialog. Remember a label is just a set of parameters which are assigned to a member by applying the label.

To apply labels to members you can either select the label and apply to members directly by clicking them OR you can select the label and then enter member numbers to which you want this label applied, then use the "Apply" button in the dialog.

Unlike other label application dialogs, the cursor will not change for single-click label application, but rather remains an arrow. As you hover over members, you will see the member highlighted to indicate which member will be selected when you click the mouse. Access this option by selecting a label and moving your mouse directly into the project window.

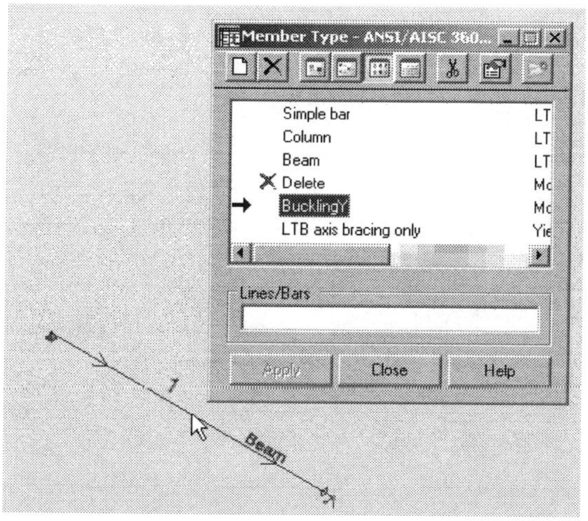

If you assign a label to a member which already has a label, the new label will simply replace the existing label.

Steel Members Design Layout

Steel member design is one area where a layout is required by the Robot interface. Access the steel member design interface (layout) either from the layout selector (Steel Design>Steel/Aluminum Design) or from the menu **Design>Steel Members Design...** Your work area will be rearranged with two fixed dialogs (non-dismissible): Definitions – ANSI/AISC 360-10 and Calculations – ANSI/AISC 360-10

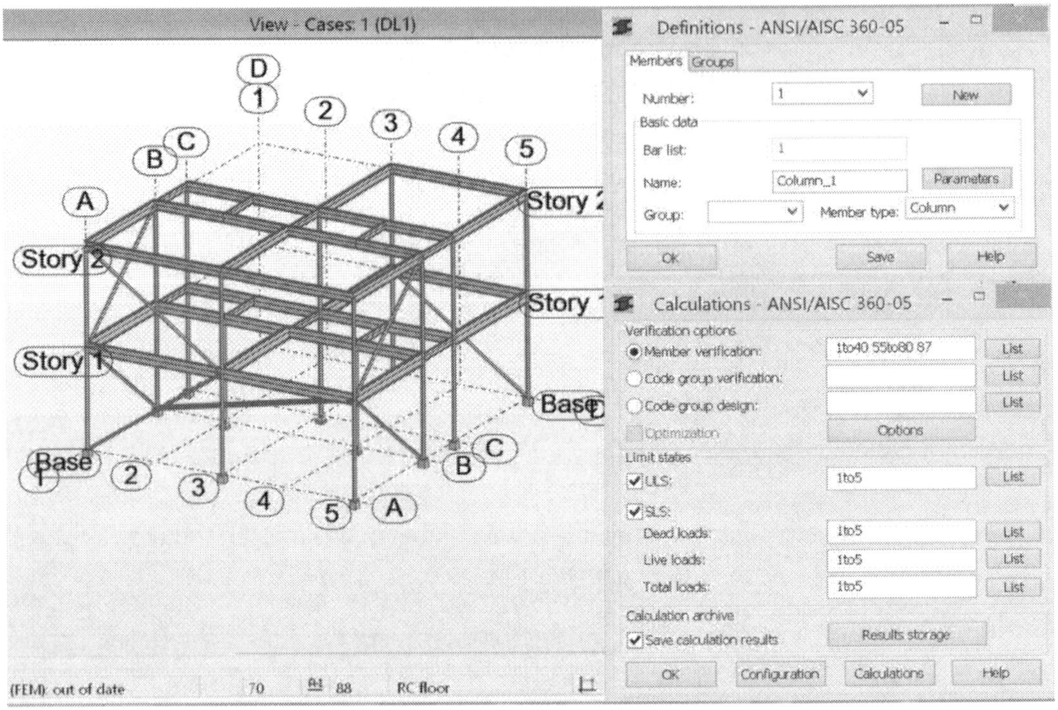

Member and Code Groups Definition:

In the steel design layout, the definitions dialog gives you member by member access to member type properties (more useful for making quick adjustments to a member or two) and furthermore access to create code groups for design.

MEMBERS TAB:

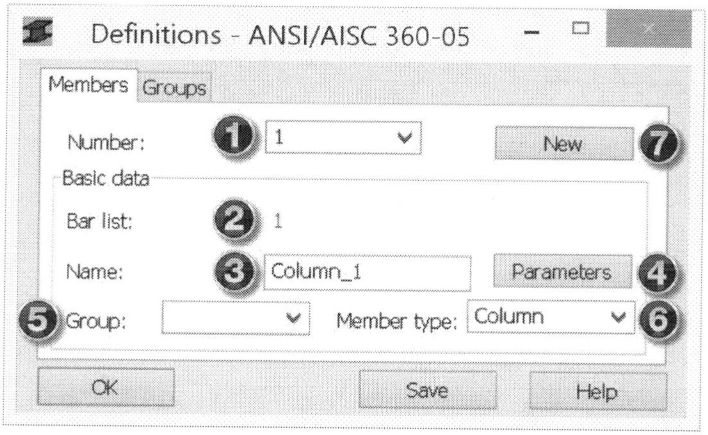

1- Access any individual member already geometrically defined in the project with this dropdown list. All current settings for the member will populate in the controls below in the Basic Data section

2- Bar List: This control is greyed out for a single member. If a super member has been defined through use of the "New" button (#7) then this list will be editable for you to define the members which will participate in this super member.

3- Name: Specify a name for this particular member. This name will be used in the tabular results of steel member verification and code group design. You can use a uniform member naming convention and apply it from **Geometry>Names of Bars/Objects** or you can specify a member number here. Sometimes Robot will overwrite your previous member numbering for the first selected member when using this dialog. Be careful to watch for any changes to the member you did not expect.

4- Parameters: Allows you access to the member type parameters for AISC 360-10 associated with this member (member type selected in #6).

5- C. Group = Code Group: In order to do member design, Robot requires that member designs be done on a code group basis where several members are considered at once. You can use this selector to switch code groups for the current member or to create a new code group for this particular member. You will still need to configure the code group settings in the groups tab.

6- Member Type Selector: Use this to adjust which member type label is assigned to the member currently selected. This list will contain all member types currently defined for your AISC 360-10 code parameters.

7- New: Use this button to create a new super member which can be comprised of several existing members. Once you select new the Bar List will become editable allowing you to specify the bars which will comprise this super member. The member will be code checked as one member. Ideally, these member's local coordinate systems should share a co-linear local-X axis for the calculation routines to be valid. Results will not be presented individually for members which are part of a super member. Note that when a super member is selected in #1 above, an additional button "Delete" is available in this dialog.

GROUPS TAB:

All member design is performed on groups of members in Robot called "Code Groups". Individual member design is not available unless you create an individual code group for each member in the model. The goal is to have similar members performing similar roles all have the same size section chosen by grouping them into "Code Groups". This ultimately helps to smooth the number of sections selected for a project and can have the effect of reducing overall project cost through reduced waste resulting from use of many different sections on one job.

Unfortunately code groups are sequentially numbered and do not follow the workflow for other types of labels in Robot. The dialog to manage code groups is a bit like working blindfolded, but once you get the hang of it, it is not bad.

Worklfow:

1- Start by pressing "New" to create a new group

2- Select the members which will comprise the group (**View>Display...** Bars section has an option to show code groups for members to give you a visual verification)

3- Use the Sections button to configure sections to be considered in design

4- Give the Group a name

5- Select the material to be used in member design

6- Save the group

Dialog Function:

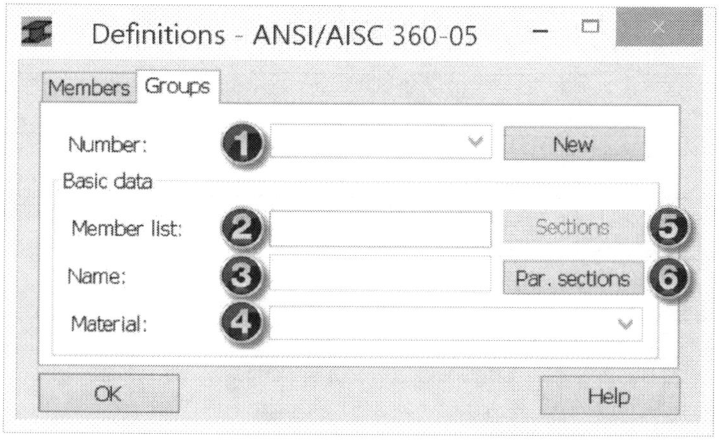

1- Code Group Number: Number of the currently selected group. Use the "New" button to create a new group. Once there is a group created, there will be additional buttons at the bottom for "save" and "delete" to manage the list of code groups.

2- Member List: Input the member numbers of members which will be considered part of this design group. You can use the typical selection tools to build this selection and copy/paste into the edit control.

3- Name: Give your code group a descriptive name. (E.g., Interior girders, Exterior wind columns)

4- Material: This is the material that Robot will use to perform design and code checking if you select code group verification in the calculations dialog. Note that the material you have configured for the member in the bars dialog will not matter in member design, *ONLY* the setting here will be used in design. If you update the member based on code group design, this material will replace your previously selected material for the bar.

5- Sections: This is the list of sections which will be considered in the design of this member. If you do not want Robot to consider W4x13s in design, simply leave them out of this list of available sections for this code group:

a. 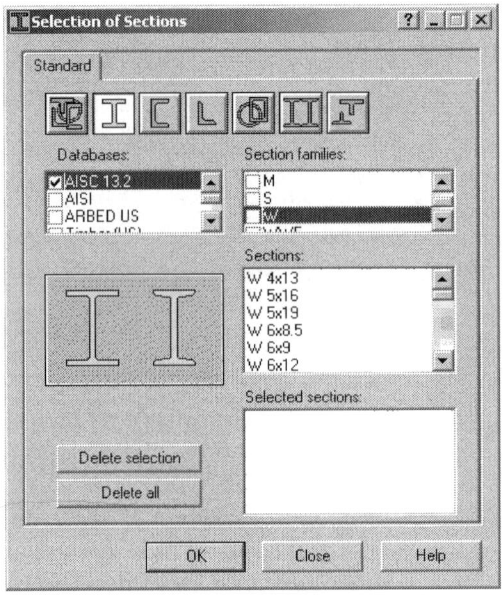

b. The selected sections portion of this dialog are the members which will be considered in design of this code group.

c. The buttons along the top will filter the section families list to make it easier to find the section family you are looking for.

d. A check next to a section family automatically populates the selected sections list with all sections from that family. If you instead only select the section family, then you can pick and choose from the sections list to populate the selected sections list.

e. Manage selected sections with the "delete" and "delete all" buttons to the left.

f. Note: Although there is no way to save preferred section lists, the selected sections will be remembered from group to group which will hopefully help facilitate group creation. It would be advisable to do group setup in batches to take advantage of Robot remembering your selected sections from the most recent group to the next.

6- Parametric Sections: Robot will perform design optimization for parametric tapered sections either by varying the section properties or by varying the member depth. Use this dialog to configure parameters to bound the design problem for this code group.

Calculations Dialog:

The calculations dialog is the heart of the design and member verification operation. Because the options are highly configurable, it is very important to check the settings carefully to make sure you have checked and designed everything.

At the top of the dialog notice that there are three main options: Member Verification, Code group verification, and code group design. To the right of each are edit controls where you select which members or code groups you will check/design. Further down, the loads to be considered also provides an edit control for you to list which load cases/combinations you intend to consider in the verification/design. Yet again, in the lower right, you further have control over whether Ultimate loads or Serviceability is checked in Member verification. (Note, that for code group design and code group verification, only one of these can be checked/designed at a time). This leaves open the possibility that you haven't selected all the members to check, haven't selected all the load cases/combinations to verify and/or have not selected all the check options for member verification. Care needs to be exercised in use of this dialog so that you get results considering all appropriate members, all appropriate loads cases/combinations and all check options.

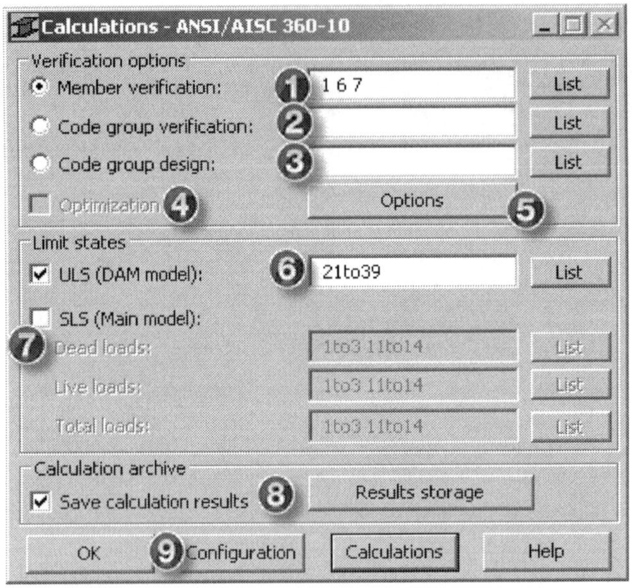

1- Member Verification: Calculations will be performed for each member listed in the edit control to the right. You can use the list button to build a selection set. Only the members included in this edit control will be checked.

2- Code Group Verification: Member verification at the code group level: individual results on a per member basis will not be presented. The list in the edit control will be the code groups to check. Only code groups listed here will be verified.

3- Code Group Design: In each code group the member with the highest forces on a considered section will represent all members of the code group and one section will be selected for this code group. Optimization options (#4 and #5) will only affect Code Group Design. Note that during code group design, only one option can be selected for limit state: either Ultimate *or* serviceability. Code group design may only be done for one of those criteria at a time. To perform a full design, you must iterate a few times to complete the design. Only code groups listed in the edit control will be designed.

4- Optimization: Unless this is checked, the list of allowable sections configured in your code group will be checked sequentially until the first section meeting the limit state is found. Robot will not check further down the list to see if there is a lighter section which also meets the criteria.

5- Optimization Options: Access to a dialog where you can configure criteria to guide the optimization. The options are mostly self-explanatory except the last option "Calculations for the entire set of sections" In the absence of other optimization criteria, this will cause Robot to not stop at the first suitable section meeting the criteria but will cause Robot to continue looking through all sections in the allowed set to find the one with the highest utilization.

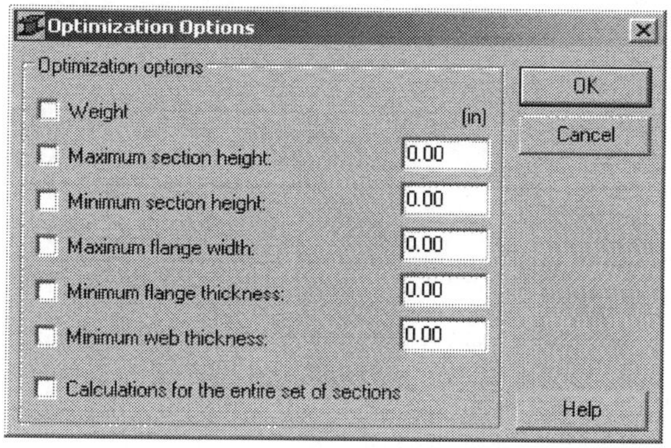

6- Limit States – Load Cases/Combinations List: This edit control should have all load cases and combinations which you intend to use in the design or verification of the members or code groups. Only load cases or combinations listed here will be used in the verification or design. It is very important that you list all appropriate load cases/combinations here. The "list" button to the right will bring up the selection set builder which you can use to prepare the selection of load cases/combinations. NOTE: Robot will automatically use the Direct Analysis Method calculated member forces for ULS design if a DAM Model has been configured and run. No additional user input is required to utilize the direct analysis forces in design. Robot will not allow you to specify any non-direct analysis method combination for ULS design if DAM is active (active meaning a DAM model has been configured and run). For service, put the proper load combinations in each combination list. Only dead loads in Dead Loads, etc. Robot will check the listed cases against the specified limiting criteria specified in the AISC 360-10 Member Type dialog "Service" settings.

7- Limit States – Limit State Checkboxes: Robot can only perform both an ultimate check AND a serviceability check, at the same time, for "Member Verification". For Code Group Verification and Code Group Design, only one or the other can be done at a time. When running a Member Verification, make double sure to check that you have both checked so that Robot will give you complete results. For design, first run a Code Group Design for the Ultimate Limit State and then after changing member sizes, check Code Group Verification for both ultimate and service. Narrow down design groups to those which fail for service. Note any code groups which fail for service and the re-do code group design for only those code groups with the service option. Finally, perform a Member Verification to check all members for both service and ultimate.

8- Save Calculation Results: Check this box to be prompted to save the results of a calculation run. Use "List" button to the right to access previously saved calculations for inspection at a later date or to screen-capture for report preparation.

9- Configuration: Access the calculation configuration options dialog (See Below)

Calculation Configuration Options Dialog:

Additional options to tune the steel design calculations are located in this dialog. Some of the settings in here are of no small importance to performing a proper set of steel design and verification calculations. The most important settings in this dialog are the Calculation Points. These are the location along the section length where the member will be checked. For instance, if you choose 2 points the member will only be checked at the ends and at no other location. Obviously this could result in very incorrect member designs. Check the settings in this dialog thoroughly to make sure that you are producing the designs that you expect.

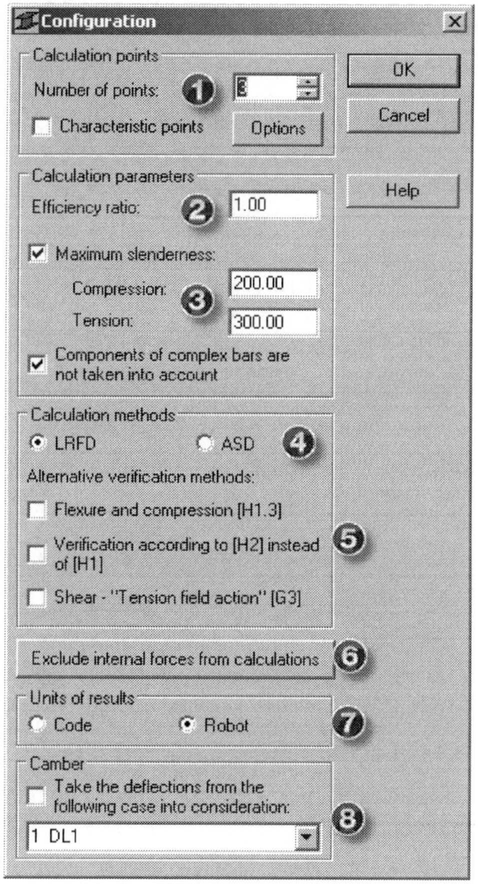

1- Calculation Points: These are the locations along the member length where the section will be checked for the forces present at that point. This is the number of divisions along the member length where calculations will be performed. Note: if you only use 2 then the member will only be checked at the ends! A larger number here will result in more accurate designs. Additionally you can opt to include characteristic points where you can indicate that Robot should perform checks at points of force maxima:

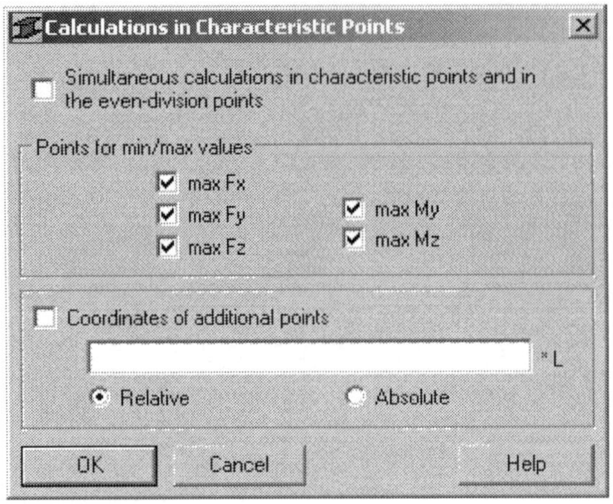

You can opt for calculations at both the evenly distributed locations as well as the points specified in this dialog with the first option. Select which forces you wish to have Robot identify maxima to generate checking locations along the member length. This means that Robot will check the forces indicated here and find maxima, then use that location along the member to perform a design check. You may additionally specify coordinate locations manually along members where checks should be performed.

2- Efficiency Ratio is the maximum utilization ratio in traditional terms. You can adjust this value to limit the maximum utilization ratio which will be allowed in the design and checking of members. For instance, a value of 1.0 would be full utilization of code determined capacities. If you wanted a 5% safety factor above the code values, you could enter 0.95 here to limit utilization to 0.95 of the full code calculated member capacity. Note that this only works with Ultimate limit state and not with Service limit state.

3- Maximum Slenderness Ratios: For compression and tension, specify maximum allowable slenderness ratios here (KL/r). Typical values are 200 for compression and 300 for tension.

4- AISC 360-10 offers options for presenting calculations in either ASD or LRFD format. Select your preferred format here.

5- Alternative calculation options which have been implemented in Robot's AISC360-10 calculators.

6- Exclude internal forces from calculations: You have the option to specify forces which should be ignored during the calculations, or ranges of stresses at which the force will be ignored (considered insignificant enough to not consider in design).

7- Units of Results: Either present results in currently configured Robot units (**Tools>Units and Formats...**) or in the units commonly used in the design code.

8- Camber: If this checkbox is checked, then the deflections due to the selected load case will be subtracted from the deflection values when checking serviceability.

Steel Design Calculations

Once you have finally configured all options, checked and double checked your settings for code parameters and calculation options you can run the member verification or code group design routines. Here are the results for running these calculations and how to understand the output.

Member Verification Dialog

Running the calculation option for member verification results in Robot checking each member you have listed and presenting the results in a tabular form. This will be the final step in your design routine and likely done several times along the way. You will find that using this tool will allow you to zero in on spots where you've under-utilized a set of members through code group design and quickly allow you to revise those designs to make more efficient use of members.

Here is a simple structure with some clear-spanning girders. We've designed all joist members in one code group, and designed the roof girders separately from the floor girders though have used the same member type (from a code properties standpoint) for both the floor girders and the roof girders (Same bracing pattern and design parameters).

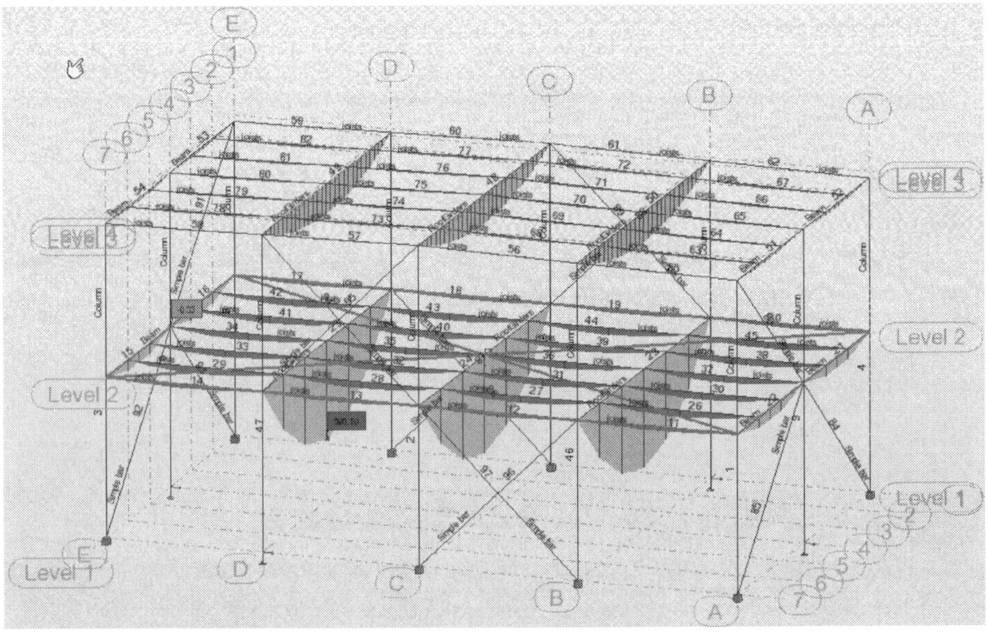

After one cycle of design here is what the member verification looks like:

ANSI/AISC 360-05 - Member Verification (SLS ; ULS) 11to14 17to20 23to45 46to50 55to82

Results | Messages

Member	Section	Material	Lay	Laz	Ratio	Case	Ratio(uz)	Case (uz)
11 joists_11	W14x26	Steel ASTM A992	42.52	222.96	0.47	12 ULS/2=1*1.20 + 2*1.60 + 4*0.50	0.51	54 SLS:STD/1=1*0.67 + 2*0.67
12 joists_12	W14x26	Steel ASTM A992	42.52	222.96	0.47	12 ULS/2=1*1.20 + 2*1.60 + 4*0.50	0.51	54 SLS:STD/1=1*0.67 + 2*0.67
13 joists_13	W14x26	Steel ASTM A992	42.52	222.96	0.47	13 ULS/3=1*1.20 + 2*1.60	0.51	54 SLS:STD/1=1*0.67 + 2*0.67
14 joists_14	W14x26	Steel ASTM A992	42.52	222.96	0.47	12 ULS/2=1*1.20 + 2*1.60 + 4*0.50	0.51	54 SLS:STD/1=1*0.67 + 2*0.67
17 joists_17	W14x26	Steel ASTM A992	42.52	222.96	0.47	12 ULS/2=1*1.20 + 2*1.60 + 4*0.50	0.51	54 SLS:STD/1=1*0.67 + 2*0.67
18 joists_18	W14x26	Steel ASTM A992	42.52	222.96	0.47	12 ULS/2=1*1.20 + 2*1.60 + 4*0.50	0.51	54 SLS:STD/1=1*0.67 + 2*0.67
19 joists_19	W14x26	Steel ASTM A992	42.52	222.96	0.47	13 ULS/3=1*1.20 + 2*1.60	0.51	54 SLS:STD/1=1*0.67 + 2*0.67
20 joists_20	W14x26	Steel ASTM A992	42.52	222.96	0.47	12 ULS/2=1*1.20 + 2*1.60 + 4*0.50	0.51	54 SLS:STD/1=1*0.67 + 2*0.67
23 RoofGirders_2	W40x149	Steel ASTM A992	38.51	44.59	0.92	12 ULS/2=1*1.20 + 2*1.60 + 4*0.50	0.88	57 SLS:STD/4=1*0.67 + 2*0.67 + 4*
24 RoofGirders_2	W40x149	Steel ASTM A992	38.51	42.07	0.92	12 ULS/2=1*1.20 + 2*1.60 + 4*0.50	0.88	57 SLS:STD/4=1*0.67 + 2*0.67 + 4*
25 RoofGirders_2	W40x149	Steel ASTM A992	38.51	42.07	0.92	12 ULS/2=1*1.20 + 2*1.60 + 4*0.50	0.88	57 SLS:STD/4=1*0.67 + 2*0.67 + 4*
26 joists_26	W14x26	Steel ASTM A992	42.52	222.96	0.93	12 ULS/2=1*1.20 + 2*1.60 + 4*0.50	1.01	57 SLS:STD/4=1*0.67 + 2*0.67 + 4*
27 joists_27	W14x26	Steel ASTM A992	42.52	222.96	0.93	12 ULS/2=1*1.20 + 2*1.60 + 4*0.50	1.01	57 SLS:STD/4=1*0.67 + 2*0.67 + 4*
28 joists_28	W14x26	Steel ASTM A992	42.52	222.96	0.93	12 ULS/2=1*1.20 + 2*1.60 + 4*0.50	1.01	54 SLS:STD/1=1*0.67 + 2*0.67
29 joists_29	W14x26	Steel ASTM A992	42.52	222.96	0.93	12 ULS/2=1*1.20 + 2*1.60 + 4*0.50	1.01	54 SLS:STD/1=1*0.67 + 2*0.67
30 joists_30	W14x26	Steel ASTM A992	42.52	222.96	0.93	12 ULS/2=1*1.20 + 2*1.60 + 4*0.50	1.01	57 SLS:STD/4=1*0.67 + 2*0.67 + 4*
31 joists_31	W14x26	Steel ASTM A992	42.52	222.96	0.93	12 ULS/2=1*1.20 + 2*1.60 + 4*0.50	1.01	54 SLS:STD/1=1*0.67 + 2*0.67
32 joists_32	W14x26	Steel ASTM A992	42.52	222.96	0.93	12 ULS/2=1*1.20 + 2*1.60 + 4*0.50	1.01	54 SLS:STD/1=1*0.67 + 2*0.67

Calc. Note | Close | Help

Ratio: Analysis | Map

Calculation points
Division: n = 3
Extremes: none
Additional: none

They are ok for a unity check: only 0.93:

Results | Messages

Member		Section	Material	Lay	Laz	Ratio
11 joists_11	OK	W 14x26	Steel ASTM A992	42.52	222.96	0.47
12 joists_12	OK	W 14x26	Steel ASTM A992	42.52	222.96	0.47
13 joists_13	OK	W 14x26	Steel ASTM A992	42.52	222.96	0.47
14 joists_14	OK	W 14x26	Steel ASTM A992	42.52	222.96	0.47
17 joists_17	OK	W 14x26	Steel ASTM A992	42.52	222.96	0.47
18 joists_18	OK	W 14x26	Steel ASTM A992	[illegible].52	222.96	0.47
19 joists_19	OK	W 14x26	Steel ASTM A992	[illegible]	222.96	0.47
20 joists_20	OK	W 14x26	Steel ASTM A992	42.5[illegible]	222.96	0.47
23 RoofGirders_2	OK	W 40x149	Steel ASTM A992	38.51	[illegible].59	0.92
24 RoofGirders_2	OK	W 40x149	Steel ASTM A992	38.51	[illegible]	0.92
25 RoofGirders_2	OK	W 40x149	Steel ASTM A992	38.51	42.0[illegible]	0.92
26 joists_26	X	W 14x26	Steel ASTM A992	42.52	222.96	0.93
27 joists_27	X	W 14x26	Steel ASTM A992	42.52	222.96	0.93
28 joists_28	X	W 14x26	Steel ASTM A992	42.52	222.96	0.93
29 joists_29	X	W 14x26	Steel ASTM A992	42.52	222.96	0.93
30 joists_30	X	W 14x26	Steel ASTM A992	42.52	222.96	0.93
31 joists_31	X	W 14x26	Steel ASTM A992	42.52	222.96	0.93
32 joists_32	X	W 14x26	Steel ASTM A992	42.52	222.96	0.93

But if we look across the table to the deflection check we'll see the problem:

Results | Messages

Member		Section	Material	Ratio(uz)	Case (uz)
25 RoofGirders_2	OK	W 40x149	Steel ASTM A992	0.88	57 SLS:STD/4=1*0.67 + 2*0.67 + 4*
26 joists_26	X	W 14x26	Steel ASTM A992	1.01	[illegible]7 SLS:STD/4=1*0.67 + 2*0.67 + 4*
27 joists_27	X	W 14x26	Steel ASTM A992	1.01	[illegible]7 SLS:STD/4=1*0.67 + 2*0.67 + 4*
28 joists_28	X	W 14x26	Steel ASTM A992	1.01	54 SLS:STD/1=1*0.67 + 2*0.67
29 joists_29	X	W 14x26	Steel ASTM A992	1.01	54 SLS:STD/1=1*0.67 + 2*0.67
30 joists_30	X	W 14x26	Steel ASTM A992	1.01	[illegible]7 SLS:STD/4=1*0.67 + 2*0.67 + 4*
31 joists_31	X	W 14x26	Steel ASTM A992	1.01	54 SLS:STD/1=1*0.67 + 2*0.67
32 joists_32	X	W 14x26	Steel ASTM A992	1.01	54 SLS:STD/1=1*0.67 + 2*0.67

Most of the members are passing nicely for both ultimate and service but some are slightly over for service. We'll go back and re-design those members for service criteria. One of the interesting things to look at here is the Ratio Analysis. This is a bar graph of each member with its utilization plotted on the vertical axis. Here is the utilization analysis for this structure:

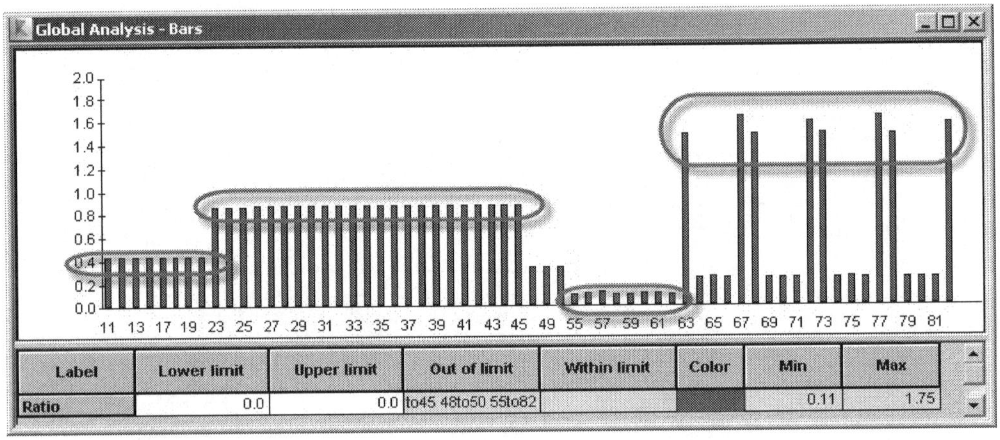

A quick glance at the utilization analysis, shows us some underutilized members which could be redesigned to smaller members (These are, not surprisingly, our exterior joists which can probably be smaller depending on load transfer from cladding to structure). The group just about a utilization of 1.0 is the set of members which need to be redesigned for deflection and off to the right we have some members which are failing due to unexpected torsion in the model. (The source of the torsion will need to be investigated and resolved. It is likely an issue with incorrect member end releases)

This graph can show additional information through right-click>table columns… or **View>Table Columns…** when this table is active, it presents some interesting options to view not only other design values (e.g., length, slenderness ratio) but also design forces:

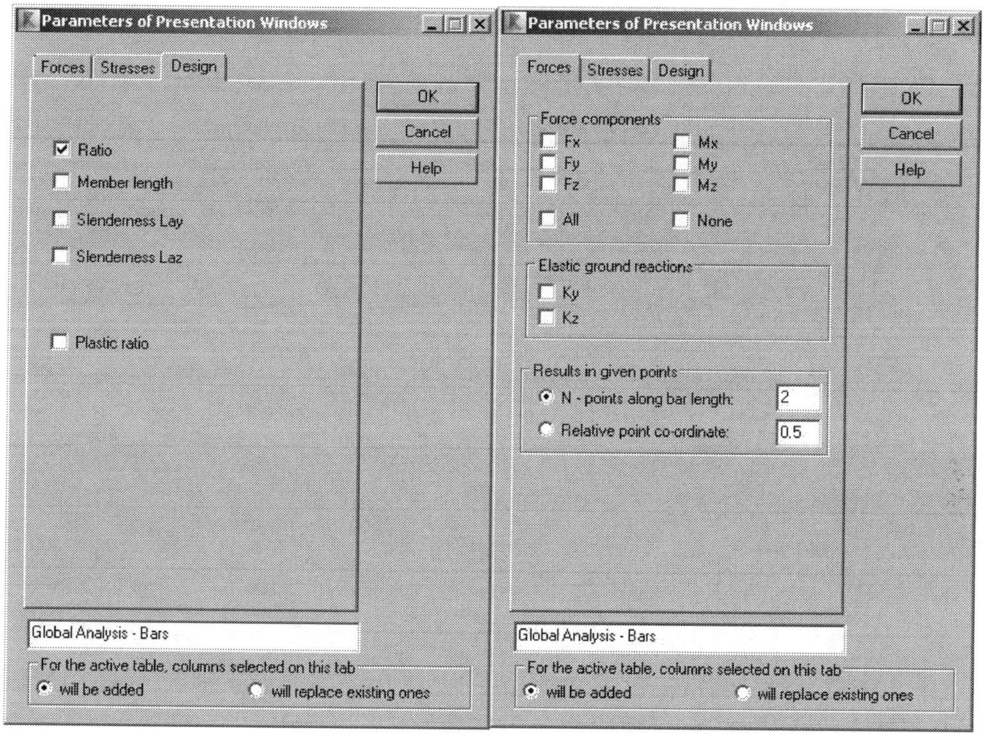

Member Verification Results Dialog:

Back to using the member verification dialog:

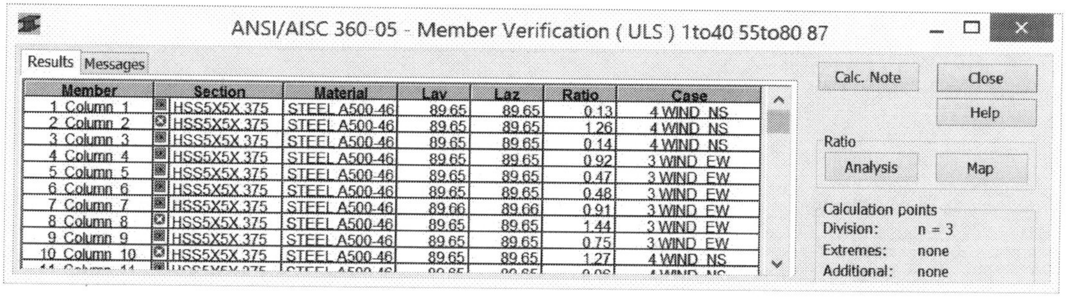

Selecting an individual line in the table will bring up the detailed results dialog:

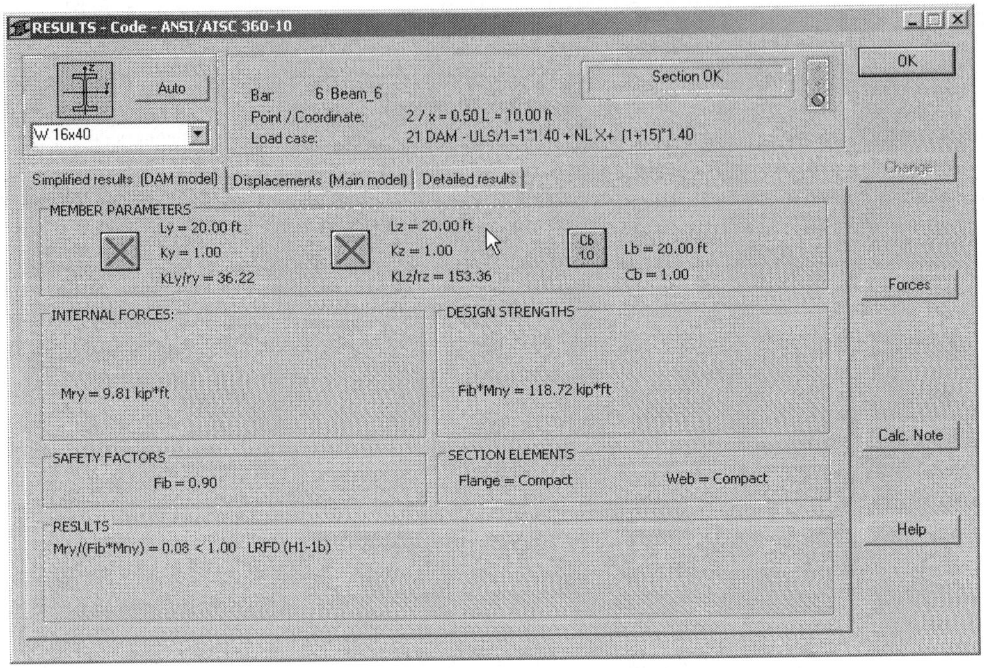

This dialog gives you quick access to the detailed design or code-checking information. Take a minute to look at this dialog and identify the member information, the controlling load combination, slenderness ratios, internal forces and nominal strengths as well as which code equations controlled the design and the results.

Switching to the displacements tab shows detailed results for displacement calculations:

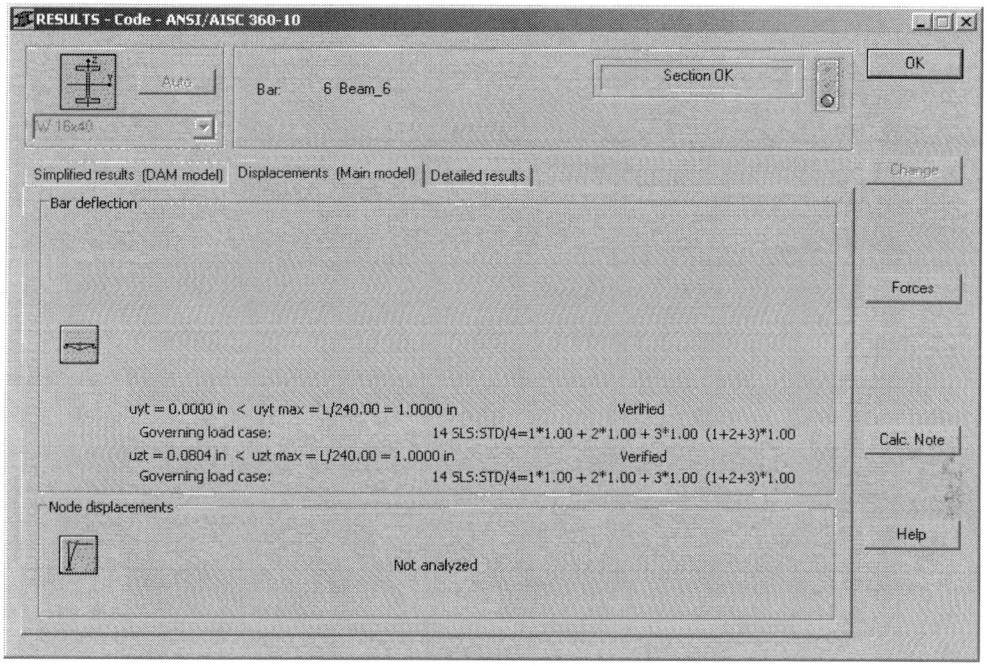

The last tab "Detailed Results" simply lists all code-checking data in tabular form.

Clicking on the Forces button will bring up a dialog that will allow you to play what-if scenarios and re-run calculations. It will allow you to change the calculation options for this member and see what the effects would be. Want to see how it would perform if you used tension field action? This is the dialog to play this what-if scenario:

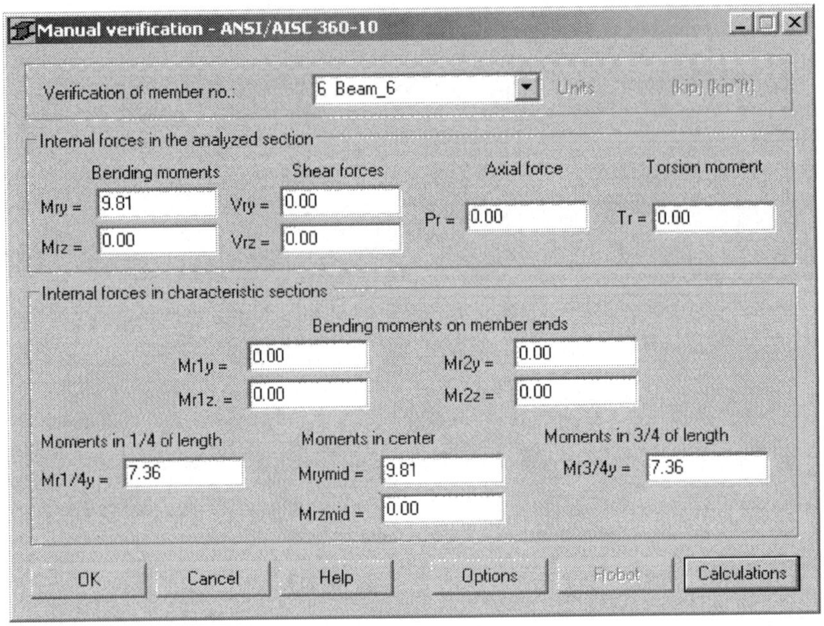

Adjust member forces here, re-run calculations or even adjust calculation configuration with the "Options" button. This has no effect on the member design or verification it is simply a what-if scenario dialog.

The last option here is to access a full calculation note. Remember that there are options for adding all of the calculation notes to the printout in the printout composition section.

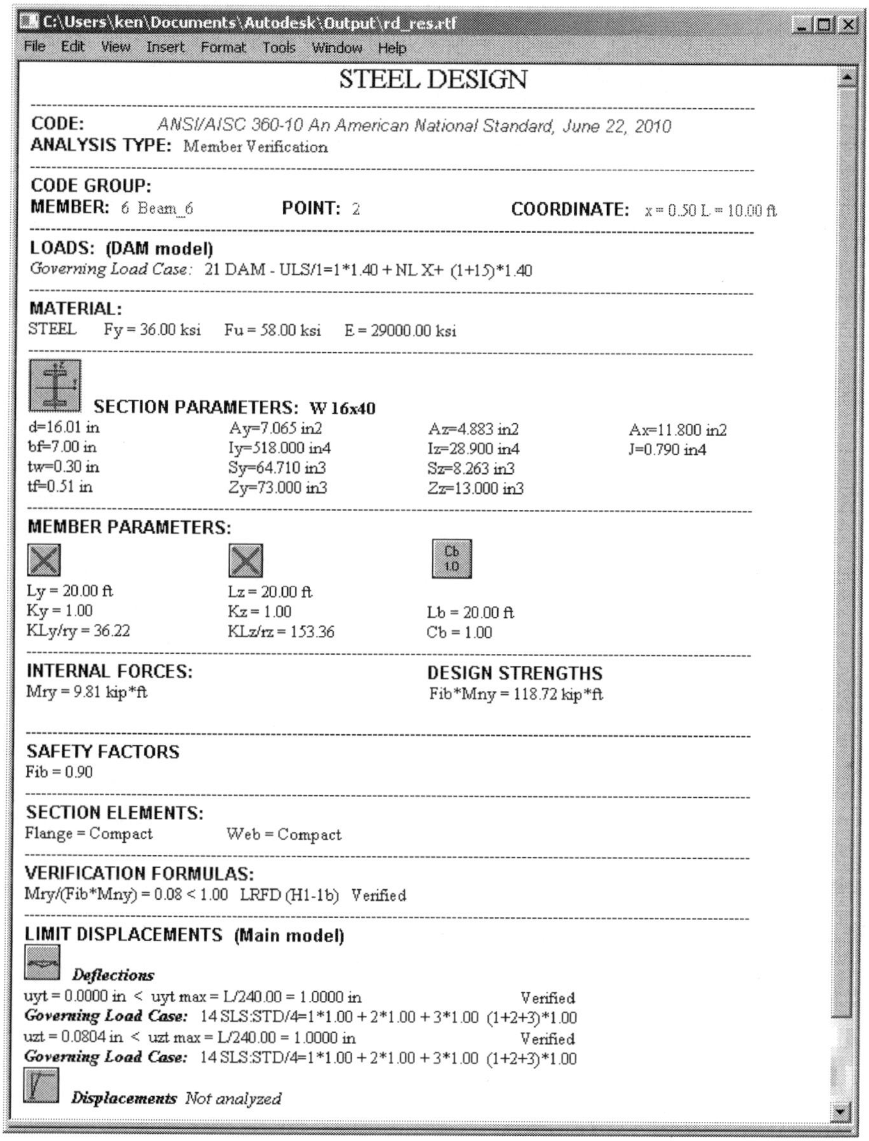

C:\Users\ken\Documents\Autodesk\Output\rd_res.rtf

File Edit View Insert Format Tools Window Help

STEEL DESIGN

CODE: *ANSI/AISC 360-10 An American National Standard, June 22, 2010*
ANALYSIS TYPE: Member Verification

CODE GROUP:
MEMBER: 6 Beam_6 **POINT:** 2 **COORDINATE:** x = 0.50 L = 10.00 ft

LOADS: (DAM model)
Governing Load Case: 21 DAM - ULS/1=1*1.40 + NL X+ (1+15)*1.40

MATERIAL:
STEEL Fy = 36.00 ksi Fu = 58.00 ksi E = 29000.00 ksi

SECTION PARAMETERS: W 16x40

d=16.01 in	Ay=7.065 in2	Az=4.883 in2	Ax=11.800 in2
bf=7.00 in	Iy=518.000 in4	Iz=28.900 in4	J=0.790 in4
tw=0.30 in	Sy=64.710 in3	Sz=8.263 in3	
tf=0.51 in	Zy=73.000 in3	Zz=13.000 in3	

MEMBER PARAMETERS:

Ly = 20.00 ft	Lz = 20.00 ft	
Ky = 1.00	Kz = 1.00	Lb = 20.00 ft
KLy/ry = 36.22	KLz/rz = 153.36	Cb = 1.00

INTERNAL FORCES:
Mry = 9.81 kip*ft

DESIGN STRENGTHS
Fib*Mny = 118.72 kip*ft

SAFETY FACTORS
Fib = 0.90

SECTION ELEMENTS:
Flange = Compact Web = Compact

VERIFICATION FORMULAS:
Mry/(Fib*Mny) = 0.08 < 1.00 LRFD (H1-1b) Verified

LIMIT DISPLACEMENTS (Main model)

Deflections
uyt = 0.0000 in < uyt max = L/240.00 = 1.0000 in Verified
Governing Load Case: 14 SLS:STD/4=1*1.00 + 2*1.00 + 3*1.00 (1+2+3)*1.00
uzt = 0.0804 in < uzt max = L/240.00 = 1.0000 in Verified
Governing Load Case: 14 SLS:STD/4=1*1.00 + 2*1.00 + 3*1.00 (1+2+3)*1.00

Displacements *Not analyzed*

Code Group Design Results

Code group design results are very similar to the member verification except that the "Change all" button is active allowing you to quickly replace all code group members with the suggested section:

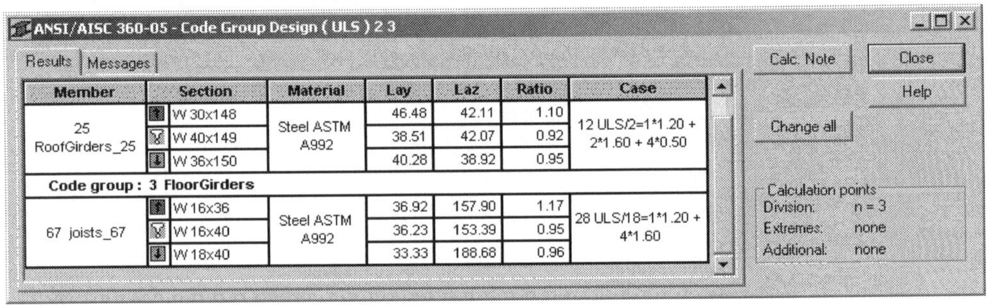

Essentially the same information is presented as in Member Verification, but sections on either side of the suggested optimal section are presented. The list is organized with icons indicating optimal section:

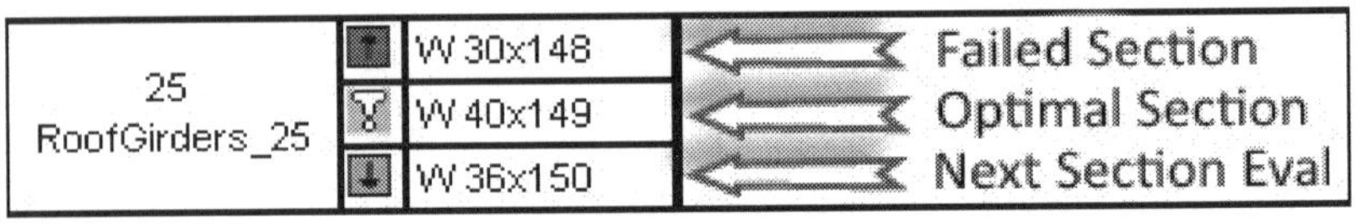

Results for each code group are presented as opposed to results for each member. Note that, even though there is a green icon on the "Next Section Evaluated" be careful to check that this is actually passing. It is simply the next section in the evaluation list and may, or may not, be a passing section. Often times it will pass but there are cases where it may not. Be vigilant.

Selecting a single line of results will bring up the detailed results dialog for that section:

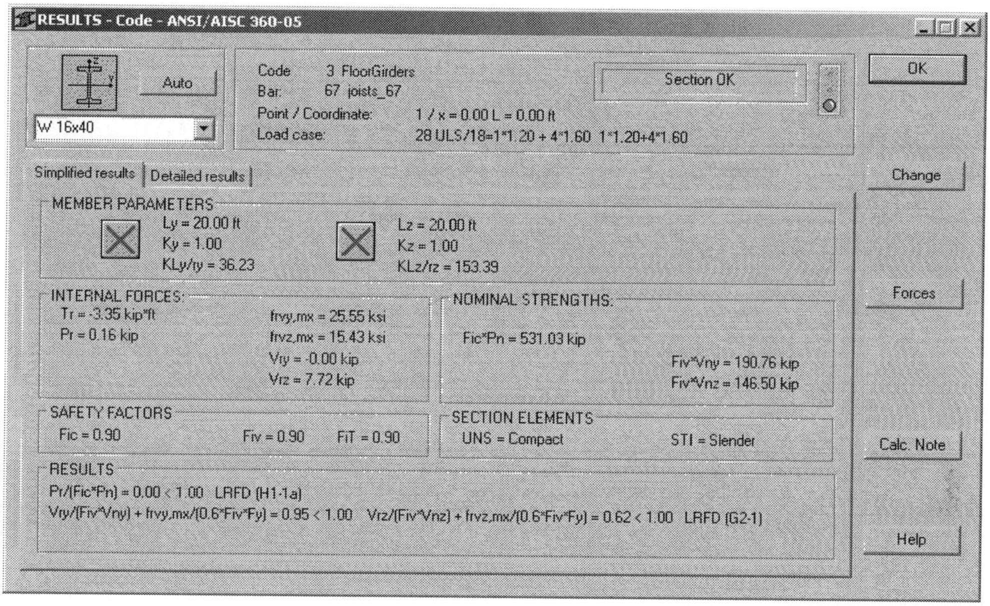

All functions are similar to the detailed results for member verification, but with the addition of a "Change" button to quickly replace all members of this code group with the current section.

Completing the Design Workflow:

Once you have performed design and changed members with optimal sections, be sure to re-run analysis to account for changes in structure mass/weight resulting from the changed sections. Do a last check of member type settings to ensure that your code parameters are still accurate (e.g., tension loads and An/Ag ratios could change based on re-sizing members) and re-run member verification with both "Ultimate" and "Service" checked for the member verification. You'll want to have a full, clean run with the final structure and member verification check completed.

Chapter 9 - Basic Wood Design Workflow

Assuming that load combinations have previously been generated, in Robot the basic wood design workflow is nearly identical to the steel design workflow:

1- Assign code parameters to all elements you wish to check/design
2- Create initial code groups (group similar members at first and refine later)
3- Configure Calculation Configuration Options
4- Perform code group design for strength with optimization
5- Update members based on strength design
6- Perform Member Verification for Ultimate and Service, note any that fail for service
7- Perform serviceability design for code groups with failing members
8- Update members based on serviceability design.
9- Perform member verification to check all members for both strength and serviceability
    a. Optionally review utilization ratio analysis and identify groups of members which can be re-designed more optimally, create new code groups for these members and iterate their design.
10- Iterate any individual members which still fail by creating a new code group specifically for each type of member which needs special attention.
11- Perform final member verification

Robot Wood Design Capabilities

Robot wood and timber design capabilities have been expanding. In addition to support for several European design codes, the 2016 version of Robot Structural Analysis has added support for the American Wood Council National Design Specification (AWC-NDS) 2012 for both ASD and LRFD design.

Here is the current list of supported codes:

USA: AWC NDS – 2012 ASD/LRFD
France: CB71/ CB71 + Kerto
France EC5: NF EN 1995-1:2015/NA:2010/A2:2014
Eurocode5: EN 1995 – 1:2004 / A2:2014
Eurocode: ENV1995-1-1:1992
Finland EC5: ENV 1995-1:1992 NAD Finland
Poland: PN-B-03150
Poland EC5: PN-EN 1995-1:2005/NA2010;A2:2014

Overall Design Capabilities per AWC-NDS-2012:

1. All materials governed by AWC-NDS 2012 Supplement tables 4A through 6b meaning Sawn Lumber including Beams and Stringers, Heavy Timbers (larger than 5"x5"), Piles and Poles, and Glued Laminated Beams/Columns of both Western Species and Southern Pine referred to in the software as (GWS and GSP respectively). Designs may be calculated in either Allowable Stress Design (ASD) or Load and Resistance Factor Design (LRFD) per the AWC NDS 2012.

2. Robot will not currently perform design or code checking of Structural Composite Lumber (SCL) including Laminated Veneer and Parallel Strand Composite Lumber.

3. Robot will not currently perform design or code checking of I-Joists per AWC-NDS

4. Robot will not currently perform design or code checking for structural panels or mechanical fasteners.

Current Limitations:

STRICT WIDTH LIMITS BASED ON SUPPLEMENT TABLES:

Current material tables have been created with consideration of the allowable member size to be designed with the values in the table. In other words if AWC-NDS Supplement Table 4A says minimum width is 2", that is what has been set for all the predefined materials. While this seems innocuous, the software does not currently take into consideration the dressed size of the members to be designed properly. Meaning that the material table says 2" min width and a sawn lumber 2x (2" wide nominally) has a dressed width of 1.5". Robot flags this member as not meeting the minimum width requirements of the design table and refuses to consider the member for design. You can work around this issue by changing the minimum width of members allowed for the material to the dressed size of the material you wish to design.

DESIGN MODIFICATION FACTORS NOT APPLIED TO YOUNG'S MODULUS

Currently Robot does not apply design adjustment factors to Young's Modulus (E) in order to calculate an E' for use in deflection calculations. Please note that for any applications where one of the applicable adjustment factors for E are required, you will want to adjust your allowable deflections to suit or run the model two times with different values of E for proper deflection calculations. Luckily it is fairly rare that we need incising, or hot/wet conditions for most members though if you do, design/code checking will require a bit more work.

ROUGH SAWN MEMBERS NOT IN STANDARD SECTION LIBRARY

The pre-populated section database has section sizes for Timbers and Beams/Stringers but they are all set to the dressed timber dimensions and there is no consideration for Rough Sawn lumber. If you wish to use Rough Sawn lumber, you will need to add a custom parametric section as shown below.

NO SUPPORT FOR DOUBLE TAPERED CURVED MEMBERS

While curved members of constant cross section can be checked/designed, as well as single and double taper (presumably compression flange only), no support is available for double tapered curved beams.

Wood Databases/Section Properties

The included database for AWC-NDS comes pre-populated with full NDS Supplement Materials (Tables 4A through 6b), and sections for standard lumber, timber, and glued laminated beams.

Adding Materials to Your Project

In order to perform code checking or design with a particular species of wood, you will first need to add that species to your project. Furthermore, you will need to add it for as many design tables as you plan to need design values. By this, I mean that if you need to do code checking for Douglas Fir Larch dimensional lumber (Table 4A) and Douglas Fir Larch Beams and Stringers (Table 4D Beams and Stringers) and you also have some heavy timbers also in Douglas Fir Larch (Table 4D but Piles and Timbers this time) You'll need to add three different material definitions to your project (and then again as many as you have different grades though most designers will work with a single grade of each type. E.g., "Select Structural").

You can access the project materials from **Tools>Job Preferences...** This will bring up the Job Preferences dialog where you can select "Materials" from the list at the left:

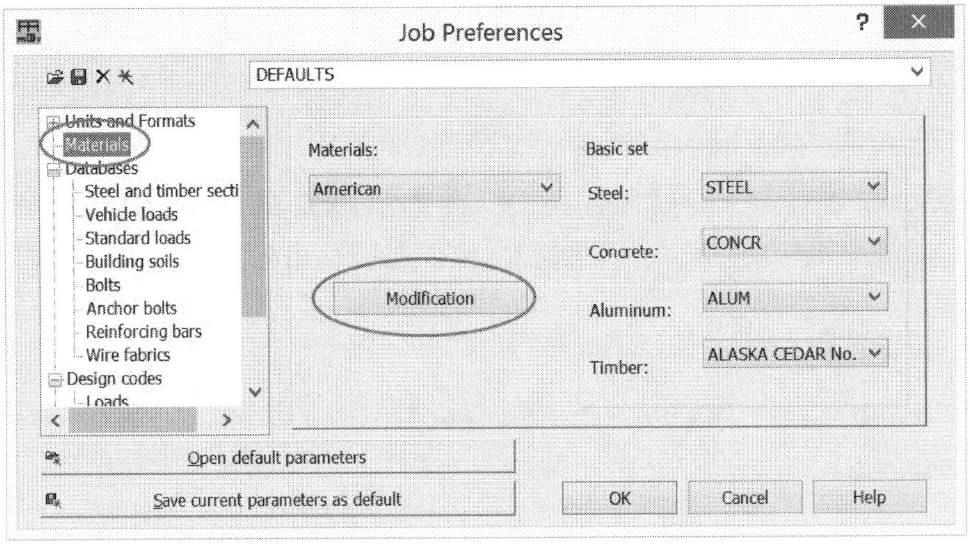

From here you can select "Modification" to access the Material Definition editor:

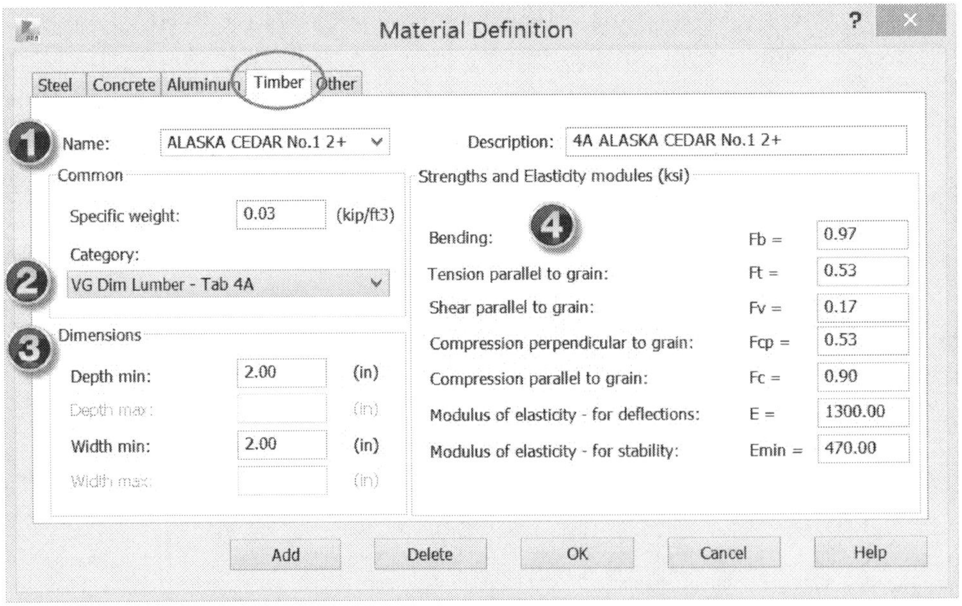

1. Name: This is actually a dual-purpose dropdown. It is the name field for a new material definition but it is also an access to the tabulated species design values in tables 4A through 6b. The tabular values can be a starting point for your material creation.

Access the tabulated design values by selecting the Name dropdown and choosing "More timbers..."

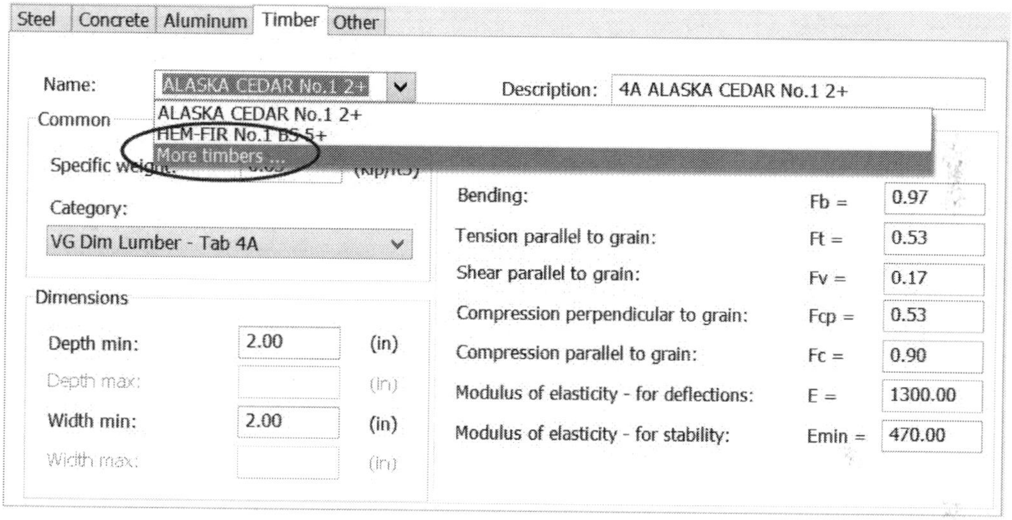

This will present you with the American Timbers AWC-NDS selection dialog:

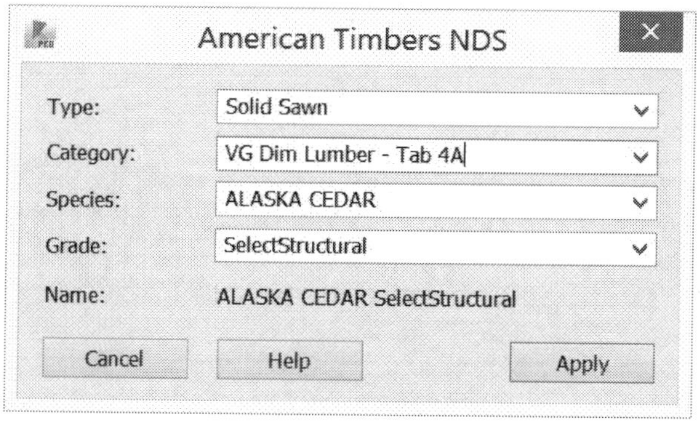

This dialog effectively gives you access to the AWC-NDS tables 4A through 6b for initially populating the fields of the material modification dialog.

Type: Choice of Solid Sawn, Glued-Laminated, and Poles/Piles. This setting controls what "Categories" are available in the next drop down.

Category: Select the desired table from this drop down. The contents of this dropdown depend on the current setting in the "Type" dropdown above.

Species: Select the species to be used in code checking/design. Note that options in this dropdown will change based on the setting of "Category" dropdown above. For instance, if you have Table 4D selected, each "Species" in this dropdown will have two entries: Species BS (beams and stringers) and Species PT (piles and timbers). You will need to select the appropriate sub-division for the type of member that you are intending to check/design.

Grade: This will be the particular grade of interest under the species sub-division. If you flip through a few of these different combinations you should be able to identify that Robot is selecting the appropriate values from the Supplement tables.

Name: This is the initial setting of the material name which will reflect several of the major properties of this "material" definition including the species name and the grade. You will likely want to give further refinement to this naming to reflect any subdivisions (BS/PT) or whether solid sawn, glued-laminated, or piles/poles.

Note that the selections in this dialog simply populate the fields of the material modification dialog with values from the AWC-NDS Supplement table. From there, you may modify as needed.

As you can see for the situation described above: We need to design 2xs, some beams/stringers as well as some heavy timber all in Douglas Fir Larch, we would need 3 different materials:

1. Soid sawn, Table 4A, Douglas Fir Larch, Select Structural.

2. Solid Sawn, Table 4D, Douglas Fir LarchBS, Select Structural

3. Solid Sawn, Table 4D, Douglas Fir LarchPT, Select Structral

If you had some glued laminated beams you would need yet another material for that.

---Back to the Material Definition Dialog ---

2. Category: This setting in the modification of timber materials dialog allows you to specify which table should be associated with the material definition. This setting is going to control which design adjustment factors are applied to the design values at code check/design time. For instance, if you select Table 4D (has no repetitive member factor Cr) and then specify that Robot should apply the Cr factor to the design/code-check, it will ignore the Cr factor based on the selection of Table 4D for the "Category" setting in the Materials Definition Dialog.

3. Dimensions: The AWC-NDS Supplement tables often specify a range or applicable member dimensions to which the table applies. For instance, Table 4A specifies that it applies to members which are Min 2" nominal thick, to Max 4" nominal thick. Robot will check member sizes in code-check/design to make sure that they fall within the allowable range set here and will not allow any members which do not fit these criteria to be designed. However, the maximum fields don't currently allow you to specify maximums.

4. Design Values: These values will initially be set to the values taken from the AWC-NDS Supplement. You may modify them as you see fit. In most cases you would ideally want to use what is suggested by the AWC-NDS Supplement but if you happen to be working with an unusual species you could specify custom values here.

Adding the AWC-NDS section catalog to your project:

If you haven't already done so, for the next section you'll want to add the AWC-NDS wood sections database to your project. Again from **Tools>Job Preferences...** Databases tab, select "Steel and Timber Sections". Use the "New Database" button to configure the AWC-NDS database for use.

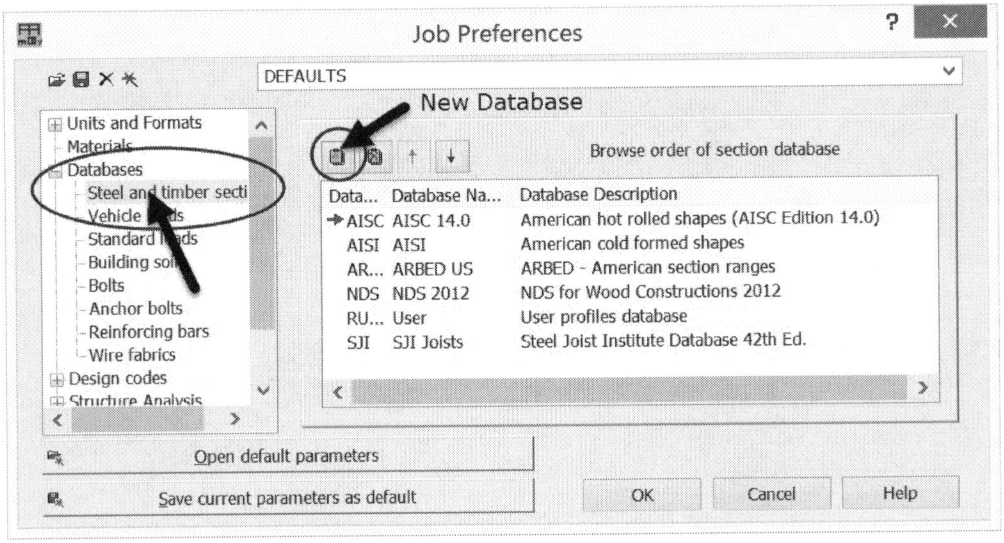

Specifying and Assigning Wood Sections to Members:

Robot has included an AWC-NDS member sections database which covers the most commonly used dressed dimensional lumber, typical beams/stringers, typical timbers, typical Glued Laminated sections, and post/pile sections. The one thing to keep an eye on is that the database covers standard dressed, it does **not** include traditional "rough sawn" sections. Should you need to use "rough sawn" sections you will need to specify them as parametric wood sections by supplying the width and height.

To begin specifying sections for use in modeling timber structures, begin by going to **Geometry>Properties>Sections...** which will bring up the Sections configuration and application dialog:

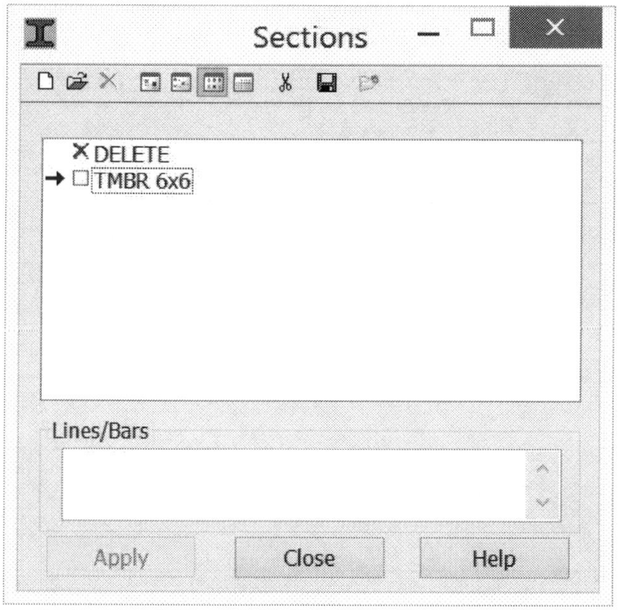

You will note that this dialog is a typical Robot label definition/application dialog.

To add a new section press the "new" button: Which will bring up the New Section Dialog:

Select "Timber" for Section type, then choose the material you'd like to use. Remember that Robot's sense of the material will incorporate a sense of the types of members for which the Supplement is intended. Therefore, if you select a Table 4A material, the "Family" dropdown will only include "LMBR" (for dimensional lumber). If you want to want to use a beam or heavy timber section with this material (not recommended), you'll need to select "ALL" in the "Family" dropdown.

This will have standard sizes of dimensional lumber, beams/stringers, glued-laminated beams, and heavy timbers. If you have a non-standard size ("rough sawn"), or need to specify pile or pole sizes you'll need to select the "Parametric" tab in this dialog to specify the exact dimensions.

In the "Label" textbox, enter the text that will show for this section type. Because the section type is material specific, you will want to include some indication of the material if you have multiple materials in use on one project.

In order to verify the member dimensions for the section (since the database sections do not show you the dimensions during configuration) you can always use the "Properties" table (**View>Tables...**) and use "Table Columns" from the table right-click menu to show the cross section dimensions of bars and you can easily verify the section properties:

Remember that only sections currently in use in the project will be shown in this table.

	Section name	Bar list	HY (in)	HZ (in)	VY (in)	VZ (in)	VPY (in)	VPZ (in)
	BEAM 6x14 DFL BS	1	5.50	13.50	2.75	6.75	2.75	6.75

To Add a Rough Sawn 6x14 select the proper material and select the "Parametric" tab of the "New Section" dialog to enter appropriate section dimensions:

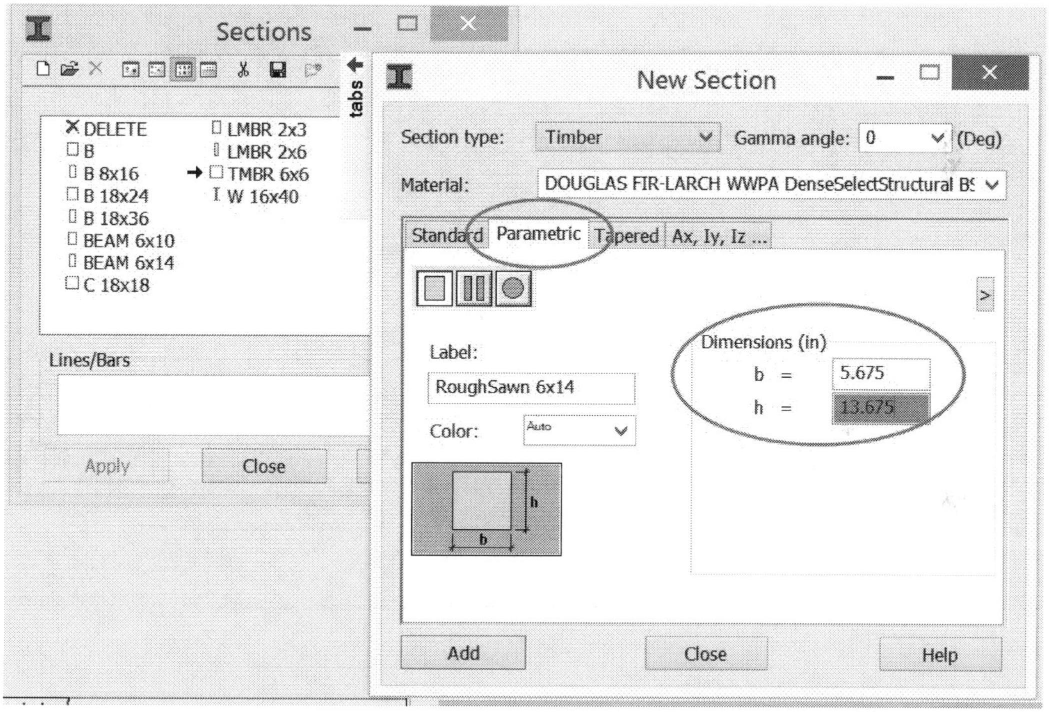

Configuring AWC-NDS 2012 Code Properties

After you have configured necessary materials, added necessary timber sections, applied them to your beams/columns, and run your analysis, you're ready to configure the Code Parameters which will be associated with each "bar" (Timber/wood member in this case). This is almost exactly the same process as for steel members. We will look at the parameters available in the AWC-NDS member settings then in the next section we'll talk about running code-checks or designs in Timber.

Configuring Code Parameters:

To launch the AWC NDS 2012 code parameters, first make sure that you have the AWC NDS 2012 code selected for timber design in the job preferences (**Tools>Job Preferences...**>Design Codes) for more information on this refer to project setup. Then from the menu, select **Design>Timber Members Design – Options>Code Parameters...** to launch the AWC NDS 2012 Member Type label dialog:

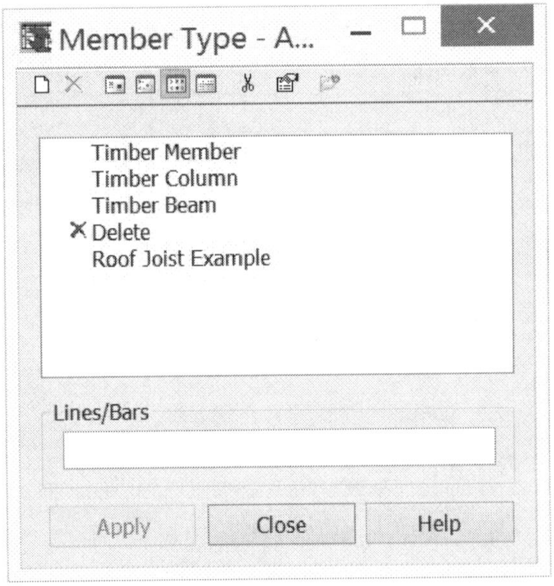

There are three default member types which cannot be deleted from the label list: Timber Member, Timber Column, and Timber Beam. You can use these as starting points for your own member types or create new ones from scratch.

Click the "New" button to start a new member type definition:  This will launch the AWC NDS 2012 Code Parameters Dialog:

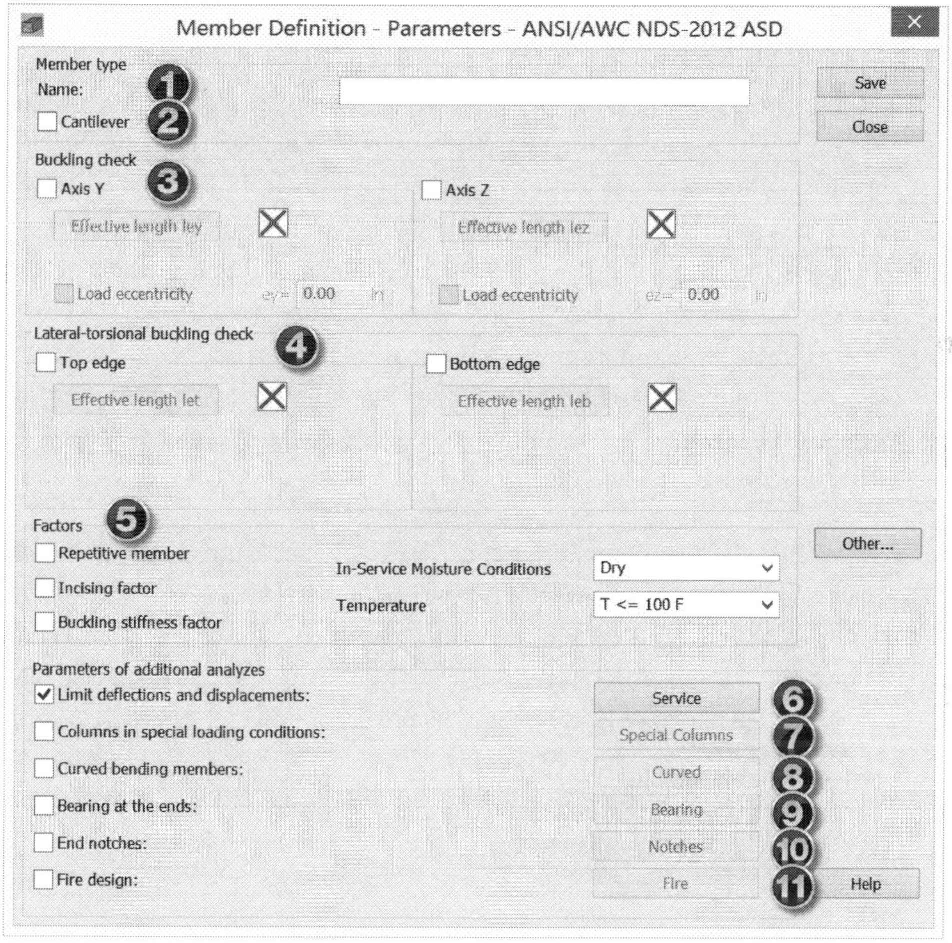

This dialog is used to configure design and code checking parameters for members you want to have checked per AWC-NDS. You will need to configure a different member type for each specific condition of use, braced length, bending/compression, or other special conditions. For instance, if you have two joists which are exactly the same except that one has allowable notches at the end, and the other doesn't, you'll need two different "Member Definitions". Likewise, if you have two stringers which are exactly the same except they have different brace points, you'll need different member definitions. Create as many member definitions as you feel you need to fully and accurately describe your member configurations for design/code-checking.

THE AWC-NDS 2012 MEMBER DEFINITION DIALOG:

1. Name: Timber Member, Timber Beam, and Timber Column are reserved names and cannot be modified. Supply your own descriptive name for the set of parameters you want to assign to members. With exception of the reserved names, saving an existing name will overwrite the definition. (You will be prompted to overwrite the definition if you do. Note that any changes overwritten will affect all members to which that set of code parameters has been assigned.) If you supply a new member type name and then save, that label will be added to the list of member type labels though there is no feedback that this has been done. You'll have to check the Member Type labels dialog to see it and apply it to elements.

2. Cantilever: If the member is a cantilevered beam, select this box. If you wish, you can break your member into two sub-members in order to apply this setting to the member definition for the cantilevered portion of the beam. Note that options available in the lateral buckling l_{et} settings (Item 4 below) will change based on whether this checkbox is ticked or not.

3. Buckling check: These settings control the application of buckling checks in the code-checking/design routines. These checks are independent of the lateral compression flange checks in the "lateral-torsional" buckling check in item 4 below.

If you leave these checkboxes unchecked, no buckling check will be performed in the direction indicated (recall that Axis Y is the Major Bending axis and Axis Z is the minor bending axis).

Checking either AxisY or Axis Z will allow you to configure the effectively length of the member for compression about the selected axis. Each axis may be configured independently. For example, the Axis Z settings:

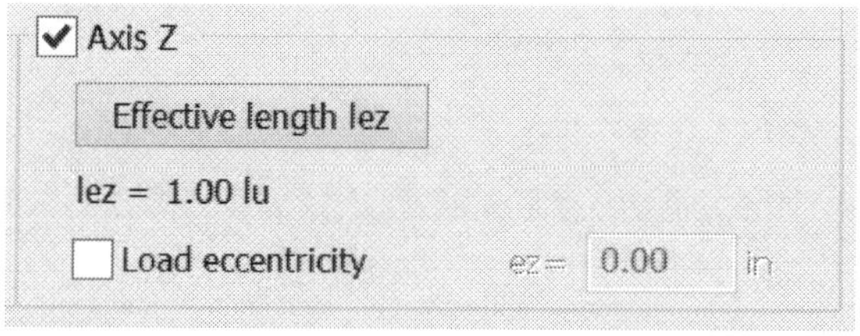

Pressing the "Effective length lez" will open a dialog allowing you to configure standard unbraced length/effective length settings or specify your own:

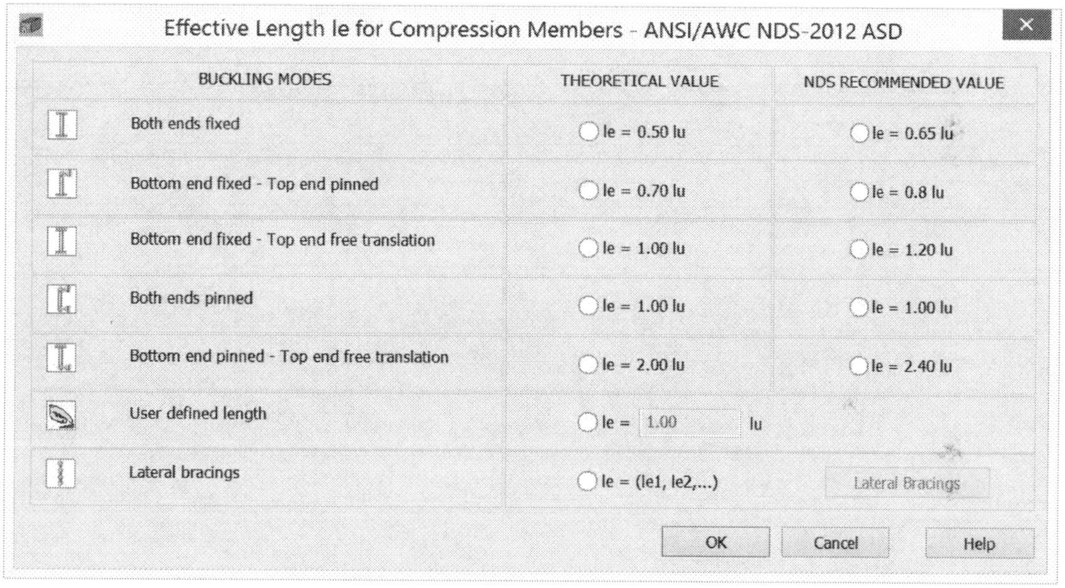

You will note that the options available here are a reproduction of Appendix G table G1 and include as standard, the theoretical and recommended effective lengths based on end conditions. You may also select your own effective length factor by selecting "User defined length".

If you have distinct points of bracing you may want to use the "Lateral Bracings" dialog to configure specific locations of bracing.

Selecting the "Load Eccentricity" checkbox will result in Robot invoking the equations in section 15.4 which require the eccentricity of the load to be specified for the calculation of combined loading with eccentricity per AWC-NDS 2012 Section 15.4.1.

--- *Continuing with the Member Definition Dialog* ---

4. Lateral-Torsional Buckling Check: These settings allow you to control the specification of lateral bracing of the compression flange with different settings available for top flange and bottom flange.

Leaving the checkbox unchecked will result in Robot not performing any lateral torsional buckling checks (C_L will be 1.0) for the flange. If the member is laterally supported on both top and bottom, leave these unchecked to have Robot use a C_L of 1.0.

Selecting the "Effective length l_{et}" button will bring up the Effective Length for Beams, between Supports dialog:

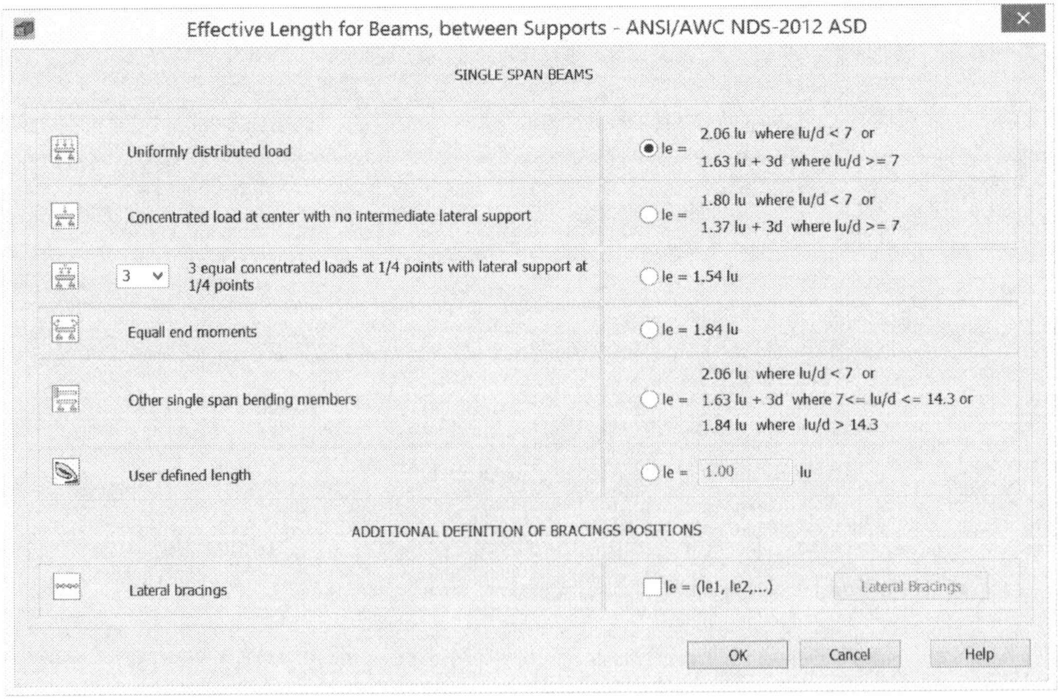

Standard options per the AWC-NDS Table 3.3.3 are available here. You will notice that cantilevered options are not available here. In order for this dialog to contain the cantilevered settings you will need to select the "Cantilevered" checkbox in Item 2 above.

Again, if you have distinct points of bracing you may want to use the "Lateral Bracings" dialog to configure specific locations of bracing.

5. Factors: Here you have the option to invoke several standard factors.

a) Repetitive Member: C_r

b) Incising Factor: C_i

c) Moisture Factor: C_m by setting "In-Service Moisture Conditions" dropdown to

d) Temperature Factor: C_t by setting the "Temperature" dropdown.

e) Buckling Stiffness Factor: C_T invoking AWC-NDS section 4.4.2 for members used as truss chords with plywood sheathing.

6. Limit deflections and displacements: This checkbox allows you to set Robot to check live, dead and total load deflections. Once you select the checkbox, you may press the "Service" button to the right and configure all settings related to limits on member deflections.

When configuring these settings, recall that "uzd" is the deflection caused by major axis bending. (The "vertical" deflection of a horizontal beam/joist element) The member local z axis is the minor bending axis of a member and therefore deflections along the z axis are perpendicular to the major axis which is the y axis.

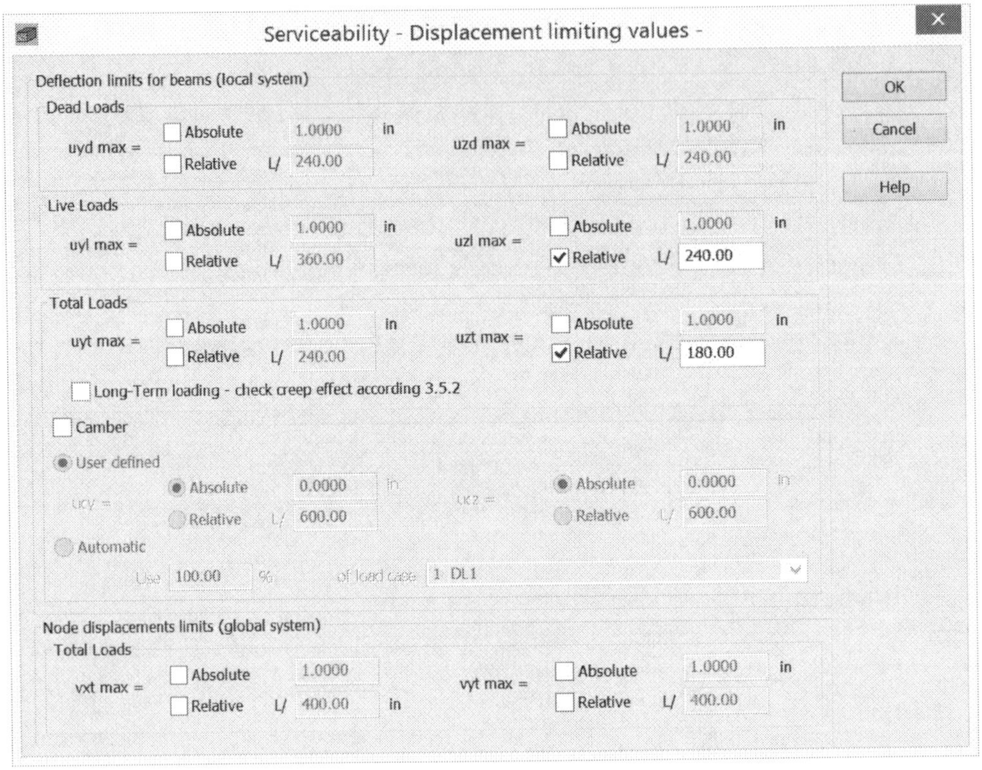

You will also notice that long term creep check (per 3.5.2 and Appendix F) can be specified here. Robot will look at the long term loading specified (based on load durations which are configured in the AWC-NDS Calculation Configuration options. See below).

You can configure both absolute and relative deflection checks for the member.

7. Columns in special loading conditions: This will allow you to specify the configuration of stitched columns or built-up columns and thereby invoke the provisions of AWC-NDS 15.2 or 15.3.

Selecting "Spaced column" will use 15.2 and allow you to select the end block condition per 15.2.2. You can also specify the center to center distance of the spacer blocks.

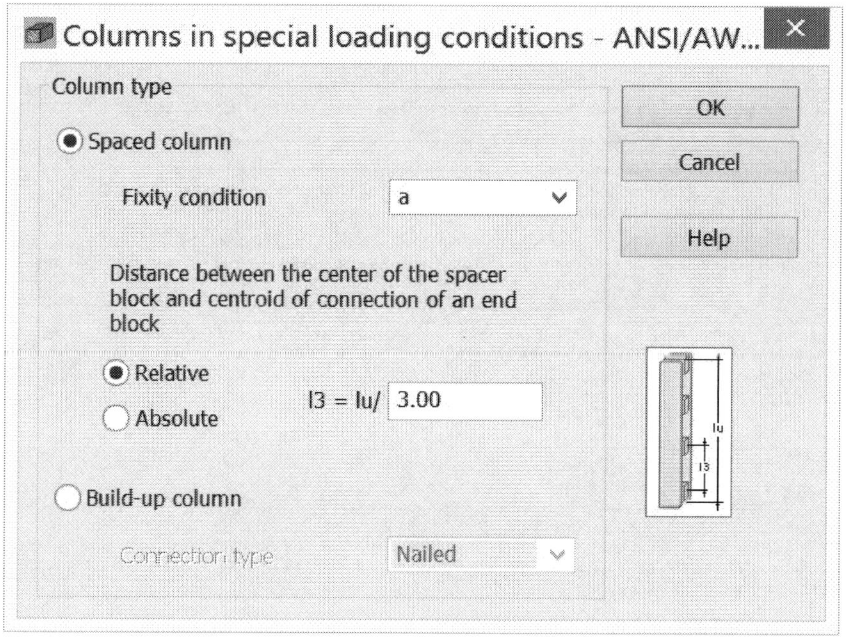

Selecting "Build-up column" will invoke the provisions of 15.3 and allow you to specify whether this member is mechanically laminated by nailing or bolting per AWC-NDS 15.3.3 or 15.3.4.

8. Curved bending members: This checkbox allows you to configure parameters for calculations of Curvature factor Cc per section AWC-NDS 5.3.8 for use with curved glued laminated timbers. Pressing "Curved" will bring up the Curved Members dialog:

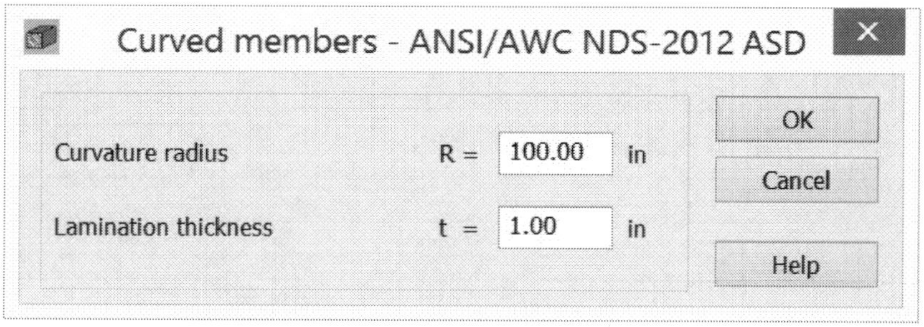

9. Bearing at ends: This checkbox invokes bearing checks per AWC-NDS 3.10 perpendicular to grain and calculates and applies Cb factor if appropriate based on the settings here. The design check basically applies the Cb factor but does not take into account the distance to the end of the member. So if you choose to use this check make sure that you have a situation where Cb applies.

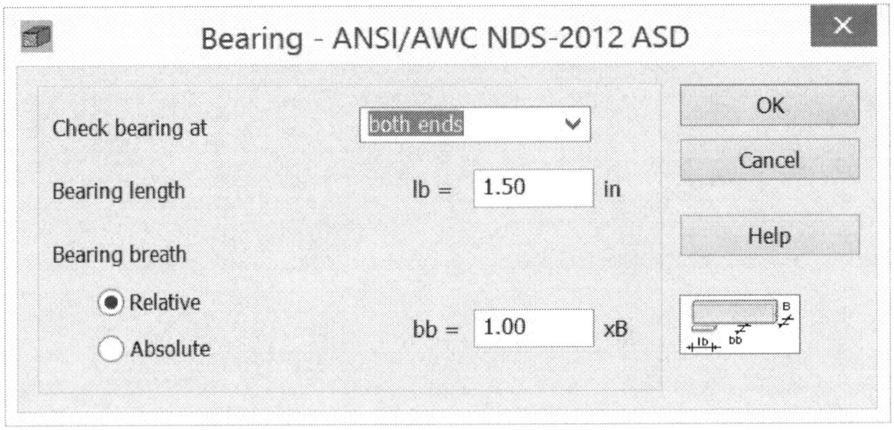

You can ask Robot to check bearing at either or both ends of the member to which this definition is assigned, you may also configure the bearing length and width. Robot will use the smaller of these values in calculating Cb factor apply.

10. End Notches: Using the End Notches dialog you can configure parameters of notches in the ends of the member. Robot will check the notches for allowable limits and apply shear calculations per section AWC-NDS 3.4.3. The End Notches dialog will allow you to configure notches at either or both ends but only top or bottom notches.

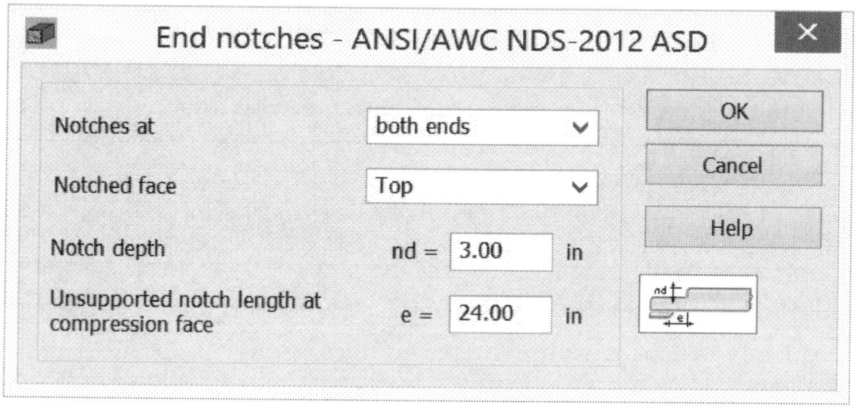

11. Fire Design: The fire design options will allow you to invoke calculations of the section reduced by fire "char". You will find option for fire duration, protection of sides, and an additional fire retardant factor.

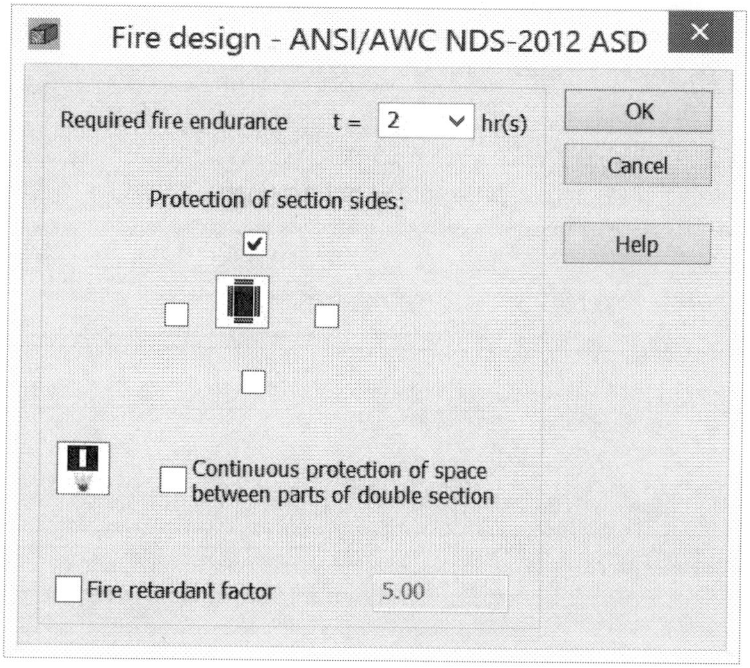

If protection is selected for one or more sides, Robot will ignore those sides in calculating the reduced section. Section reductions are calculated per the char rate equation given in AWC-NDS 2012 16.2-1. The fire retardant factor will be applied to design values as a direct multiplier of the design value and is assumed to be provided by the fire retardant manufacturer.

Wrapping up Timber Design Setup:

Once you have configured and applied all required AWC-NDS member types to the members you wish to check/design, if you want to perform design, you will need to further configure design groups. This is standard functionality in Robot and applies to Steel/Aluminum, Required Reinforcing, and Timber design. All design in Robot is performed on design groups.

Wood design layout

The wood design layout may be accessed from the layout manager dropdown or from the menu **Design>Timber Members Design...**

You will notice that the layout includes the Members/Groups dialog where you may configure super members (comprised of smaller calculation members if you have this situation) and/or Design Groups.

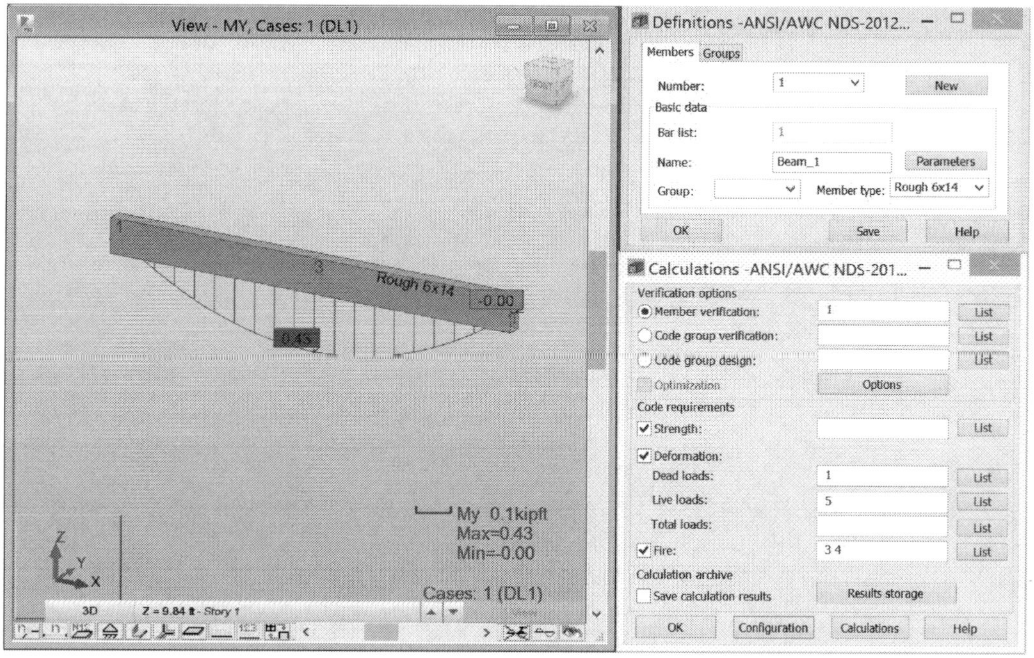

Calculations Dialog

The calculations dialog is very similar to the steel design dialog with the exception of the addition of Fire load cases/combinations as an additional option.

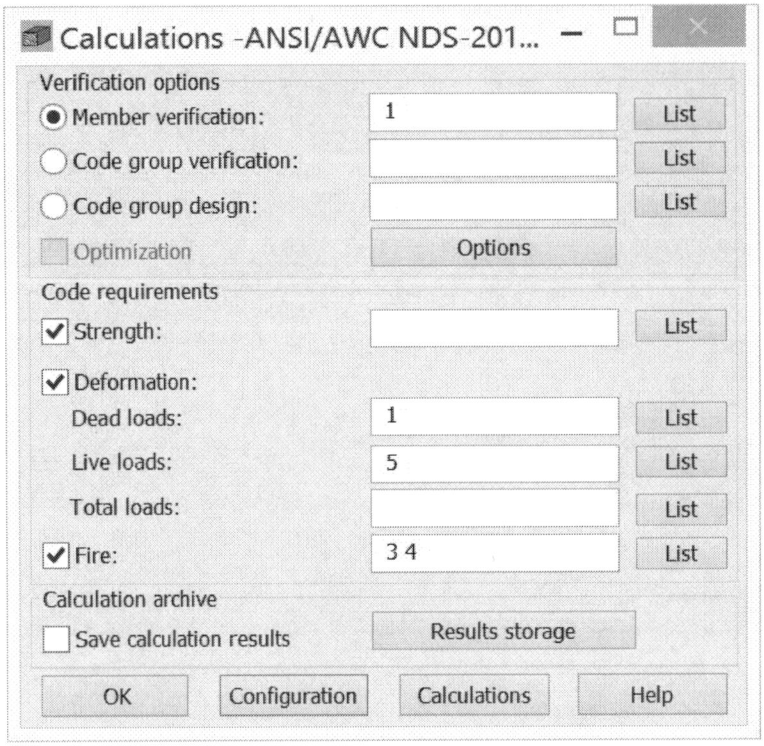

The "Configuration" options provide some of the standard options as well as one very important aspect of wood design: Load Durations!

Here is the configuration options dialog:

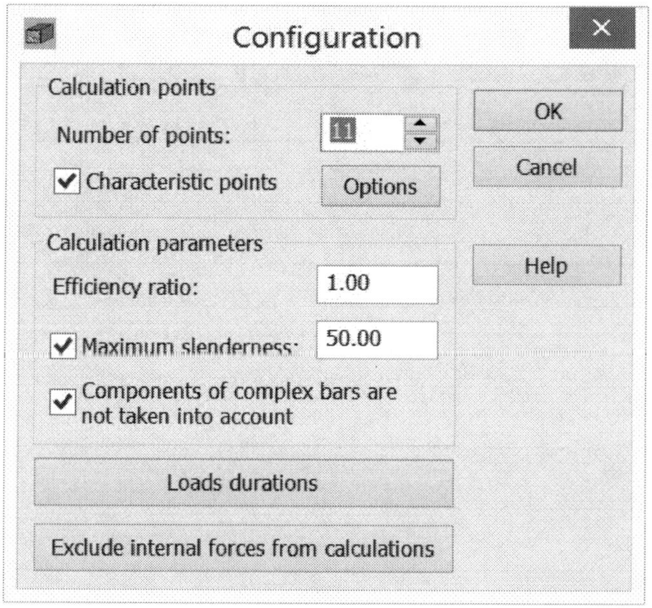

As always with Robot design you will want to make sure that you have selected an adequate number of calculation points along the member and also configured any characteristic points or locations at which you would like to force design calculations to be made. (e.g., points of max member force, specific locations of interest along members).

Most importantly for timber/wood design, the "Load Durations" button will give you access to configure loading durations associated with the load cases you have previously defined:

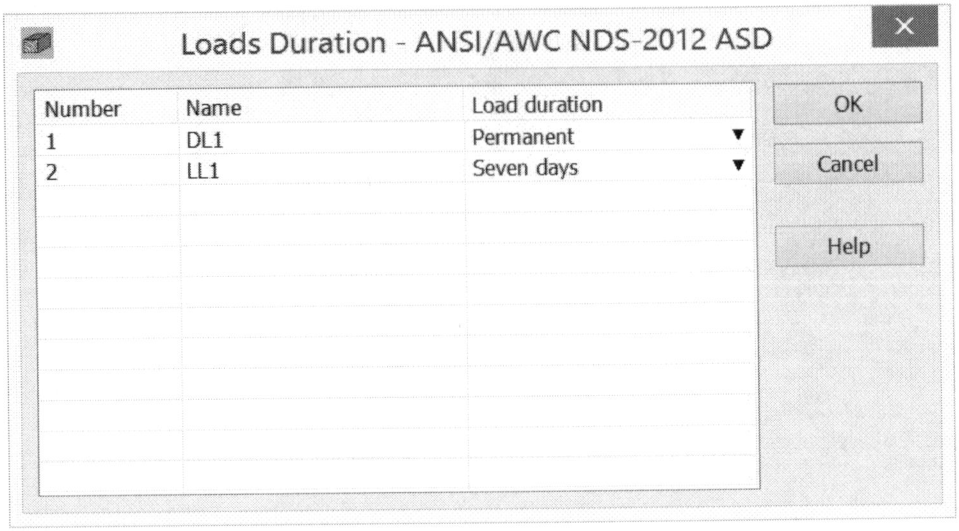

All configured load cases will be shown here and the dropdowns to the right of the case name offer access to AWC-NDS code load durations:

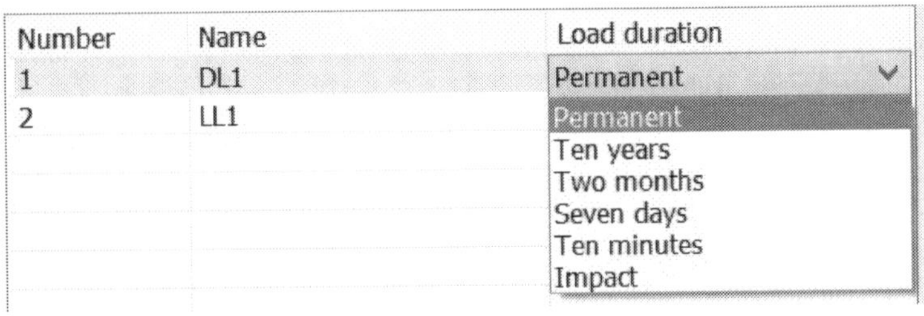

Once these settings have been properly configured, you are ready to start design calculations.

Chapter 10 - Structure Modification

Once you have created a model, sometimes you'll find the need to change it to adapt to design changes or perhaps, if the model was brought in from Autodesk Revit, you may need to adjust or correct some nodal positions to get the structure fully connected.

Modifying Axes/Levels

Axes and levels cannot be modified in the dialogs where they were created. Short of deleting and re-adding them, the only way to modify levels or grids is to access their object properties via the right-click menu:

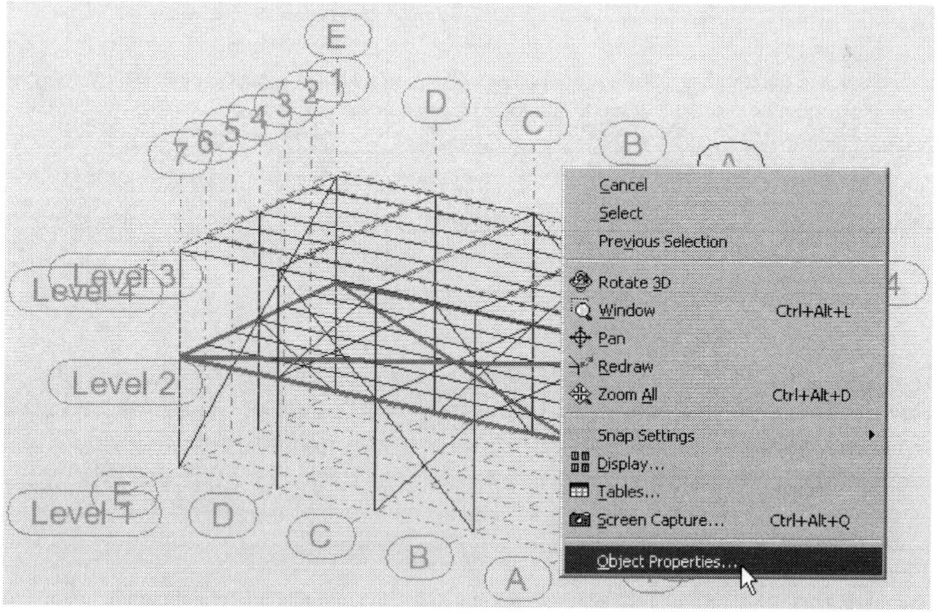

Selecting Object Properties will bring up the Structural Axis Modification dialog:

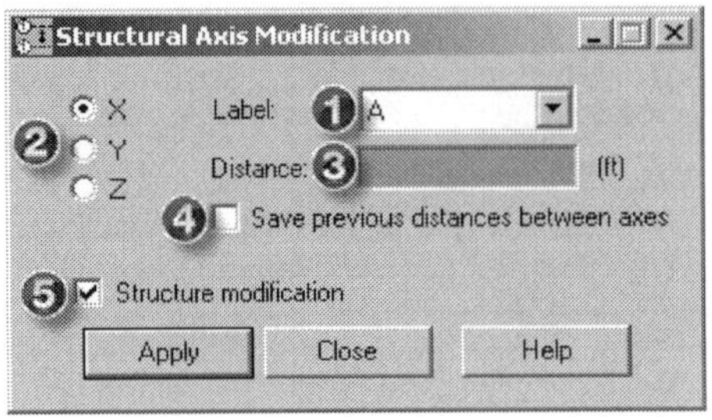

1- This dialog can adjust any structural axis, not only the one you right-clicked. The label selector allows access to any of the axes which are of the axial persuasion (X,Y, or Z) selected in #2 below. To access other axes, select a different axis direction on the left. This should be set to the axis you selected to get to this dialog.

2- Direction selector. Changing this setting will populate the label control with axes perpendicular to this direction.

3- Distance to move grid. This can be positive or negative but cannot be farther than the distance to the next grid line in the direction of movement.

4- Preserve the distance between all other axes. If you want to increase or decrease the space between interior grids, but leave all other spacing of grids after this grid as they are, then select this option. For instance, if you have grids A through F and change position of grid C with this option checked, Grids C,D,E, and F will be relocated as well to preserve the prior spacing. With this option off, only Grid C will move relative to all other grids.

5- All elements and nodes which lie on this axis will be moved along with the axis. Members will be shortened as necessary to adjust to the new grid position.

The Edit Menu

All modifications to the physical model will be manifested through the Edit menu. Not only movements of the structure but the edit menu can also be used to copy, rotate, mirror elements during creation and even be used to divide existing members and correct small offsets in the model.

Edit>Edit>

This edit menu offers options useful in modifying or even in just building the structural model itself. Move/Copy is fairly obvious though the usage is not necessarily straight forward, Vertical mirror is to mirror across a vertical plane (parallel to the Z axis), Horizontal mirror is to mirror about a plane parallel to the X and Y axes. Planar Symmetry is to mirror about any user defined plane and axial symmetry is a double mirror (about the Y axis plane through the axis and then about the X axis plane through the axis). The last option "scale" is exactly what it sounds like.

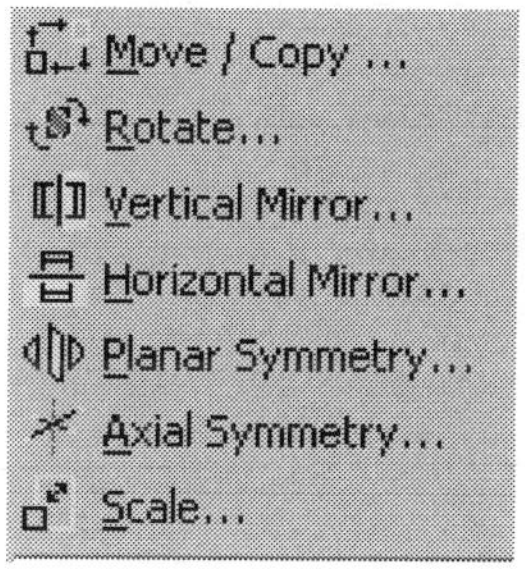

General Workflow for Edit Tools:

1. Unless you want to edit the entire structure, select the elements you want to edit before starting the command.

2. Start command from **Edit>Edit>**<command-name>

3. Choose edit mode (move or copy)

4. Specify number of repetitions

5. Specify translation vector (or other parameters to control the operation) either by clicking once in the edit control and then selecting points in the project environment or by typing coordinates directly into the edit controls. The behavior is different for each.

6. If you use the cursor to select points in the project, Robot automatically executes the edit after specifying the last point (this is slightly different for each tool but there is one parameter that, if specified in the project will automatically execute the edit). If you manually specify points/vectors by typing into the edit control, then you must use the "Execute" button after entering the last point.

7. To re-execute the operation on other elements with the same settings, right-click in the project environment and choose "Select", select elements and, then use the Execute button.

8. To use the same tool with different settings, right-click in the project environment and choose "Select", select the elements and then configure the new parameters. Again, if you select parameters by clicking in the project environment (specifying a translation vector for instance) the operation will be executed automatically. If you manually enter the parameters, you will use the "Execute" button to complete the operation.

MOVE/COPY...

This dialog seems very straightforward, but the operation is a bit unintuitive. In the dialog below, notice the Execute button. This button is only used if you have already moved something and, then want to select something else to move by the same move vector or if you have manually entered coordinates of the translation vector. Otherwise, this dialog function is automatic: selecting the second point of the vector in the project environment will automatically execute the tool. Most of the edit dialogs work this way, there is one parameter which, if selected in the project by clicking, will automatically execute the tool. A little confusing at first but easy to get the hang of it.

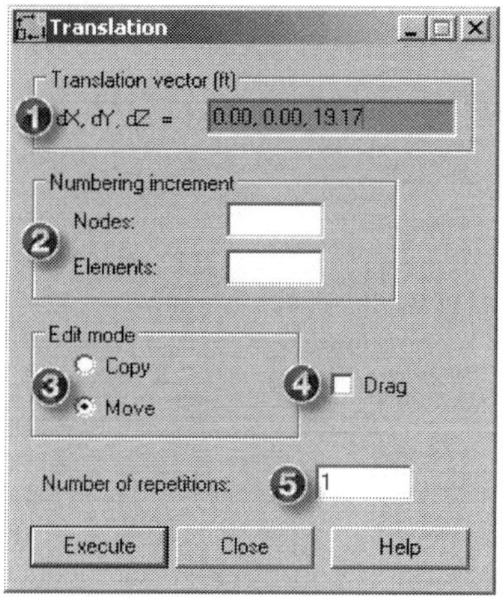

1- Translation vector. Specify this last. You have two options, either enter it manually by typing into the edit control (use "Execute" button to execute the move/copy) or by left-clicking once in the edit control (it will turn green) and then selecting two points in the project environment which represents the translation vector. The move will occur automatically after selecting the second point in the project.

2- If copy is selected, new nodes and members will be automatically numbered. If you wish to specify an increment for them from the original numbers, enter that value here. For instance, if you specify 1000 and copy member number 51 twice, the new members will be 1051 and 2051 respectively.

3- Move or Copy. Copy creates new elements, move moves the existing elements.

4- Drag. Only for copy. This actually means: create new members between new nodes and existing nodes parallel to the translation vector. The last selected bar element in the sections dialog will be the member created on drag. To get a quick look at what that is, look at the status bar along the bottom of the Robot window. There is a rotating info section cycling through each of the current last configured elements.

5- Number of Repetitions: Note that for move, this must be 1! Or it will not move. If you select a number greater than one, then the elements will be moved that number of times. This is more obvious for Copy where (this number of) additional copies will be created.

ROTATION...

The workflow for rotation is similar to Move/Copy... familiarize yourself with the move/copy workflow. Here the specification of the rotation angle on-screen will cause automatic execution; otherwise manually entering the rotation angle will require use of the "Execute" button.

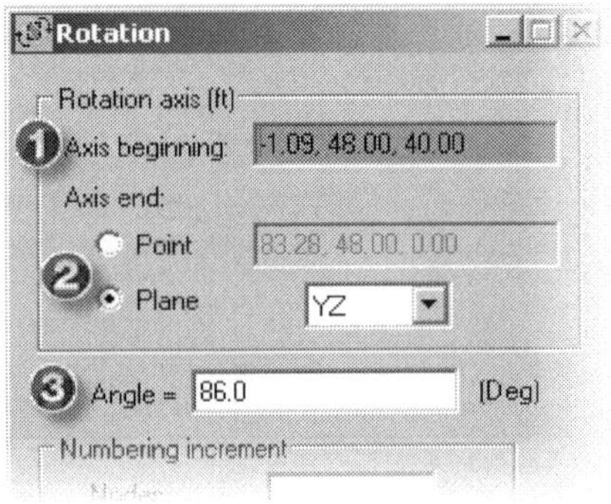

Focusing only on the differences between Move/Copy and Rotate:

1- Specify the beginning point of the axis of revolution (the axis about which you intend to rotate the selected elements.

2- Either specify the end coordinate of the axis of revolution or select the plane perpendicular to the axis to fully specify the axis of revolution.

3- Angle of rotation: Either enter this value manually or select graphically in the project environment. Selecting graphically in the project environment will cause automatic execution of the rotate command. Entering it manually will require the use of the "Execute" button.

VERTICAL MIRROR... (HORIZONTAL MIRROR...)

This is for mirroring about either the XZ or YZ planes (mirror in plan). Most functions of this dialog are similar to Move/Copy or Rotate above. Focusing only on the differences:

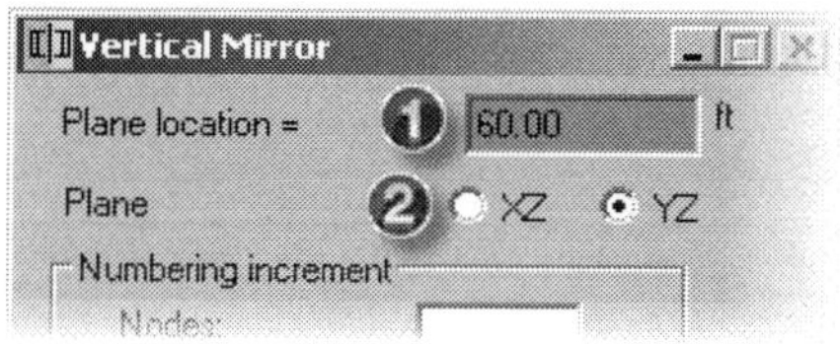

1- Specify the plan location (in the XY plane) of the mirror plane. The mirror plane will either be the XZ plane or the YZ plane

2- Select whether the mirror plane is the XZ plane or the YZ plane. To mirror about an arbitrary plane use the "Planar Symmetry" tool.

PLANAR SYMMETRY...

This functions exactly like other edit tools, but allows you to specify three points to define a plane about which to mirror elements:

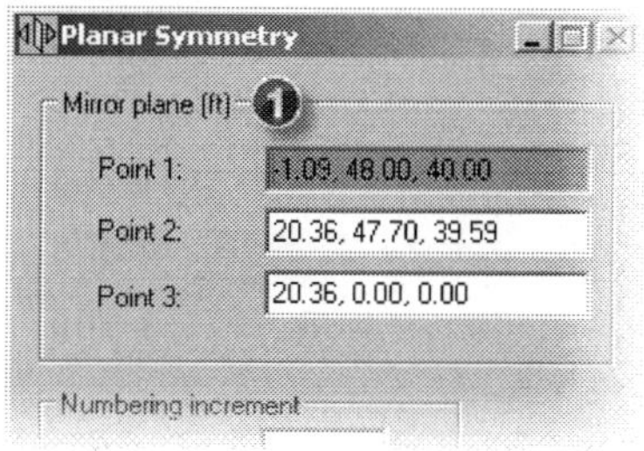

1- Either manually enter 3 points to define the plane of symmetry or select points in the project environment by first clicking in the first edit control (it will turn green), and then specifying three points in sequence in the project environment. Upon selecting the last point in the project, the command will be automatically executed. If you choose to manually enter the coordinates of points, then use the "Execute" button to effect the mirror.

AXIS SYMMETRY...

This is effectively a double mirror. Here is an example of mirroring through the vertical axis (parallel to Z) at grid 7A:

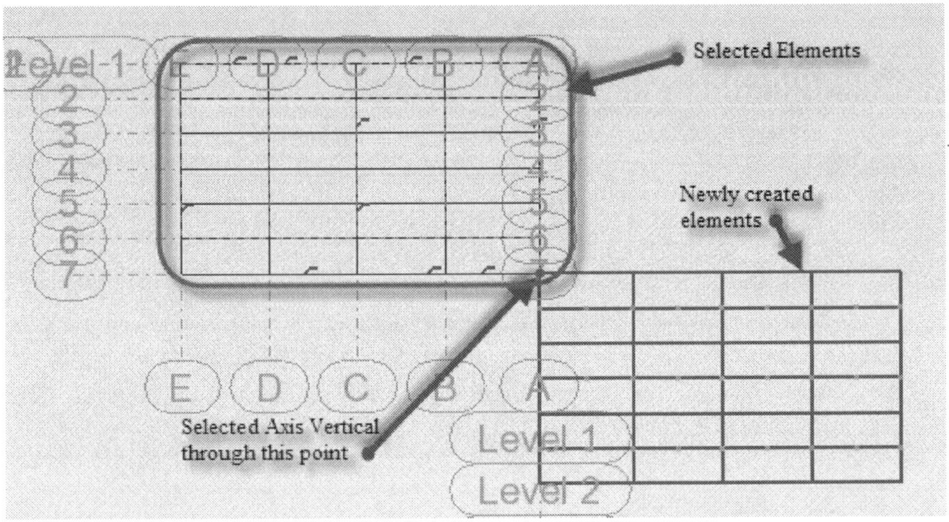

SCALE...

Scale is probably the most straightforward of the edit tools, but the in-project selection of the center point and ratio should be avoided mostly because it is not the most intuitive. Select the elements you want to scale up or down, select a center point and enter a scale factor (ratio) then press "Execute". The scale is applied to the selected elements relative to the selected center point.

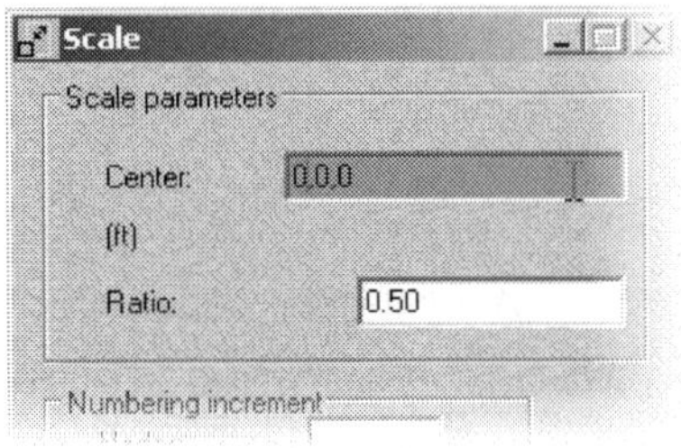

The Edit Menu Modify Commands

Often times a model will need to be broken into smaller elements, calculation nodes will need to be added, or even corrections to the geometry of the model may be required. The modify commands below will help make these kinds of adjustments to your model.

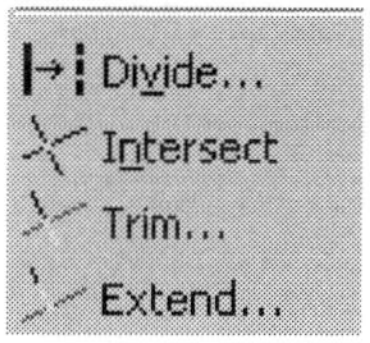

DIVIDE...

Divide is used to either add calculation nodes along the length of a member or to actually break members into smaller members. With a member (or members) selected, launch the Divide dialog from **Edit>Divide...** The divide action will apply to currently selected members. As with other dialogs, once you have divided members you may change the selection by right-clicking in the project environment and choosing "Select" from the contextual menu.

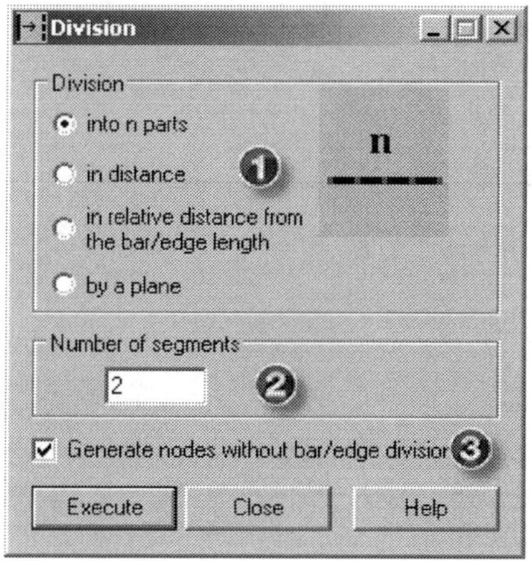

1- Division: "Into n parts" is the only option which will multiply divide a member in one go. The other options you will need to execute multiple times to divide a member more than once.

2- This area of the dialog will change based on the setting selected above in the "Division" section. Configure the parameters of each type here prior to executing.

3- Generate nodes without bar/edge division: If checked, the bar or edge itself will not be divided but rather calculation nodes will be added at the division points. Calculation nodes give you the ability to control calculation points along a member, results presentation can be shown at added calculation nodes and calculation nodes will be used as one of the design sections during member code checking and design. This can be important if force maxima occur at locations other than the standard divisions of the elements. (Refer to member point loading, in Chapter 4, illustrating the addition of a calculation node at the time of load application to see an example of how this can be advantageous).

Example:

Selecting this bar and dividing it into an equal number of segments generating a calculation node without bar/edge division:

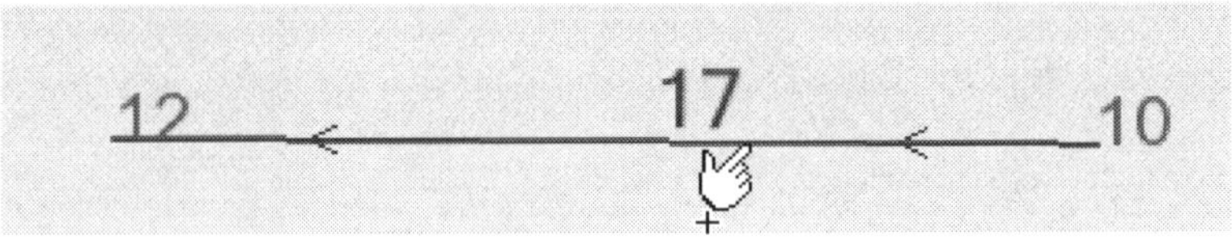

Will result in this:

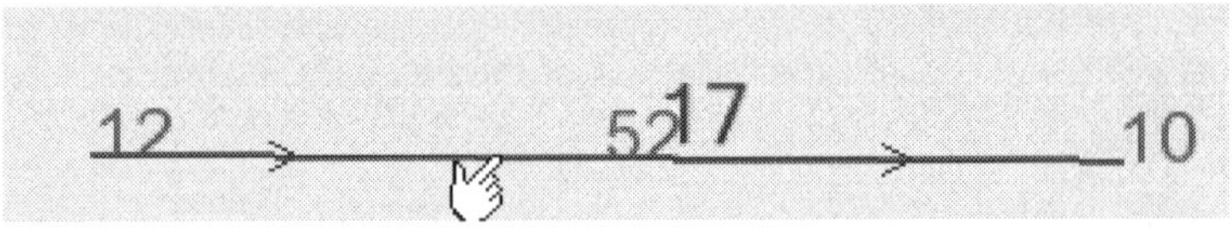

Note that node 52 has been added yet member 17 is still the same as it was before. However, if the option is not used, then we'll wind up with two different members:

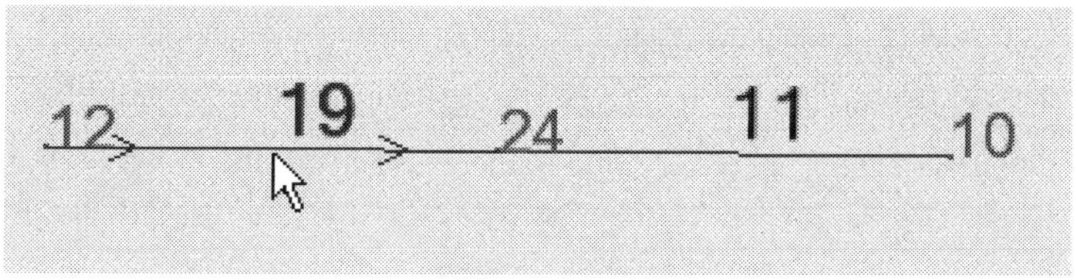

Notice that we now have new members 19 and 11 where we only had 17 before as well as having a new node (24). Dividing members may create extra work where simply generating a calculation node might suffice for the purposes of the results presentation and forcing consideration of a particular point in the design and member checking (a section check will be performed at this point). Extra work is required for proper design of a member which has been "physically" broken into multiple members in terms of making super members of the individual elements.

INTERSECT

Note that Intersect is a command and does not launch a dialog. First select the members for which you want to perform the intersection, then start the command. Any place where members cross each other or shell finite elements touch a bar member will be divided at that point.

Example:

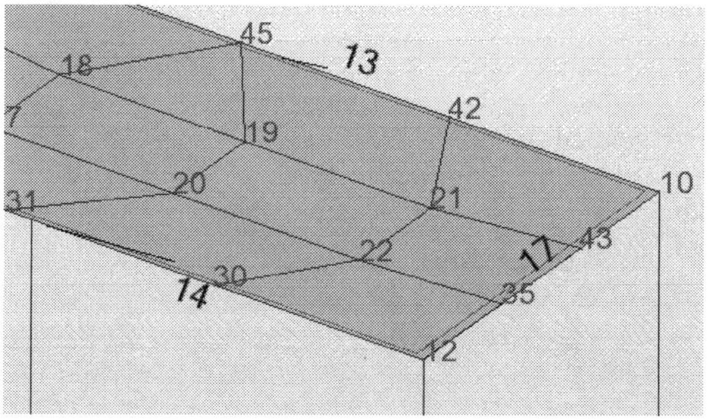

Note that the shell element is supported by three members 13, 14, and 17. Also, note that there are finite elements which have a dimension less than the members which support them. At the time of calculations, these calculation nodes will force a division of the bars in the calculation model (generated at time of calculation) and the results will all be re-assembled at the super-bar level after calculations are finished. This allows you to interact with the bars and meshed shells without the extra work of dealing with highly divided supporting members. However, you can use the Intersect command to divide these bars and any intersecting bars in the model. The result will look like this:

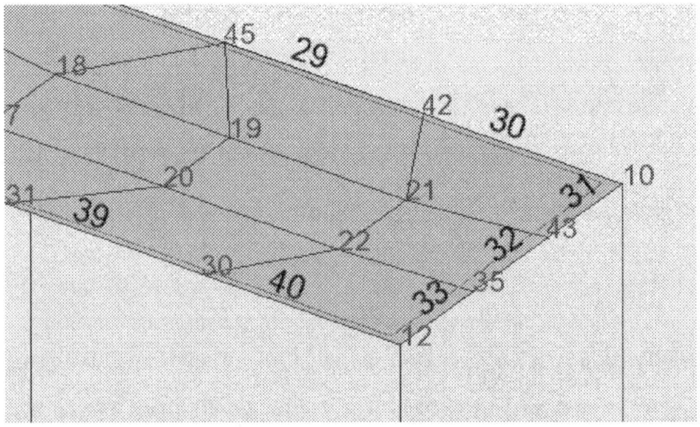

Notice that all the nodes are the same but now we have members 29to33, 39, and 40 (7 in all) supporting our shell instead of just 3.

This technique can also be used for troubleshooting model problems. Take a look at the troubleshooting section to see how this tool is used to find issues in a model.

TRIM.../EXTEND...

These are, more or less, your standard trim and extend tools. Select a cutting or extension boundary (or boundaries), select the element you wish to cut or extend, and then select the part you wish to remove.

Trim example:

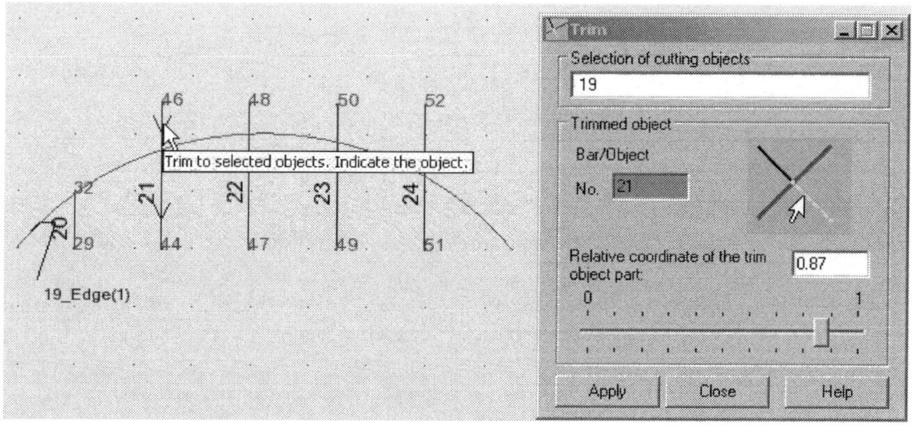

Here we have selected object 19 as the cutting object. Hovering over successive members will automatically populate the Trimmed object "Bar/Object" number as you hover over members. The Relative coordinate of the trim will also automatically update as you hover. However, if you would prefer, this is a relative distance along the member to indicate the part (measured from local x) of the member you wish to trim off. Graphically however, simply select the part of the bar you wish to trim off.

Extend Example:

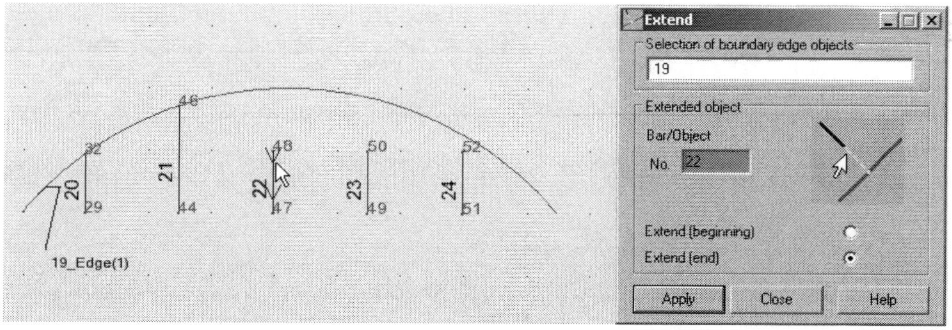

Hovering over the next bar to be extended, will automatically populate the Extended object "Bar/Object" edit field. Click near the end of the member you wish to extend as opposed to using the beginning/end option in the dialog. You can always use the beginning/end option along with the Apply if you prefer to enter the data manually.

Edit Menu Drawing Correction Tools:

If you start a model by opening a DXF file for your member axes or perhaps import a model from Autodesk® Revit® Structure, you might find that some of the members are not intersecting or that members do not end at a connection with another supporting element. The drawing correction tools can help you find and merge duplicate nodes, fix nodes which are close (but not connected) by moving them and making the connection, allow you to adjust nodes in very specific ways with several different tools, like adjust to point, adjust to line, etc. and even add rigid links between members which are slightly out of plane yet were intended to be connected.

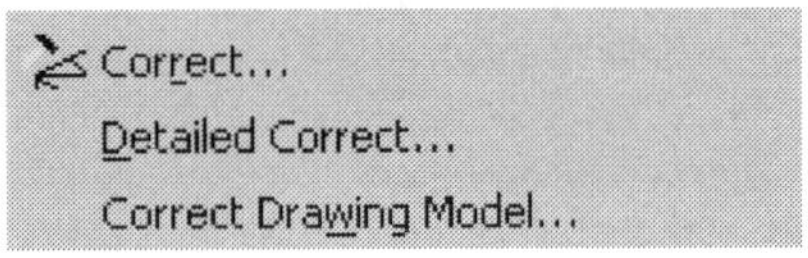

CORRECT...

This tool can be used for basic model adjustments. It has two main purposes: find nodes which are close and merge them and find locations where bars intersect each other and split them. In general, finding close or overlapping nodes and merging them is quite useful. Splitting bars which intersect may or may not be the right idea for a particular situation and caution should be exercised in using the "entire structure" option with this tool, unless you are simply locating close nodes and merging them.

Start the tool from the edit menu **Edit>Correct...**

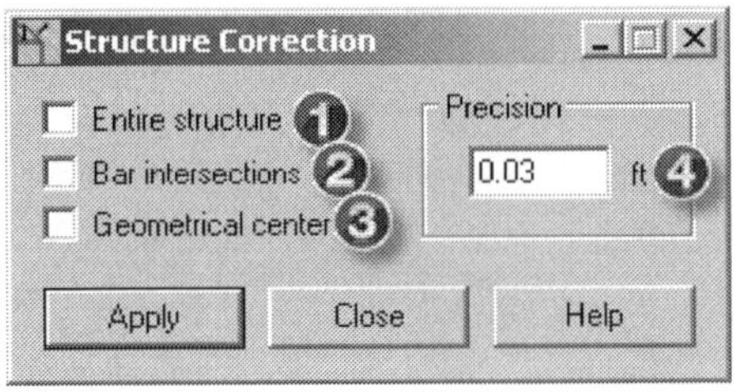

1- Entire structure: Perform node merging and/or bar intersections for the entire project. Unselect this to perform this operation for only the selected nodes. Use right-click>select… to create a new selection set.

2- Bar Intersections: Check this to have Robot also break bars at intersections and create new members as needed. This is usually not desirable unless you actually intend the members to be connected at that point. In the case of diagonal braces where you intend to have them connected at the midpoint, you can alternatively use the options in the structure model tab of the analysis types dialog to add nodes at these points. Use this option carefully and make sure that what you want is actually two or more separate elements at all bar intersections. Note that bar intersections will not only break members which intersect, it will also break them and reposition if within the precision sphere. This can inadvertently result in more disconnections than you had before. See below for an example.

3- Geometrical center: Controls how Robot merges nodes. If checked, Robot will move both nodes to a point representing the geometric mid-point between the two nodes and create a new node there. Otherwise, Robot will choose one of the two nodes and use that as the new node for both member ends.

4- Precision: Is the volumetric radius within which to search for nodes which are considered "close". Nodes which are within a sphere of this radius will be joined. Members which are separated by this distance and theoretically intersect each other will be broken and joined together (often changing the member orientation).

Examples:

-Performing nodal correction with the geometric center option:

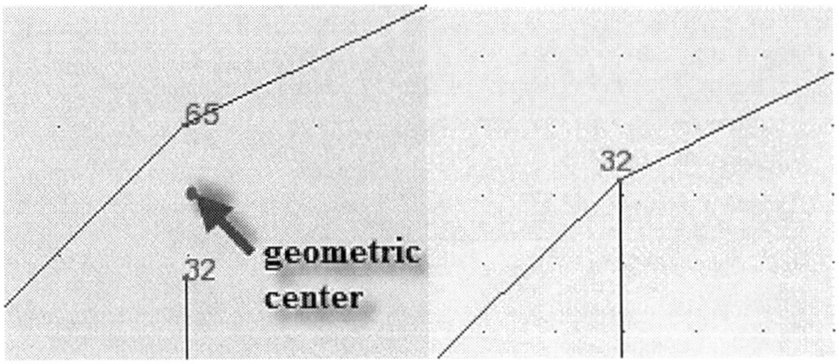

Before adjustment we had two nodes 65 and 32, after the correction operation we now have only one node and all members frame to this node. Note that the orientation of the upper members has changed and that the vertical member has been extended to the new nodal position.

-The same adjustment without the geometric center option:

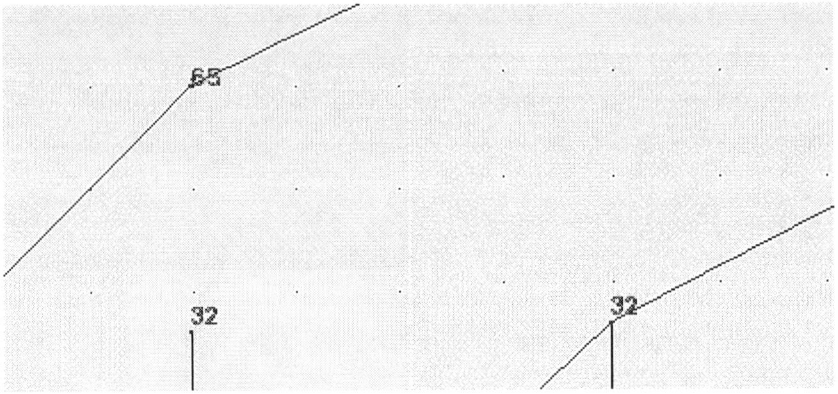

In this case, node 32 was chosen (at random) as the final location of the nodes and the extraneous node (65) was removed.

-Bar intersections:

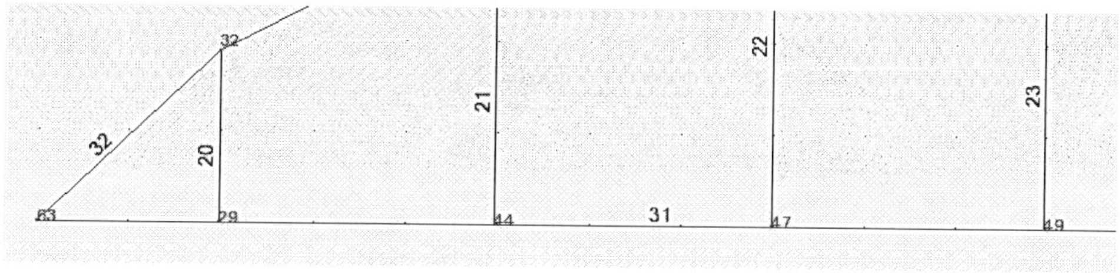

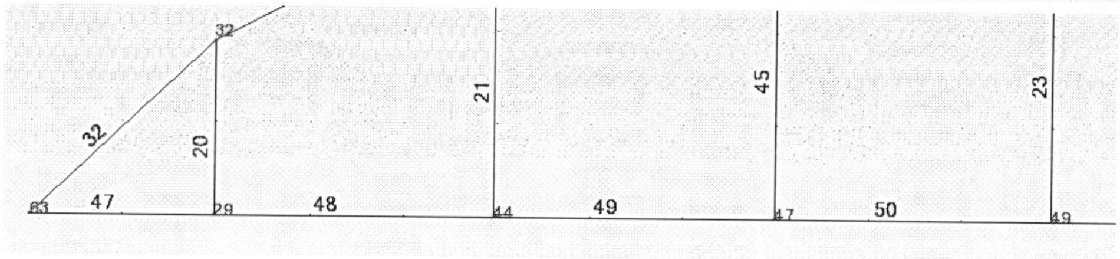

Previously, we had one single member along the bottom of this truss (member 31) and after running the correction tool with the bar intersections option, we now have 4 members (47 to 50) along the bottom chord which was split at every intersection with another member (in this case the vertical webs). This may, or may not be the change you had intended. It is best to use the bar intersection option with a selection, and then only in cases where you really want to break up one member into several members.

<u>-Bar intersections (unintended consequences)</u>

In this 3D perspective of a truss, Bar 24 and node 51 are beneath the bottom chord (member 31). However, the bars, at their theoretical intersection point, are within the selected precision. Here is what happens if only the bar is within tolerance:

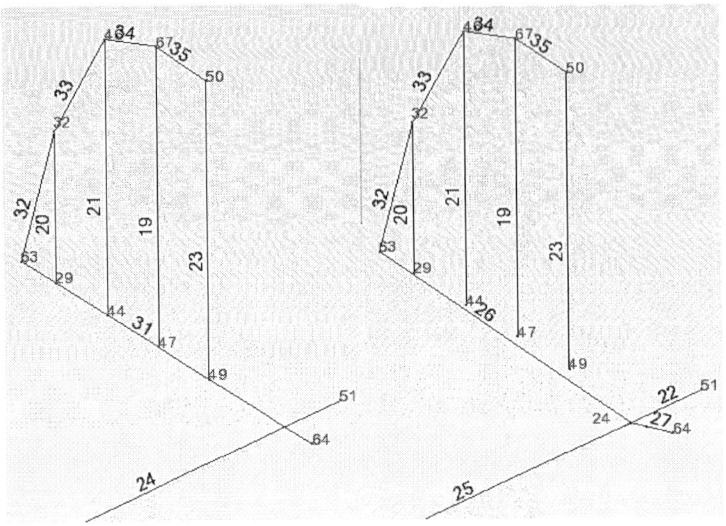

Notice that the bottom chord and also member 24 have been broken into two members and the new (middle) end-points joined together at new node 24. In general, you will be using a precision value which will not result in this type of behavior, but it is important to be aware of what Robot can do with this command so that you can spot areas which might be complex (geometrically) and pay close attention to what Robot is doing as a result of this command.

In the case that node 51 also falls within the precision value, here is what happens:

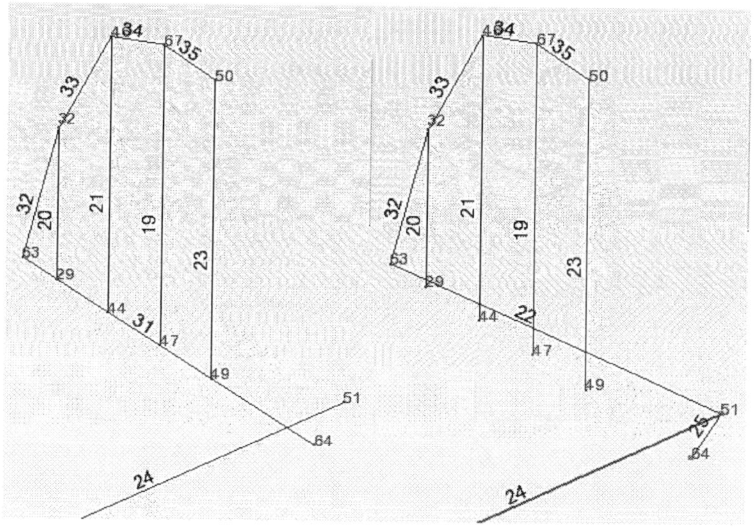

In both cases, the resulting bottom chord member is now less well connected than it was before. We can now go back and fix this with the detailed correct tool, but it is probably more likely that the result of using bar intersections on this model with the specified tolerance (unusually large for demonstration purposes) was not the desired result of this tool.

Here is one final look at another potential outcome. With the same situation in a 3D perspective on this truss with a potentially unrelated member passing underneath. If you select whole structure and bar intersections, here is the result:

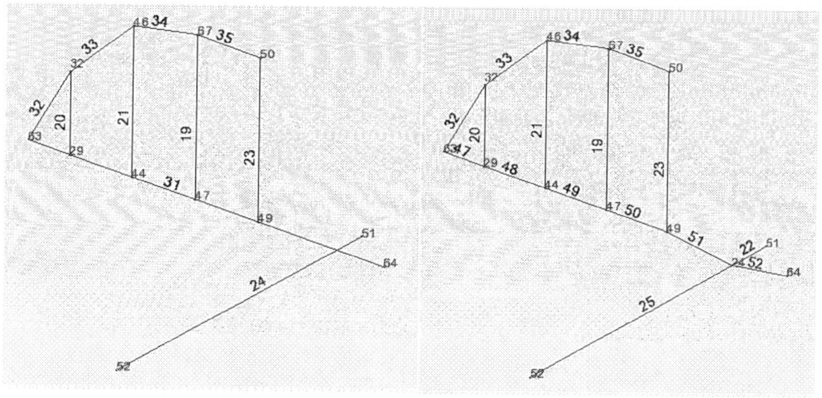

Note that the bottom chord has been broken into multiple sections as we have seen before and as a result, has reduced the potentially strange member re-arranging to only the part of the bottom chord closest to the nearby member.

DETAILED CORRECT…

Detailed Structure Correction (**Edit>Detailed Correct…**) is a really powerful tool. It is important to understand how it works in order to not create more problems through its use. With this tool, you can move nodes into perfect alignment with either points, lines, planes, structural axes or other Robot objects. For instance, if you want the end-points of your straight members to lie on a perfect arc, you can use this tool for that. You can also use this tool to do very precise relocations of nodes to put the structure into alignment geometrically speaking. There are cases where the Detailed Structure Correction tool will not give you the results you are probably after and the following examples should illustrate the issues as well as suggest alternatives.

Start the detailed correct tool from the edit menu **Edit>Detailed Correct…**

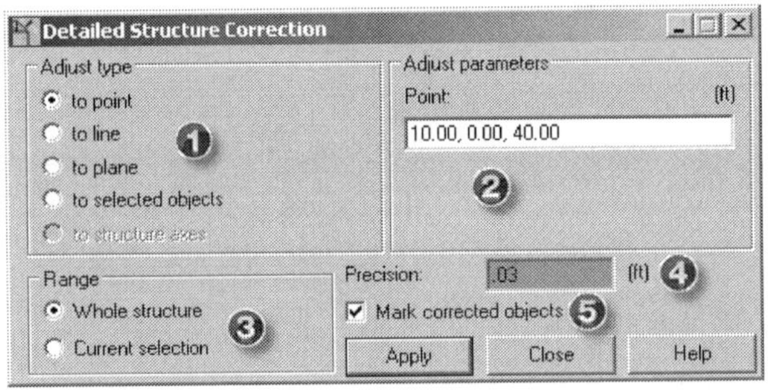

1- Adjust type: Here are the different options you have for making structure adjustment:

   a. To Point: Adjust nodes within the precision value to the point you select

   b. To Line: Adjust all nodes within the precision to the line specified in Adjust parameters. NOTE: Adjustment will be done as a projection perpendicular to the selected line. See examples below.

c. To Plane: Adjust all nodes within the precision of a plane to that plane. As with "to line" above, nodes are moved along a projection perpendicular to the plane.

d. To Selected Objects: You can create geometric objects in Robot from the Geometry menu such as arcs, spheres, cones, rectangles, cubes, etc. You can use these objects for adjustment of nodes which are within the specified precision of the object. See example below.

e. To Structure Axes: Adjust nodes that are within the specified precision to the specified structure axes. Nodes will be adjusted according to a projection perpendicular to the axes.

2- Adjust parameters: Depending on which adjustment type you specify in #1, the options here will update to allow you to configure appropriate options. For instance, for "to point" only one edit control will be available for selecting the point to which you want to adjust nodes within the precision value. For "to plane" you will have three edit controls to specify three points which lie on the plane to which you want to adjust nodes.

3- Range: Apply adjustment to the entire structure or only to selected nodes/bars.

4- Precision: This is the tolerance value for determining whether a node should be moved or not. If a node is farther away from the configured node, line, etc., then it will not be adjusted.

5- Mark corrected objects: After adjustment the elements which have been adjusted will be selected for you to investigate further. Use of the edit in a new window tool (**Edit>Substructure Modification>Edit in a New Window**) can be helpful in understanding exactly which elements have been adjusted.

Examples:

<u>-To a point:</u>

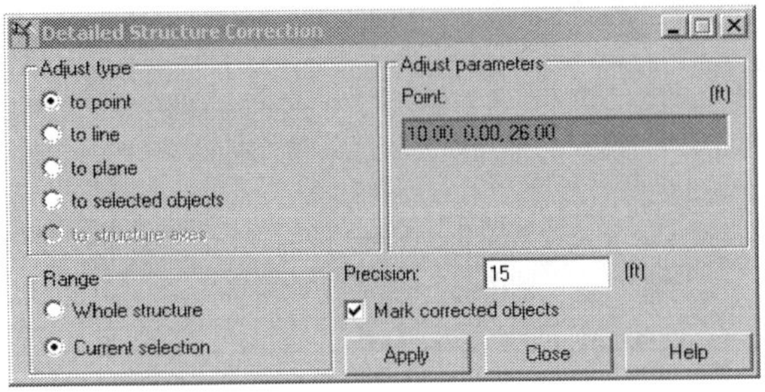

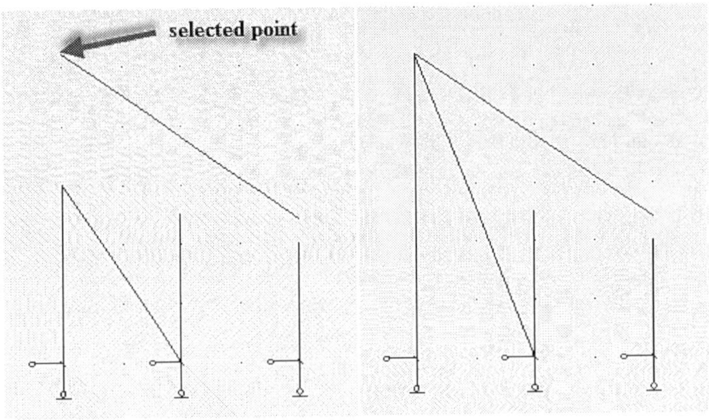

Note that the node within the tolerance (exaggerated here) has been moved to the specified point (10, 0, 26). If any other nodes had been within the specified precision, they would also have been moved to the specified point (10, 0, 26).

<u>-To a Line:</u>

This time we will select the upper line as the adjustment line for the "to line" option.

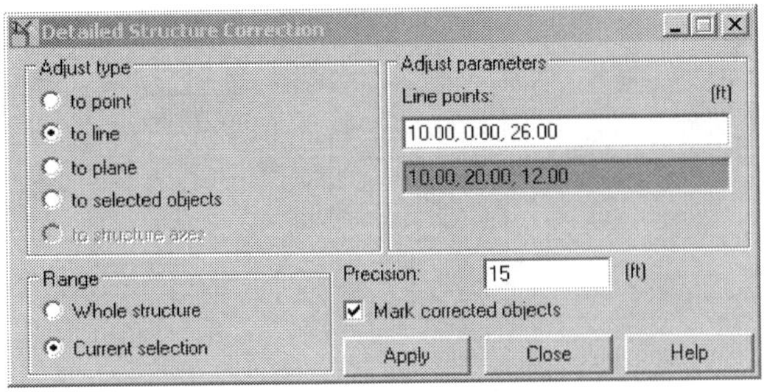

Notice what happens when we apply this tool now:

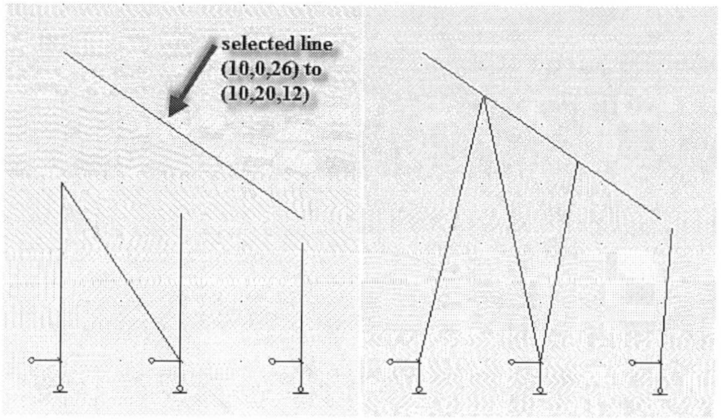

The projection has been performed perpendicular to the selected line, resulting in inclination of previously vertical members. While this might be the intended result, you can actually cause changes you might not have intended as a result of using either the "to line" or "to plane" which has a similar projection behavior.

Here is the basis of what is happening in the "to line" and "to plane" projections:

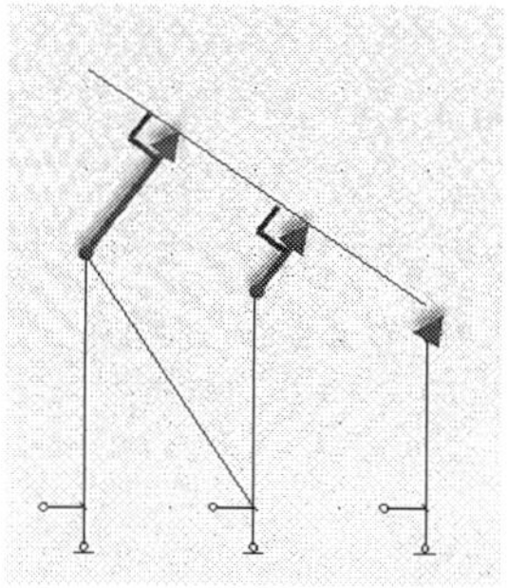

Also noteworthy is that the definition of the line is infinite, the new location of points does not necessarily need to lie on the previously specified line.

It is worth mentioning that these examples are somewhat contrived, but are intended to show you in large scale, what is happening when you use these tools. They can be very powerful and can help you quickly get a model lined up and ready for analysis, but they can also have unintended consequences which should be considered.

CORRECT DRAWING MODEL...

The correct drawing model tool was intended to help clean up the geometry of models which were either begun from DXF files or imported from Autodesk Revit Structure. It has some very cool functionality, but if used indiscriminately, it can have some unintended consequences. It has the ability to add rigid links between elements, move and merge nodes as well as to specify actions to be taken at bar intersections for specified bar pairs. Unlike the "Correct" tool, Correct Drawing Model operates on member to member relationships, not simply nodal positioning. This tool is looking for ways to connect members together such that the structure can be analyzed.

Here is an example of a structure model with a typical set of issues. In this case, the joists are not extended fully to their supporting member on one side and on the other they are out of plane with the supporting member. We can use the Correct Drawing Model tool to fix this model, but only if we are selective about how we apply the tool.

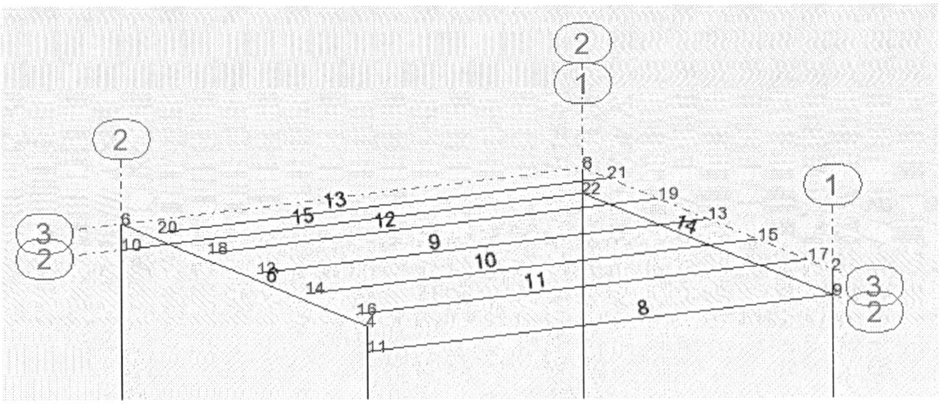

What we'd like to do is to extend the joists at one end and add rigid links at the other so that the joists remain horizontal yet are connected to the supporting structure. If we first select the members which are in the same plane to extend and execute the tool:

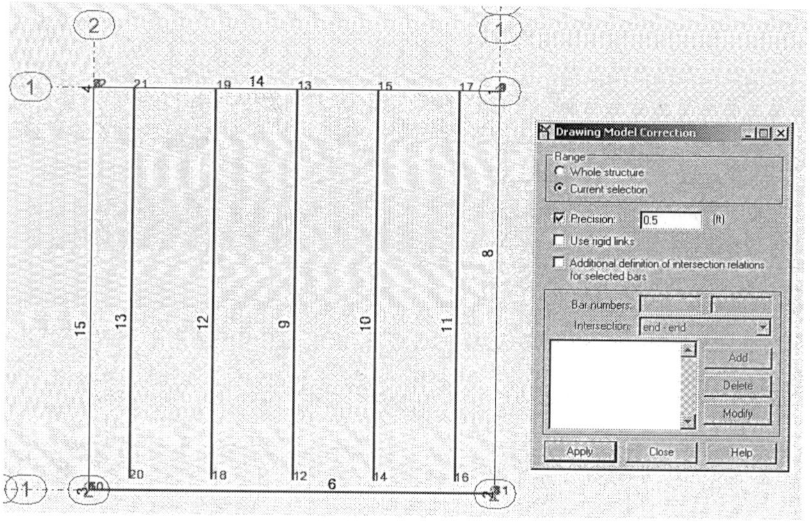

The members will be extended to intersect the supporting member.

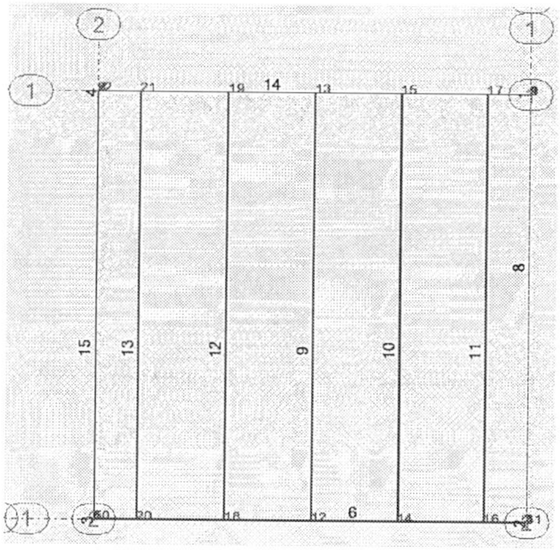

To deal with the other side, you can use the rigid links option to connect these out of plane members. Note that we have only selected the members on which we wish to operate.

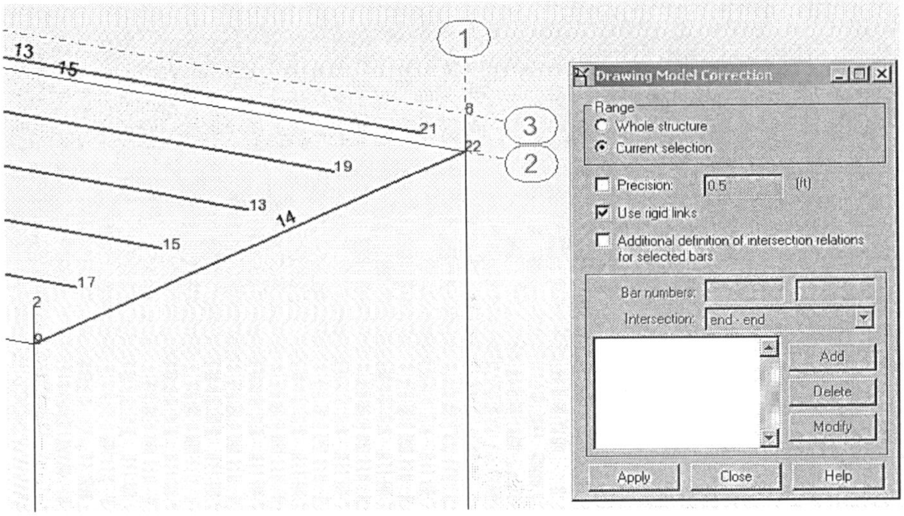

The result is the creation of rigid links between the ends of the members and the supporting beam:

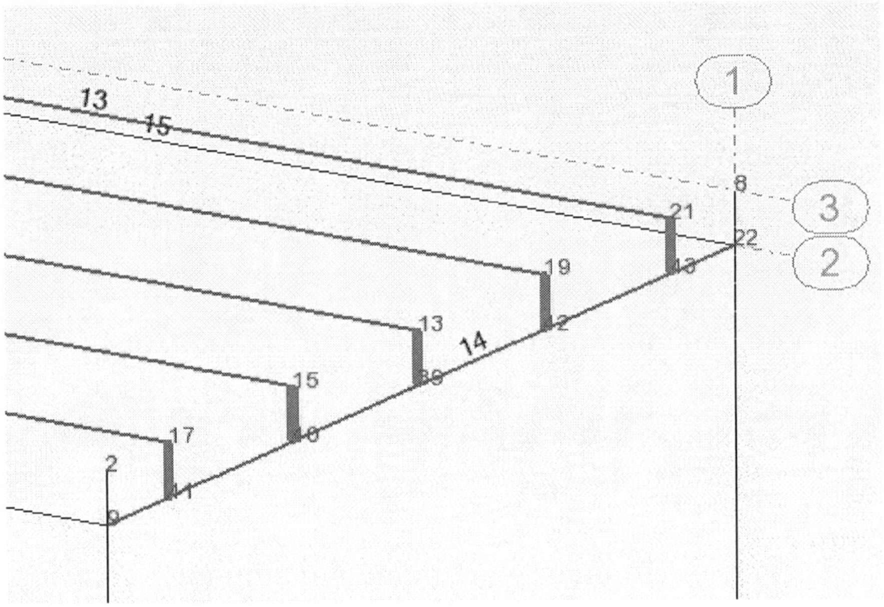

You can also use the "Additional definition of intersection relations for selected bars" to define behavior for bars which intersect each other. The most common use of this option would be to tell Robot to not add an intersection point between crossing members such as diagonal x-bracing. Another use of this option is to join member ends which are within tolerance of each other. However, it is a tedious process to add all these relations and is probably best done with model editing tools trim/extend/correct.

Changing Member Orientation

When you have a member which needs to be rotated about its local x axis, you can use the gamma angle tool to make this adjustment. The gamma angle is the angle between the member major axis and global Z axis. Conversely, you can also specify a gamma angle in the sections definition dialog. The advantage of specifying it in the definitions dialog is that if you have a wind-girt for instance, by specifying it in the section definition dialog you can simply place the member and it is already at the proper gamma angle as opposed to manually adjusting each one with the gamma angle tool (or operating through the bars table to adjust it).

Gamma Angle Tool

Launch the Gamma Angle Tool from the geometry menu: **Geometry>Properties>Gamma Angle...**

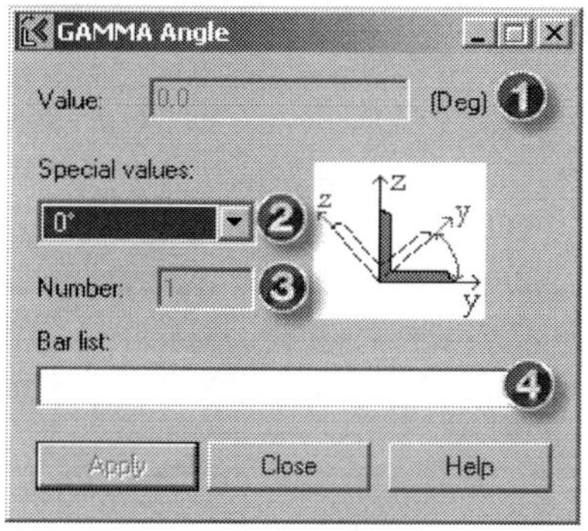

1- Value: This edit control will not be active unless "definition" is selected in the "Special Values" dropdown. When "definition" is selected, enter the specific gamma angle for the member(s) in this edit control. Otherwise, this may simply show you the value which will be applied or be blank.

2- Special Values:

a. 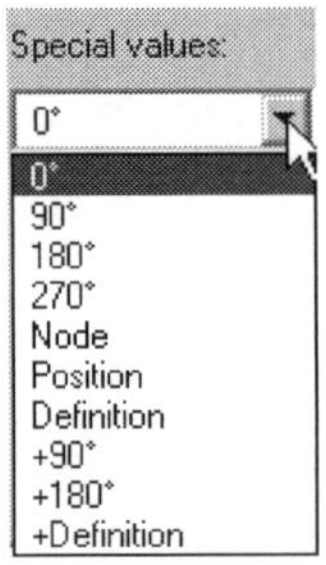

b. 0, 90, 180, and 270: should be self-explanatory.

c. Node: Specify a node in the project which defines a plane with the member and Robot will align the member local z-axis with this imaginary plane. (The plane is defined by the two end points of the member and the node you specify)

d. Position: Intended to make it easy to align member's gamma angle perpendicular to an inclined supporting member. Choose position, then enter the member number to be used for the orientation of the member local y-axis. (z-axis perpendicular to this member).

e. Definition: Enter a specific value for gamma angle in the "Value" edit control #1

f. +90,+180,+definition: increment the current gamma angle by 90, 180 or a custom value you enter in the "Value" edit control above (#1)

3- Number: Node number or member number for Node and Position options respectively

4- Bar list: List all members which should be adjusted to this gamma angle. You can select manually in project or use the member selection tools from the selection toolbar.

Creating a section with the gamma angle predefined:

From the Sections dialog (**Geometry>Properties>Sections...**), create a new section label or modify an existing one. Note that in the New-Section dialog, you have the option of setting a gamma angle (only standard 0, 30, 45, 60, 90, 180 are available). The option is in the lower left hand corner of the dialog and it's probably best practice to give it a name that indicates it's a rotated member like so:

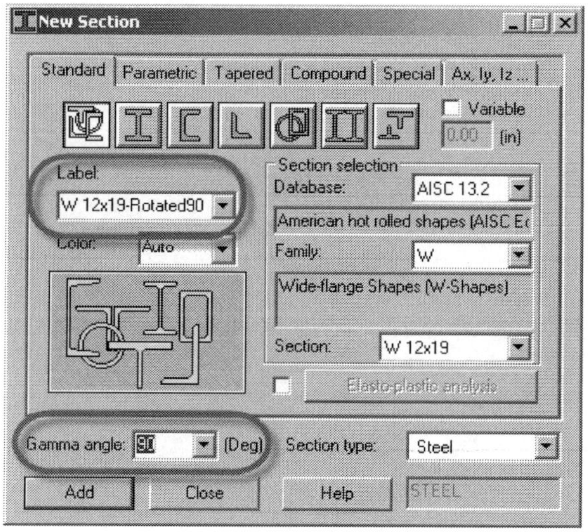

Once you add this section, notice that the icon in the section labels dialog helps you differentiate section labels with gamma angles specified:

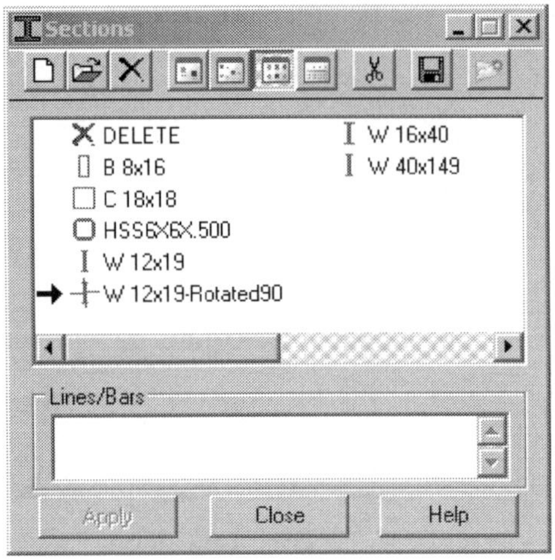

When you place this section in the project, it will automatically be oriented at the specified gamma angle. This is a great option for wind-girts.

Chapter 11 - Beginning Decks and Walls

The following section should give you a great start on analyzing structures with shell elements. We present the basic creation, definition and results analysis. Prior to heading down the path of full shell analysis for floor slabs, consider whether a rigid slab approach is appropriate as this could save time in working out the issues with full shell element analysis. Considerations such as mesh density, mesh refinement, relative stiffness, and influence of the slab in load sharing between supporting members and the slab are important.

Basic Workflow:

1- (Optional) Configure floor/wall thickness types (**Geometry>Properties>Thickness...**)

2- (Optional) Configure shell calculation model (**Geometry>Properties>Panel Calculation Model...**)

3- (Optional) Configure shell reinforcement model (**Design>Required Reinforcement of Slabs/Walls – Options>Code Parameters...**)

4- Define Geometry with either classic method or floor/wall method

Geometric definition

We will begin with Geometric Definition, as Thickness, Panel calculation model, and Code Parameters are all labels which can be assigned at any time to a panel, either at creation time or after the panel has been created.

The method of defining your floor or wall elements will vary depending on which structure type you currently have configured. Not all Structure Types will offer options for creating shell (floor/wall) elements so if you do not see the options you were expecting, check your structure type. Only the following Structure Types allow you to configure plate/shell elements:

1- **Building Design** (Functionality tailored to building analysis and design)

2- **Shell Design** (Very similar to Building Design with access to more functionality)

3- **Plate Design** (Environment tailored to flat plate design, reduced functionality)

We will primarily discuss the options common to both Building Design and Shell Design structure types. There are two major options for defining floor and wall elements. They may be defined classically or with new tools tailored to creating floors or walls. Robot originally provided the "Panels..." tool and later added the "Floors..." and "Walls..." tools which offer a streamlined creation experience.

The major difference between the two workflows is that to create shell elements with the "Panels..." tool, you must first create a contour (outline) using the **Geometry>Objects>** functionality, and then select that contour as the basis for the element boundary. In the "Floors..." and "Walls..." tools (**Geometry>Floors...** or **Geometry>Walls...**) the boundary definition is part of the tool and not only does not require contour objects, but offers no way to use previously created contour objects in selecting the boundary for the element.

Any of these three methods for creating shell elements result in shell elements in Robot. The method you choose is largely up to you and as you get more familiar with the tools, you will find that some offer distinct advantages over others depending on the task at hand. For example, the "Walls..." tool is lightning fast for standard wall layout done in plan view.

Classic Definition (Panel... tool)

This method can be used for floors or walls or actually any flat or warped situation you encounter. Panels do not need to be flat by any stretch though the degree to which the meshed elements will fit the surface contour will depend entirely on how fine the mesh is. This workflow requires first constructing a geometric shape (contour), and then using the "Panel..." tool to create the shell element. In addition to polylines and contours, you can also use closed loops of beams or other objects to define a panel outline.

NOTE: A CONTOUR OR COLLECTION OF LINES AND POLYLINES DEFINING A BOUNDARY FOR A PANEL WILL BE CONSUMED IN CREATING THE PANEL. I.E., THEY WILL NO LONGER EXIST AS ELEMENTS IN YOUR MODEL ONCE YOU USE THEM TO CREATE THE PANEL.

Start creating a contour or boundary outline from the geometry menu: **Geometry>Objects>Polyline – Contour...** this opens the Polyline – Contour dialog:

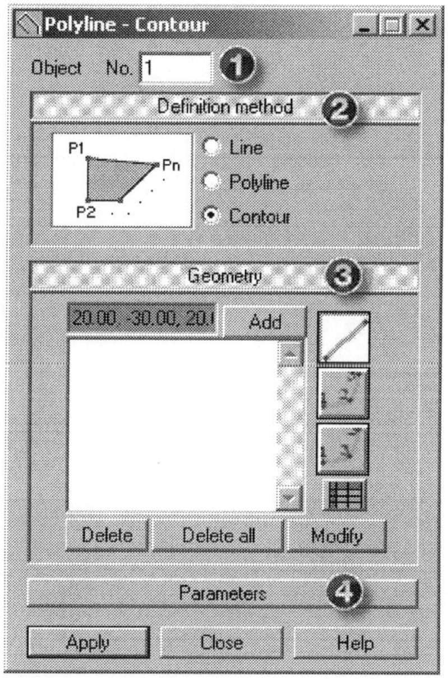

1- Object Number: Robot will automatically pick the next available element number. If you choose to specify one and that element already exists in the project, Robot will ask you if you want to redefine it.

2- Definition Method: Choose the style of object you wish to create:

a. Line: Straight line segments

b. Polyline: Line segments chained together which do not need to close. Lines may be straight or arced.

c. Contour: Freeform closed polygon. Edges may be straight lines or arcs

3- Geometry: Section expands to allow configuration of the vertices of the object. It will change depending on the style of object geometry you have selected in #2 above.

4- Parameters: Contains advanced settings regarding curve approximation and filleting at corners. Curve approximation will replace any arc you specify with a number of linear segments according to the settings configured in the Parameters section.

a. 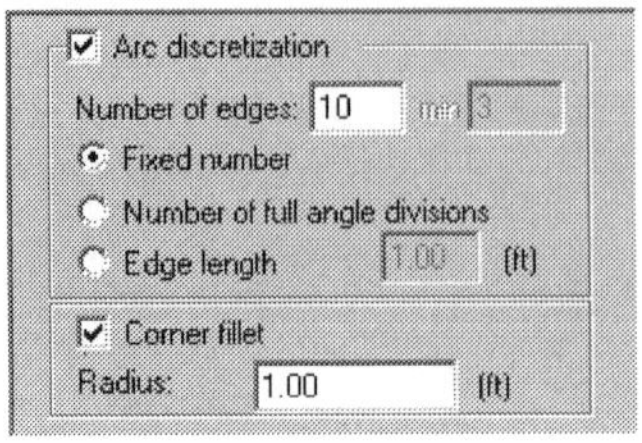

DEFINING THE GEOMETRY:

First select the type of object you want to create: Line, Polyline, or Contour.

<u>Line</u>: The Geometry area of the Polyline - Contour dialog will show the line definition configuration:

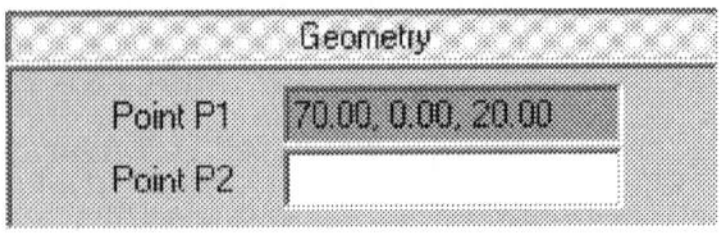

Either manually enter the coordinates of two points to define the line or select two points in the project with the mouse. Left-click in the Point P1 edit control to highlight this edit control green, then select the point in the project. You may select the next point immediately following the first.

<u>Polyline or Contour</u>: The Geometry area of the Polyline - Contour dialog will show the contour definition configuration: (The only difference between polylines and a contour is that the contour is by definition closed.) For polylines, if you double-click on the last point of the polyline or single-click on the first point it will finish the polyline definition.

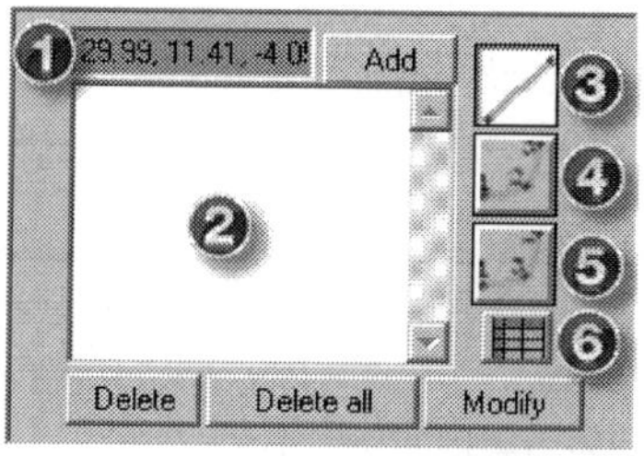

1- Coordinate entry area. You may type coordinates manually and use the Add button to add them to the current list of vertices or you may click in the edit control and move your mouse out into the project environment where you may indicate vertices directly by clicking in the project.

2- Current vertex list: as you either manually Add points or click them in the project, you will see them populate this list. (While you can modify them, you cannot reorder them in the list.)

3- Create linear segments: Between successive clicks, Robot will create a linear edge for the cladding. For buttons 3-4 as you click in the project environment, you can select back and forth between these options for each edge you wish to create thereby stringing together linear and arc segments in order to specify exactly the contour configuration you desire.

4- Arc Beginning-Center-End: Create an arc by specifying three points in the project or by manually entering the coordinates

5- Arc Beginning-End-Center: Create an arc segment by specifying the start, the end and then a point that the arc should pass though

6- Spreadsheet Input: Opens a spreadsheet interface for indicating contour vertices in a spreadsheet style. To use this interface, right click in the spreadsheet to insert a new row and configure the parameters. Use "Table Columns…" from the right-click menu to add columns in order to specify arc center points and other parameters.

Use the "Apply" button at the bottom to create the contour (or the polyline if you have not finished it by double-clicking).

IDENTIFYING OBJECTS:

Panels and contours look almost identical with one small exception. Here are both a contour and a panel side by each. The contour is on the left and the panel is on the right:

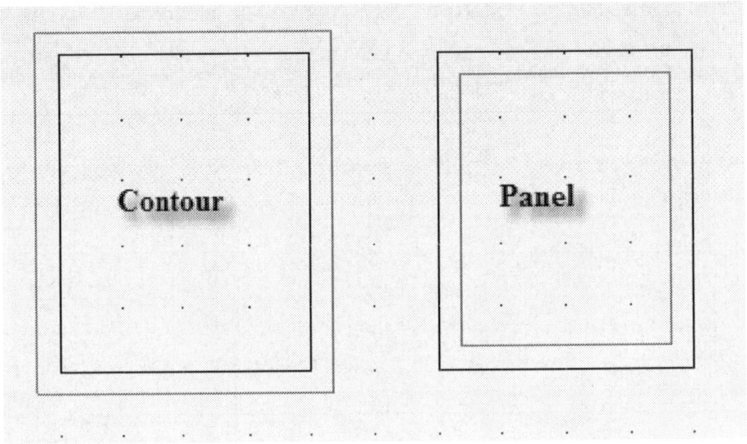

Note the location of the light line. In the Contour, the lighter line is around the object and in the Panel the light line is inside the panel boundary. In both instances, the boundary is exactly the same size and is defined by the heavier line as illustrated here:

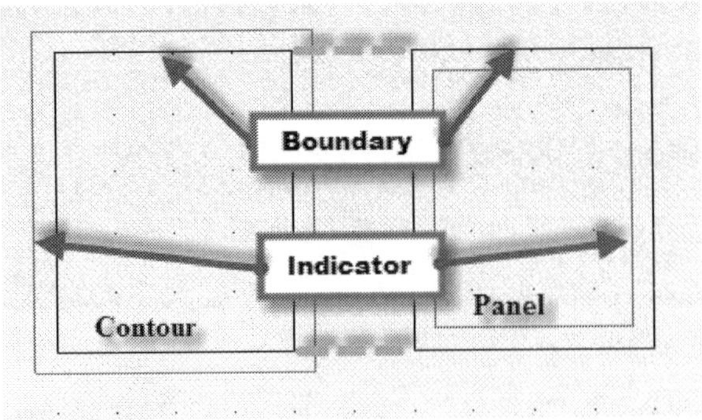

We'll call the lighter line the "indicator". When selecting edges of panels or contours, you can hover over either the boundary line or the indicator line. Remember, once used to create a panel, contour and polyline/line objects are consumed in the process.

CREATING THE PANEL

Once you have either defined objects which define the panel outline or have a closed loop of beams or walls which define the boundary of the panel, start the panel tool from the Geometry menu or, if in the Shell structure type, the icon from the modeling toolbar on the right. This will launch the panel dialog:

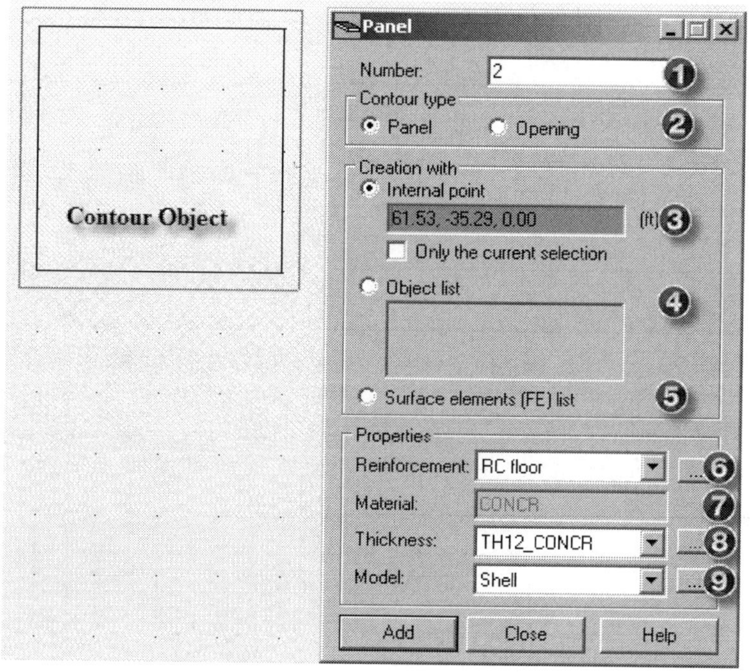

1- Number: Robot will automatically pick the next available element number or you can specify one yourself.

2- Contour Type: Are we selecting an object to make an opening in a panel or to create a panel?

3- Creation with Internal Point: Simply specifying a coordinate location within a closed loop of beams, walls, a contour, or polylines will automatically create a panel (or opening) filling the area enclosed by the objects around the selected point. NOTE: If you intend to configure other panel properties (reinforcement, thickness, etc.) specify these before selecting the internal point.

4- Creation with Object List: Simply enter the numbers of a set of objects which form a closed loop. You can also select them manually in the project environment. You will use the "Add" button with this option.

5- Creation with Surface Elements (FE) list: Once a panel has been meshed you may find that edge elements or elements around openings need to be thicker. You can use this option to specify surface element numbers (to display use **View>Display...**>Panels/FE>Finite Elements>Finite Element Numbers). This will create an additional panel, but at meshing time only the properties of the new panel will take precedence.

6- Reinforcement: ACI code parameters assigned to this panel. Select from your predefined label list of code settings or adjust/create new labels by using the ellipsis button to the right. See Panel Properties below for more information.

7- Material: This is a read-only property which is derived from the thickness setting (#8)

8- Thickness: The thickness and other properties of the section. Select from your predefined label list or adjust/create new labels by using the ellipsis button to the right. See Panel Properties below for more information.

9- Model: This is the calculation model label you would like applied to the panel. Select from predefined labels or use the ellipsis button. See Panel Properties below for more information.

CREATING PANEL OPENINGS:

Panel openings are created in very much the same way that a panel is created. You first create boundary objects (polylines, rectangles, contours, etc.), then select "Opening" in the Panel dialog and either select an internal point of the opening boundary or specify a list of objects which comprise the opening boundary. You can also use the Openings tool from the geometry menu: **Geometry>Openings...**. We'll cover that after the floor and wall tools.

Using the Floor and Wall Tools

The floor and wall tools were added later in the development of Robot and if you use Autodesk® Revit®, you will find that they are very similar in functionality to the Revit floor and wall tools which is intentional. The elements created by the floor and wall tools are identical to elements created with the panel tool from Robot's viewpoint. These are simply different ways to achieve the same end result.

FLOORS:

Launch the floor tool from the geometry menu: **Geometry>Floors...** or from the modeling menu (if in shell type, then modeling bar>structural elements>Floors icon) with the  icon. This will launch the floor dialog. You will note that it looks almost identical to the cladding dialog we discussed previously.

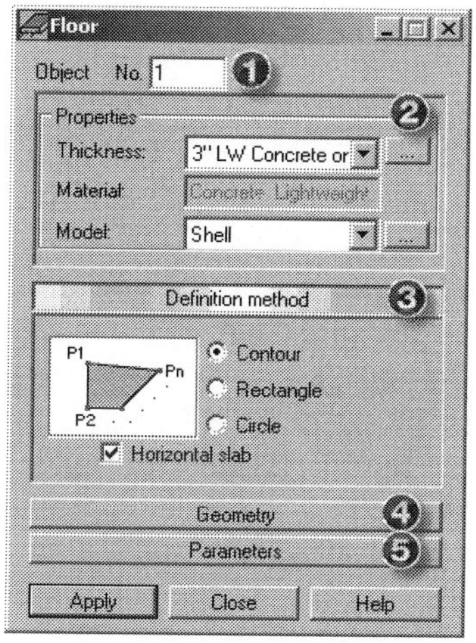

1- Object Number: Robot will automatically pick the next available element number. If you choose to specify one and that element already exists in the project, Robot will ask you if you want to redefine it.

2- Properties:

   a. Thickness: The thickness and other properties of the section. Select from your predefined label list or adjust/create new labels by using the ellipsis button to the right. See Panel Properties below for more information.

   b. Material: This is a read-only property which is derived from the thickness setting.

   c. Model: This is the calculation model label you would like applied to the panel. Select from predefined labels or use the ellipsis button. See Panel Properties below for more information.

3- Choose the style of floor boundary you wish to create:

   a. Contour: Freeform polygon. Edges may be straight lines or arcs

   b. Rectangle: Simplified definition by three points

   c. Circle: Simplified definition by three points

   d. Horizontal Slab: This option will force the created floor panel to be horizontal in the project. Effectively ignoring Z coordinate of selected points.

4- Geometry: Section expands to allow configuration of the vertices of the floor. It will change depending on the style of cladding geometry you have selected in #3 above.

5- Parameters: Contains advanced settings regarding curve approximation and filleting at corners. Curve approximation will replace any arc you specify with a number of linear segments according to the settings configured in the Parameters section.

a. 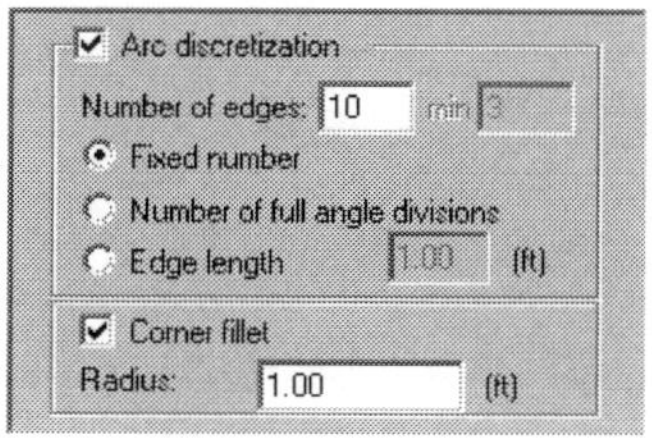

Defining the Geometry: First select the type of floor boundary you want to create: Contour, Rectangle, or Circle.

<u>Contour:</u> The Geometry area of the Claddings dialog will show the contour definition configuration:

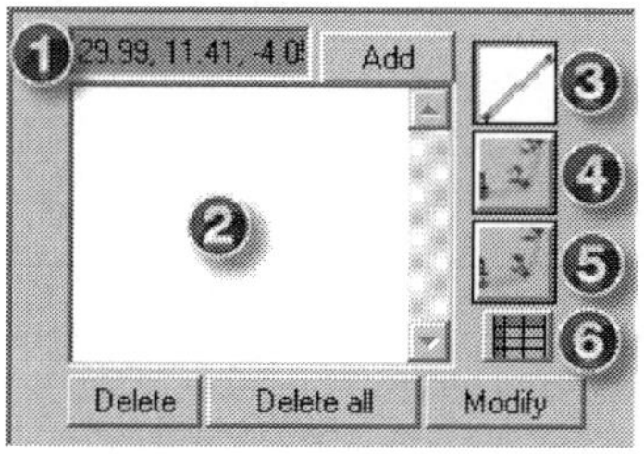

1- Coordinate entry area. You may type coordinates manually and use the Add button to add them to the current list of vertices or you may click in the edit control and move your mouse out into the project environment where you may indicate vertices directly by clicking in the project.

2- Current vertex list: as you either manually Add points or click them in the project, you will see them populate this list.

3- Create linear segments: Between successive clicks, Robot will create a linear edge for the floor. For buttons 3-4 as you click in the project environment, you can select back and forth between these options for each edge you wish to create thereby stringing together linear and arc segments in order to specify exactly the boundary configuration you desire.

4- Arc Beginning-Center-End: Create an arc by specifying three points in the project or by manually entering the coordinates

5- Arc Beginning-End-Center: Create an arc segment by specifying the start, the end and then a point that the arc should pass though

6- Spreadsheet Input: Opens a spreadsheet interface for indicating floor vertices in a spreadsheet style. To use this interface, right click in the spreadsheet to insert a new row and configure the parameters. Use "Table Columns..." from the right-click menu to add columns in order to specify arc center points and other parameters.

<u>Rectangle</u>: This is a simplified version of the contour definition above and will likely be the most frequently used. Changing the Definition Method will discard your current vertex. Selecting Rectangle will reconfigure the Geometry section as shown here:

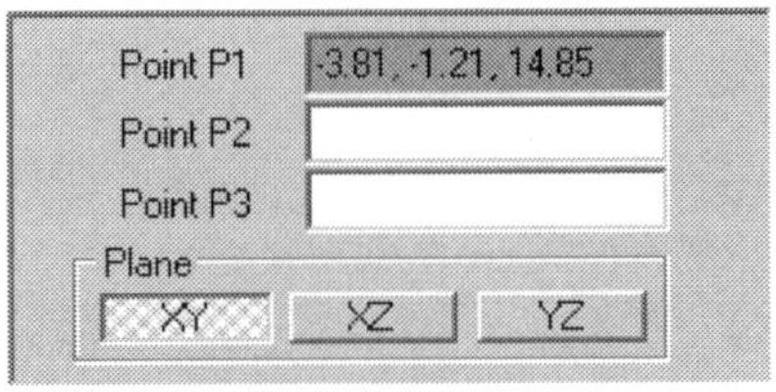

You can type in the coordinates of three points defining the rectangle the first two points define the first edge and the last point merely defines the distance perpendicular to the first selected line (and potentially the inclination if a z-coordinate magnitude is specified by manually typing in coordinates or snapping to elements in the project)

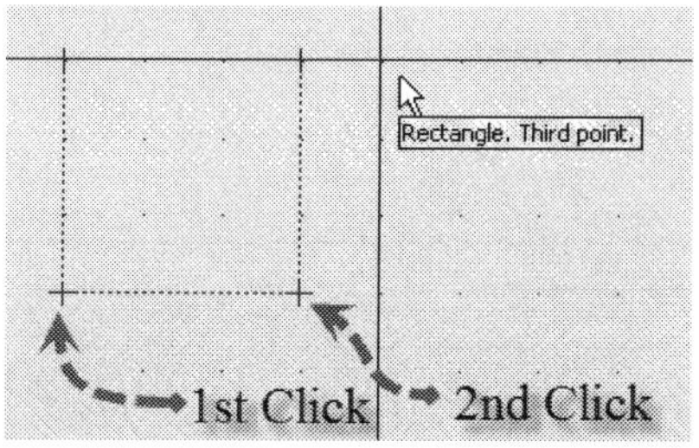

Circle: Requires the definition of three points that lie on the circle. You may manually type in the coordinates or select the points in the project environment. You may create a sloped circle by either manually specifying a z-coordinate or snapping to an elevated piece of geometry in the project.

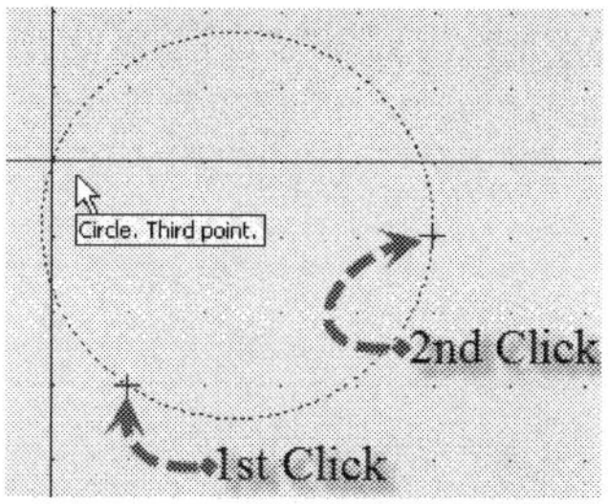

Finishing the floor definition: If you have manually entered any of the coordinates you will need to use the “Apply” button at the bottom of the Floor Dialog in order to create the floor in the project. If you are clicking in the project to define points, simply selecting the last point for rectangle or circle will finish the definition and add the floor. If you are using the contour option, you may either click again on the last vertex to complete the definition and add the floor or use the Apply button in the Floor Dialog.

WALLS

Launch the wall tool from the geometry menu: **Geometry>Wall...** or from the modeling menu (if in shell type, then Modeling Bar>Structural Elements>Floors icon) with the icon. This will launch the wall dialog.

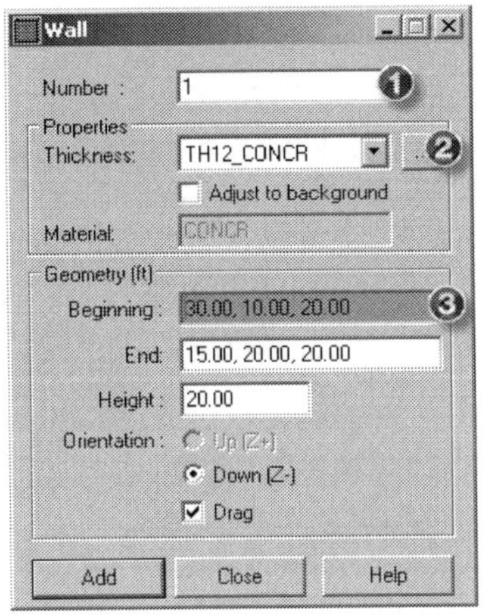

1- Number: Robot will automatically pick the next available element number. If you choose to specify one and that element already exists in the project, Robot will ask you if you want to redefine it.

2- Properties:

a. Thickness: The thickness and other properties of the section. Select from your predefined label list or adjust/create new labels by using the ellipsis button to the right. See Panel Properties below for more information.

b. Adjust to background: This tool is supposed to pick up the thickness from a background DWG or DXF file. It seems to make more sense to create your thickness labels deliberately as opposed to trying to pick them up from a background drawing which may or may not be accurate.

c. Material: This is a read-only property which is derived from the thickness setting

3- Geometry:

a. Enter coordinates defining the member beginning and end or place the cursor in the edit control (it will turn green), then select a location in the project window directly. The edit control focus will switch to the "End" once you have selected a start point in the view.

b. Height: Simply the height of the wall. From your current placement plane, this will either be up from that plane or down based on the setting under Orientation below:

c. Orientation: Will this wall being created extend upward from the current placement plane or downward? (Look at the bottom of the view to see which plane is the current placement plane:

d. Drag will automatically set the next start point to the current end point when selecting points in the project view. This allows you to quickly place walls without needing to select the start point again if your start point for the next wall was the end point of your last one.

As you can see, placing wall elements is significantly faster than using the Panel tool, granted that your walls are vertical and otherwise well behaved.

OPENINGS TOOL:

This is a special tool dedicated to putting openings in shell elements. It is so similar to other tools that it doesn't seem worth an in depth description. Here is the dialog and the workflow:

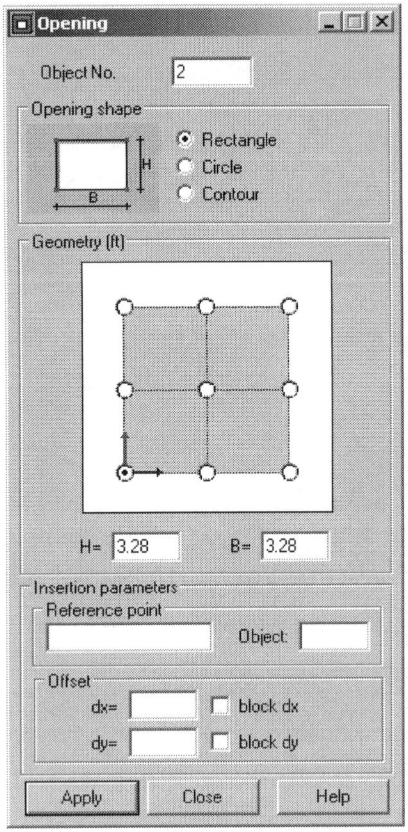

1- Select the opening shape type

2- Select the insertion point location on the opening boundary

3- Set the parameters of the opening shape and/or create a contour with the ellipsis button.

4- In Insertion Parameters: Enter the number of the object into which you intend to put an opening and select the reference point on that object.

5- Enter offset values or select location of opening directly in the project environment.

Panel Properties

You can specify panel properties either at time of creation or at a later time by creating and applying labels for Thickness, Calculation model and Reinforcement parameters (Code Parameters).

Thickness

This label specifies panel thickness as well as orthotropy and initial material. Launch the thickness labels dialog from the geometry menu: **Geometry>Properties>Thickness...** or from the modeling toolbar . This will launch the Thickness dialog:

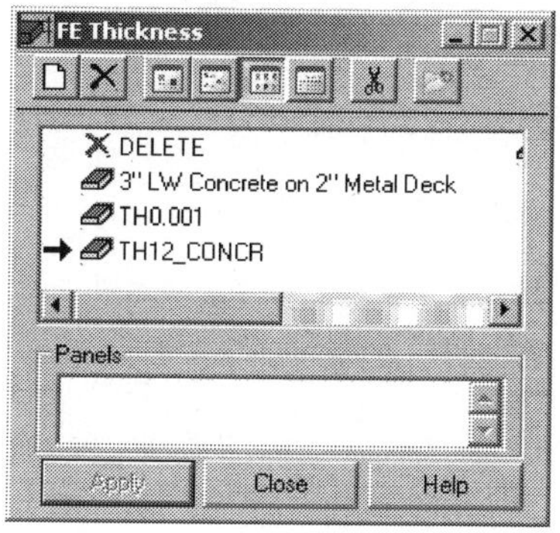

This dialog behaves exactly like other label dialogs: New page button will create a new label, double-clicking on an existing label will open its properties for editing. In the label properties dialog:

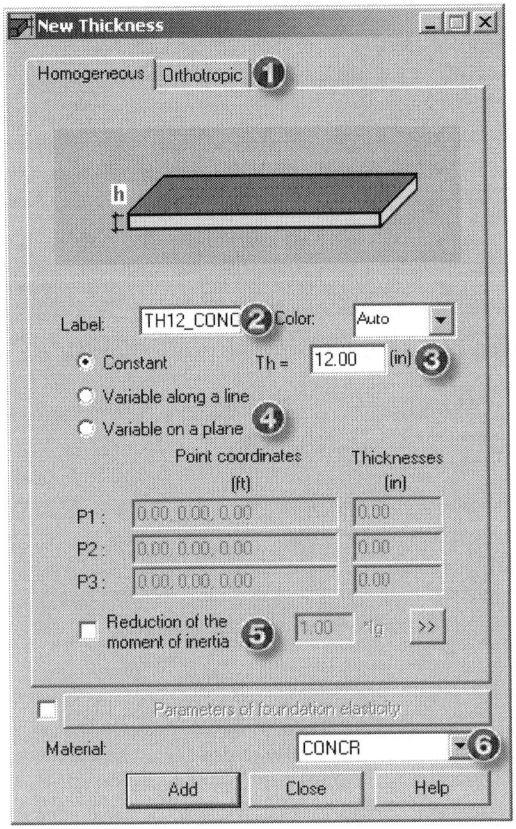

1- There are two tabs here for configuring different types of slabs/decks. Choose the Homogeneous or Orthotropic tab and configure the options. The tab selected at the time you press "add" is the type of element which will be created, regardless of settings in the other tab. Select the tab for the type of element you plan to create.

   a. Orthotropic thickness parameters affect the bending stiffness of the element in the different panel directions. It does not affect the overall membrane stiffness or in-plane shear/tension/compression forces.

2- Label Name: This is the name of the label you are either modifying or creating. Enter a new name here to create a new label or use an existing one to modify the properties.

3- Th: Thickness of homogenous panel.

4- Panel may be variable thickness. Select the style variation here.

5- Reduction of the Moment of Inertia. Use this for checking cracked sections for deflection in concrete.

6- Select the material of the panel.

Orthotropic:

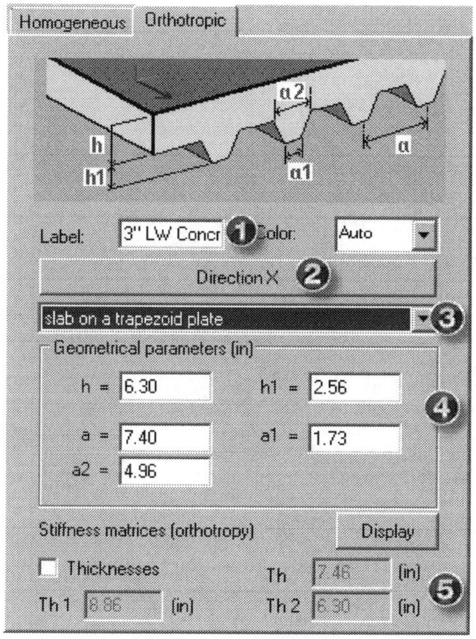

1- Label Name: This is the name of the label you are either modifying or creating. Enter a new name here to create a new label or use an existing one to modify the properties.

2- Direction X: Since this thickness style is orthotropic, it is important to decide which direction is the major axis. Use this button to access the Orthotropy Direction dialog

a. 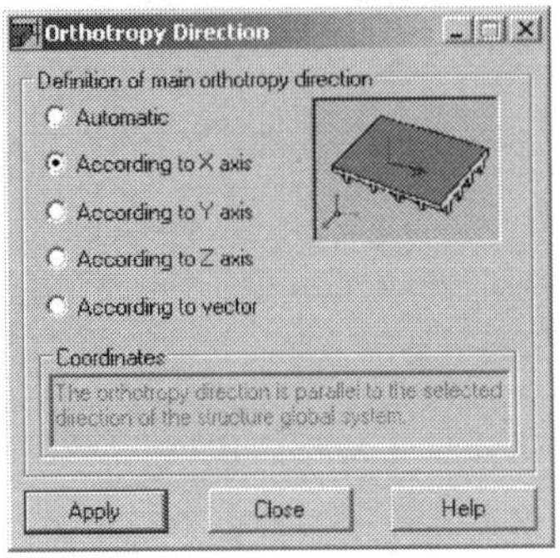

b. Automatic is the best choice for most situations here. Automatic sets the major axis of the panel orthotropy to the local x direction of the panel. NOTE: The other options are global axes or global vector which will be projected onto the panel to determine the major axis direction of the panel orthotropy.

3- Style: Many typical styles of orthotropic panel are provided here. Selection of style will affect the options in the geometric parameters below (#4). Options are:

a. 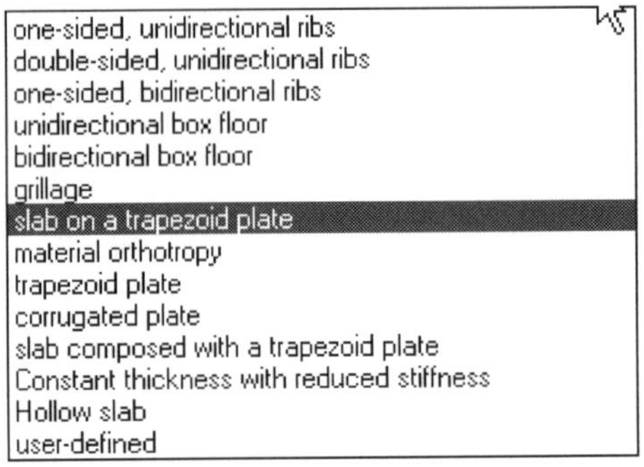

b. See the help file for specifics on defining geometric parameters for each style.

4- Geometric parameters: Based on the style selected in #3, options here will allow you to configure the panel geometry which is used to develop the stiffness matrices for the panel.

5- Thickness: These are automatically calculated, but may be manually specified. The first "Th" is the thickness used to determine dead weight of the panel, the next two are used only in thermal loading calculations as effective thickness.

Panel Calculation Model

The Panel Calculation Model Label is where we configure settings to tell Robot exactly how we want the panel to be treated in terms of analysis. We can have panels treated as claddings to merely distribute loads and not participate any further in the analysis or we can achieve full shell element treatment with these settings and several different options in between.

Launch the Panel Calculation Model Label dialog from the Geometry menu: **Geometry>Properties>Panel Calculation Model...**

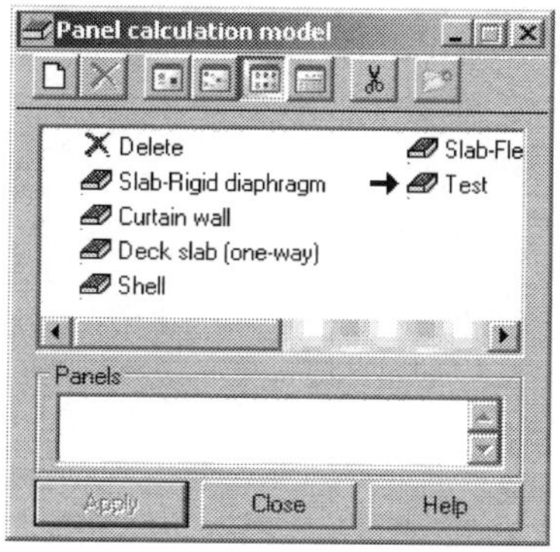

This is, once again, your standard label manager, this time for panel calculation models. Double-click one of the existing labels to edit it or use the new label button in the upper left corner to start a new label definition:

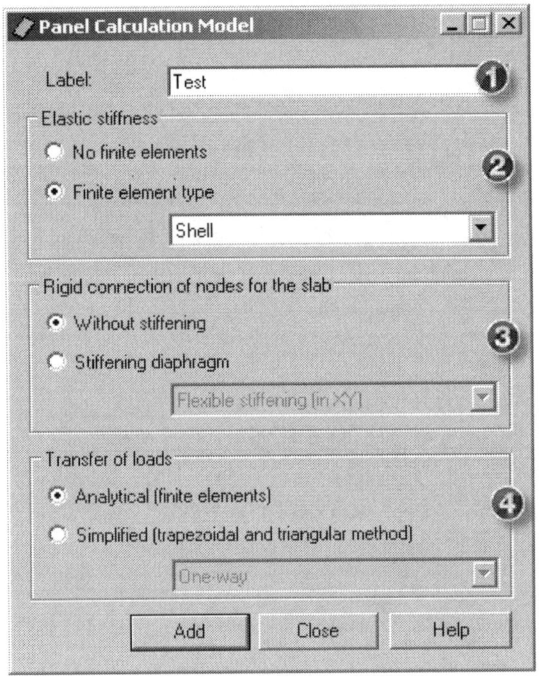

1- Label Name: Give this calculation model a recognizable name so that it will be easy to identify later. Leave this name the same as an existing label to overwrite the existing label. Give it a new name to create a new label.

2- Elastic Stiffness:

a. No Finite Elements: This element will either be a cladding element (or a rigid diaphragm depending on rigid connection settings below) no results will be generated for this plate.

b. Finite element type: Choose the type of finite elements which will be used in meshing this plate.

3- Rigid Connection of nodes for the slab

a. Without stiffening: No additional forcing constraints will be applied to this shell element. It will be calculated as a finite element if elastic stiffness (above) is set to a finite element type; however, if "no finite elements" option is selected above, it will be completely ignored in analysis beyond distributing loads to the supporting elements (act as a cladding) and elements it connects will be independent of each other.

b. Stiffening Diaphragm: Apply additional constraints to the nodes connected to this shell. (NOTE: only horizontal slabs can have flexible or full body stiffening applied)

   i. Flexible Stiffening (in XY): Apply a constraint that all xy (in-plane) movements (and z rotations) of the nodes connected to this slab must be remain fixed in relation to each other. This is the typical rigid-diaphragm assumption. The shell will be able to translate, slope and take bending moments/beam shears, but no xy movements or z rotations of the nodes will be allowed relative to each other. Therefore, no diaphragm forces will be generated in this slab, all lateral loads pass through this slab via the rigid diaphragm and into the lateral system.

   ii. Full Stiffening (rigid body): Entire shell element will be treated as completely rigid; therefore, no bending moments, no diaphragm forces. It will transfer loads through itself but will not take any bending, shear, or membrane forces. The shell will displace along with the rest of the structure, but will not, itself, sustain any loads. (Results will be zero for all quantities except displacement) NOTE: This option will have the unexpected effect of over stiffening your structure and will cause issues with modal analysis in terms of calculating a period which is much lower than you might expect.

4- Transfer of loads:

a. Analytical (finite elements): Use the finite elements to distribute loads on the shell element to other supporting elements. You could call this "full finite element analysis" in terms of letting the elements themselves transfer loads via their nodal stiffness relationships.

b. Simplified (trapezoidal and triangular method): Do not load the shell elements directly with perpendicular loads but rather apply them to supporting elements by assuming a triangular or trapezoidal distribution. (Will result in no bending moments developed in the shell element resulting from applied perpendicular loads.)

EXAMPLES OF BEHAVIOR:

Using the following basic model, we will look at the effects of these different settings on the behavior of the deck. The wall elements are all set to shell type (Finite Elements, No Stiffening, and Analytical Load Transfer). Only the deck will change type. The structure is a long plate with a horizontal load of 800k applied at mid-point in the plane of the deck and a uniform load of 0.3ksf down applied over the entire surface of the deck.

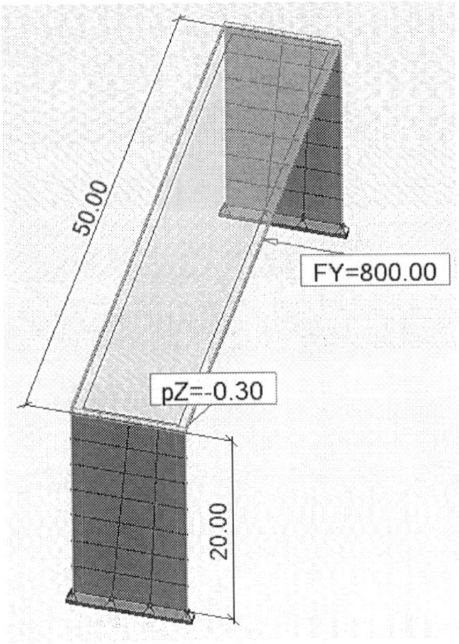

We will look at Tension/Compression in the deck and walls (Membrane Nxx tension compression in the local x direction in plane), bending moments in the deck and walls (Moment Mxx bending between the walls along the deck) and a visual display of the deflected shape of the model.

First: Full Finite Elements, No Stiffening, Analytical distribution of loads Results:

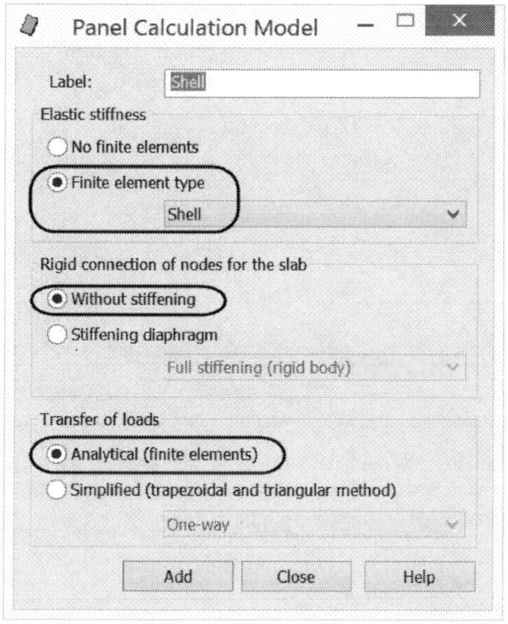

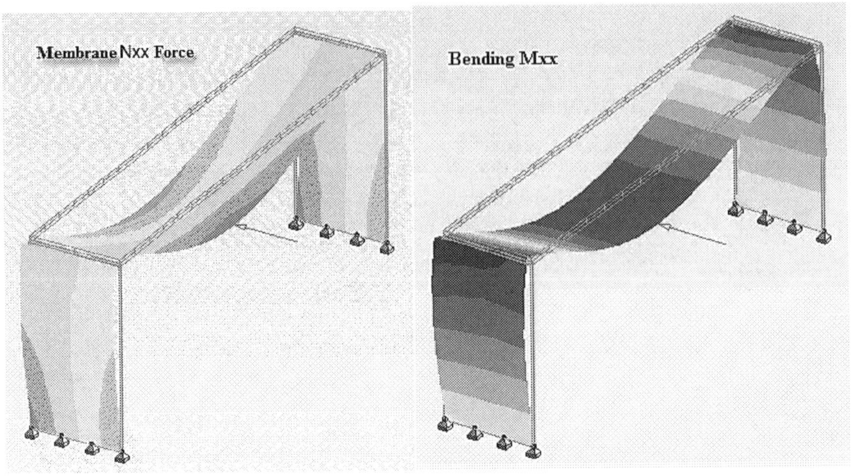

Note that deck takes membrane and bending forces and also displaces.

Second: Finite Elements, Flexible Stiffening (XY), Analytical Transfer

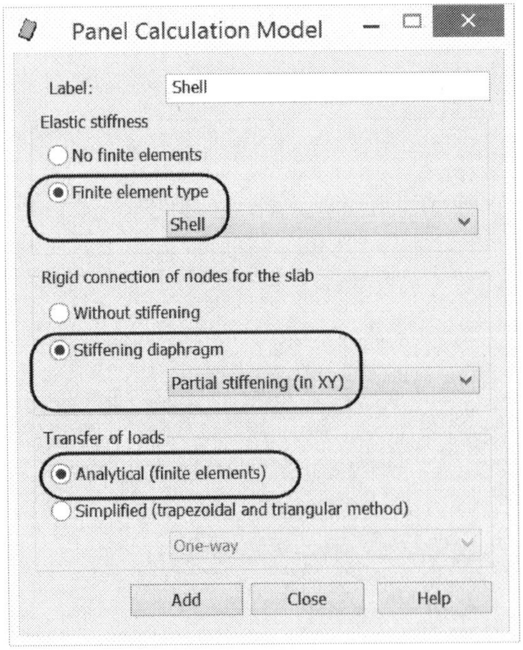

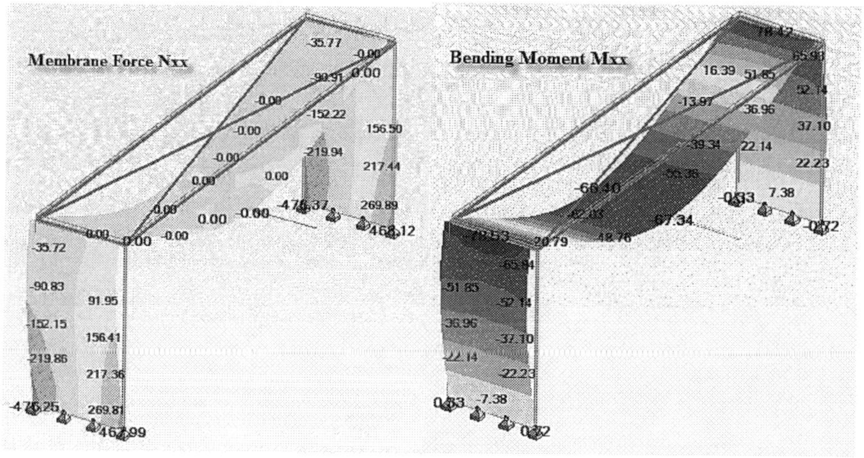

Note that bending forces are developed in the deck but no membrane forces.

Third: Finite Elements, Full Stiffening (rigid body), Analytical Transfer

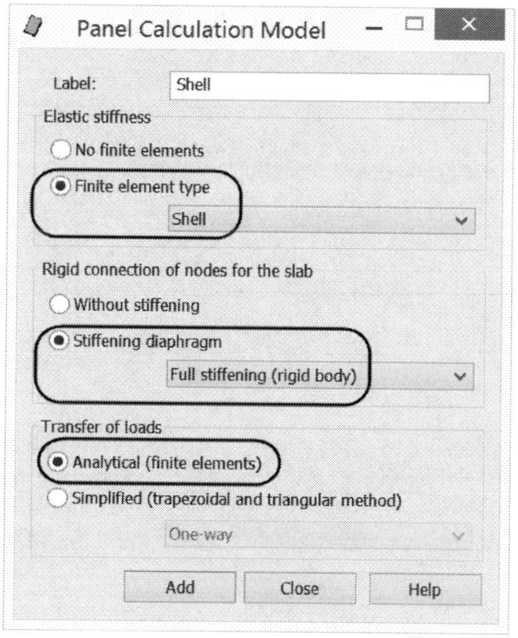

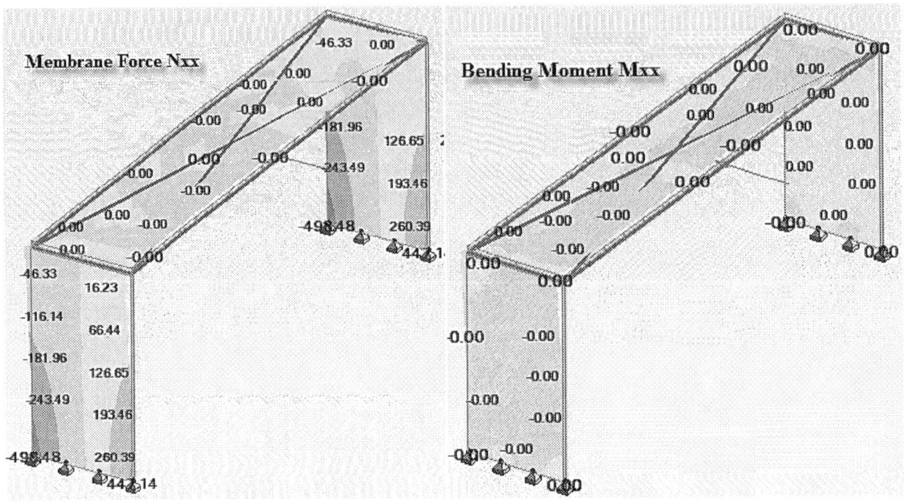

Note that the deck is acting like a completely rigid body, it still translates but no membrane nor bending forces are generated in the deck itself. It behaves exactly like a complete pass-through element and behaves like a full rigid body. You can see the impact of selecting full rigid body: this is typically **not** the desired method of stiffening a slab to consider it a rigid diaphragm.

Fourth: Finite Elements, No stiffening, Simplified Transfer

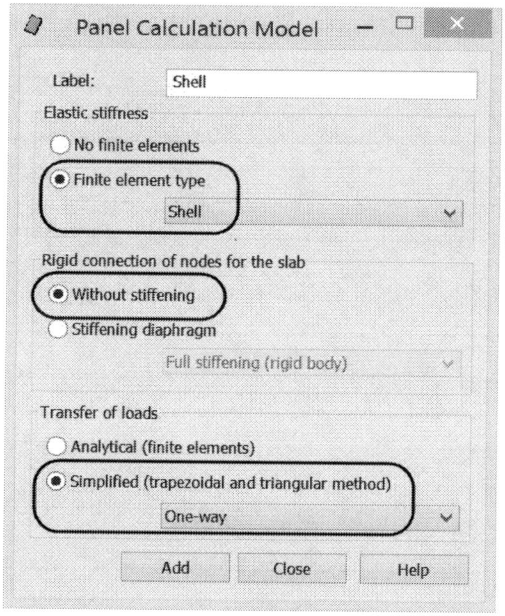

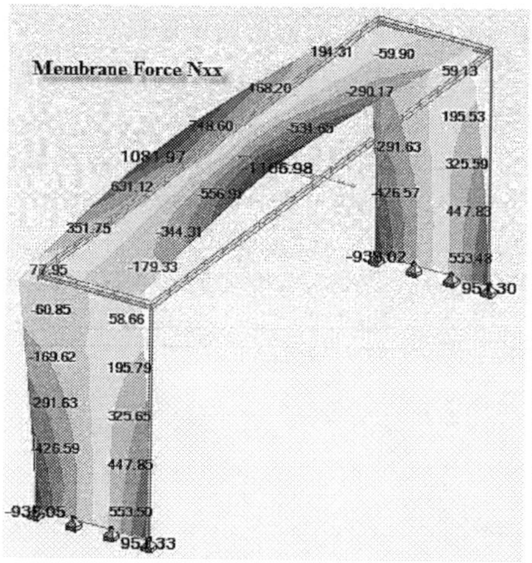

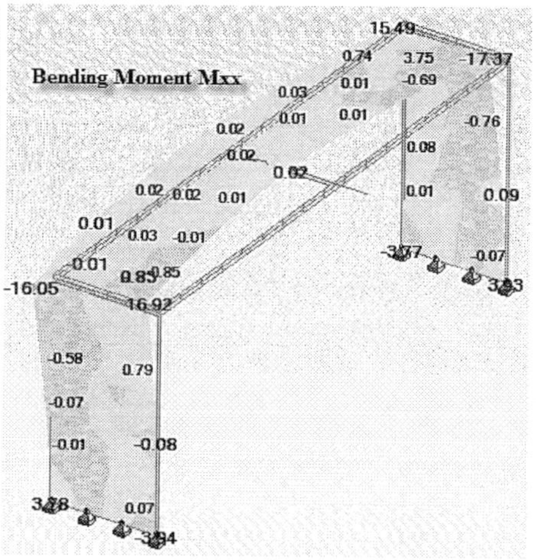

Notice that we do get membrane tension/compression in the deck, but we're not getting any bending in the deck!! This is because the simplified transfer of out-of-plane panel loads by-passes the slab and distributes the loads directly to the walls themselves as compression loads! Be careful using simplified transfer if you want to account for bending forces in the slab.

Fifth: No Finite Elements, Flexible Stiffening, Simplified Transfer

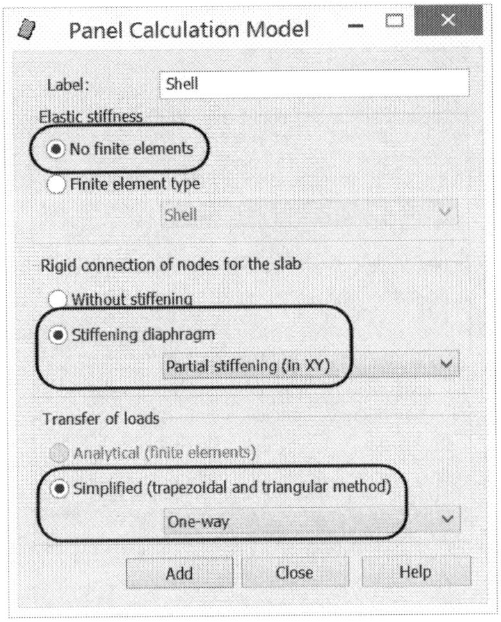

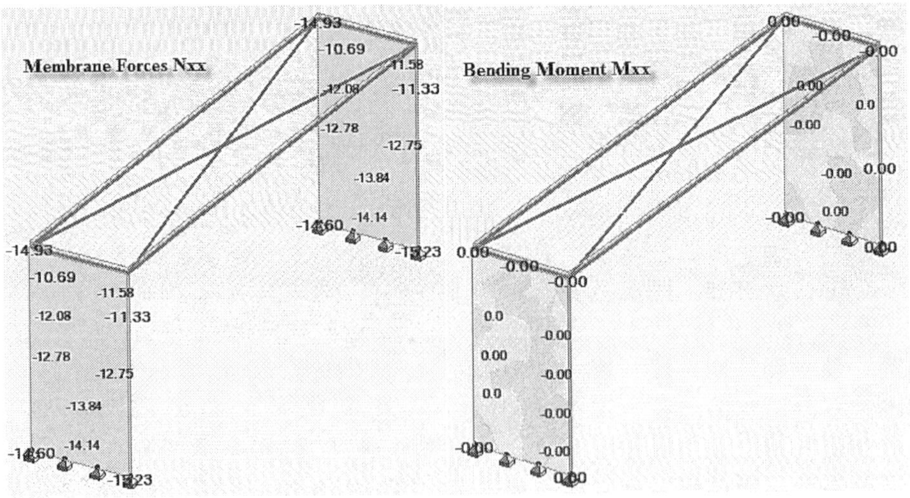

Notice that the in-plane 800k load has been ignored! The vertical load on the deck has been transferred to the walls, but you'll notice that there are no results reported for the deck and the in-plane load has been ignored by Robot. Be very careful when modelling panels with no finite elements. Make sure you are getting the results you are expecting. If you didn't use flexible stiffening, you will get a warning that you have two separate structures (each wall) because the panel effectively does not exist in your project other than to distribute loads to the supporting elements.

Panel Supports

Panels can be supported either at their edges (Linear Supports) or over their entire surface (Planar Supports). Both types of supports are applied and managed in the Supports dialog accessed via the geometry menu: **Geometry>Supports...** or from the modeling toolbar with the supports icon: . This opens the supports dialog:

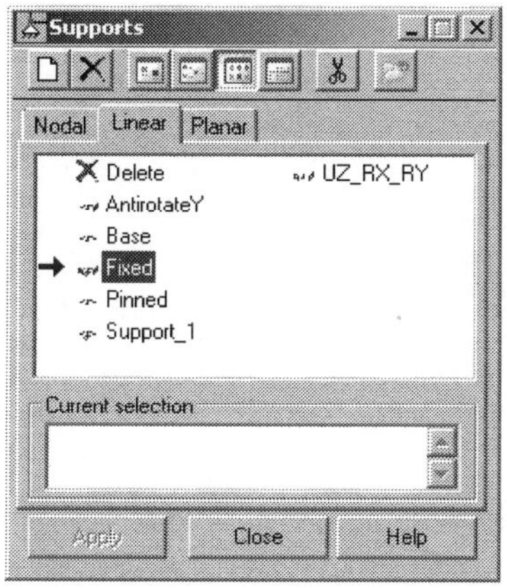

This functions exactly as the supports dialog for nodal supports. Notice the tabs across the top allowing you to place linear and planar supports with this label manager. The only difference between nodal and linear/planar supports is that the support label has the added option of being aligned with the global coordinate system or the local coordinate system of the element being supported. See Modeling Nodal Supports above.

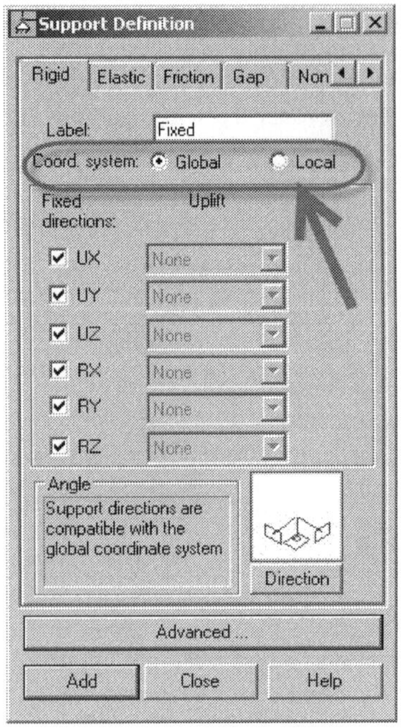

What happens when Robot applies linear or planar supports is that it uses the edge as a support definition and when the panel is meshed in generating the calculation model, the support is actually applied to each node of the meshed elements that lie along that edge or on that plane. Here is an example from the supports table (**View>Tables...**>Supports) before and after meshing:

Before generating the calculation model:

Support name	List of nodes	List of edges
Base		1_EDGE(3) 32_EDGE(3) 31_EDGE(3) 32_EDGE(3) 31_EDGE(3) 32_EDGE(3) 31_EDGE(3) 32_EDGE(3) 31_EDGE(3) 32_EDGE(3) 31_EDGE(3) 32_EDGE(3)

After generating the calculation model. Notice that the list of nodes which have this "Base" support applied now includes all the nodes of shell finite elements which lie along that line, which were created during generation of the calculation model.

Support name	List of nodes	List of edges
Base	458 459 474 475 482 483 0 458 459 474 475 482 483 458 459 474 475 482 483 58 459 474 475 482 483 58 459 474 475 482 483 58 459 474 475 482 483 474 475 482 483	1_EDGE(3) 32_EDGE(3) 31_EDGE(3) 32_EDGE(3) 31_EDGE(3) 32_EDGE(3) 31_EDGE(3) 32_EDGE(3) 31_EDGE(3) 32_EDGE(3) 31_EDGE(3) 32_EDGE(3)

Basic Panel Meshing

Panel meshing is another deep subject with tons of options. The way you mesh your panel elements will have a huge effect on the results you obtain from analysis. It is highly recommended that you visit your finite elements text for a thorough review of the effects of mesh density and mesh proportion on results of finite element analysis. In general, a mesh that is not fine enough (elements not small enough) will tend to give you very inaccurate results. Many people have worked to find methods of meshing that give optimum results accuracy yet do not tax computational capabilities beyond reason. Typically, engineers have tended to use less dense meshes in broad regular areas of panels and increased the mesh density in complex areas or areas where the geometry changes drastically (notches or other geometric discontinuities).

Mesh Settings

Panels are meshed automatically during the generation of the computational model. You have quite a bit of control over how panels are meshed, how fine the mesh is for each panel and which algorithms are used in generating the mesh. Mesh settings are accessed from the Analysis Menu: **Analysis>Meshing>Mesh Options...**

You can also access them via the meshing toolbar , then choose Mesh Options icon 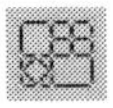 to launch the Meshing Options dialog:

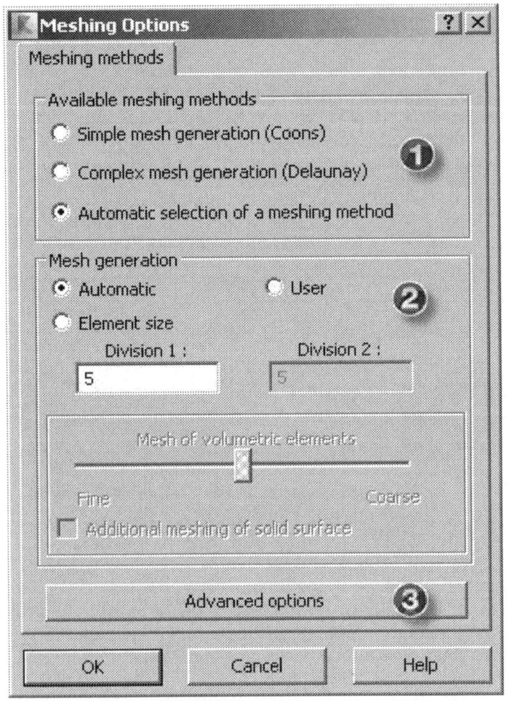

Default settings allow Robot to automatically configure the mesh by selecting a generation algorithm and a mesh density. These settings will typically give you a computable model, but are unlikely to give you the optimum mesh density/configuration for your model. Starting with the automatic options is a great way to get a quick calculation in, then look at refining your mesh either with the refinement tool or by adjusting settings here or in the advanced options.

1- Mesh Method: If you wish to be more specific with your meshing you can select a meshing algorithm here. An additional tab will appear at the top of the dialog with options specific to either the Coons method or the Delaunay method.

a. Coons: Regular mesh of either quadrilaterals or triangles produced by dividing opposite edges of the panel an equal number of times and connecting those divisions to create a mesh. This method may have issues with irregularly shaped geometry (other than a rectangular panel). In order to better control this option, you can specify base points for the mesh with the Base Points tool from **Analysis>Meshing>Base Mesh Points...**

b. Delaunay: A mesh of triangular or quadrilateral elements which is tuned to avoid triangles which have very acute or very obtuse angles. The ideal mesh is one where triangular elements are as close to an equilateral triangle and quadrilateral elements are as close to a square as possible. The Delaunay algorithm minimizes deviation from equilateral triangles and squares. Also, refinement options for points of interest are only available with the Delaunay + Kang algorithm. Local mesh refinement at points of interest or geometric discontinuity, e.g., notches, supports, and corners. Emitters will only be used in mesh generation if this option is selected and either Characteristic points (default) is selected and/or user defined emitter points (User) is selected:

c. 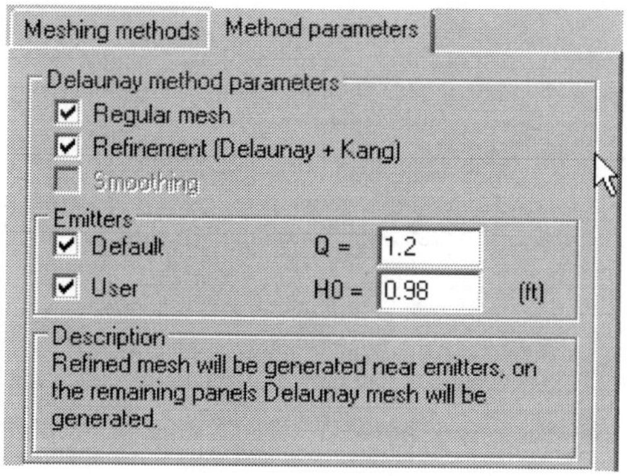

d. Emitters (Kang Refinement): From each emitter or characteristic point (default emitters), a theoretical wave pattern is generated. The parameters H0 and Q control the "wavelength" and relationship of the first wavelength to the next successive wavelength. In each complete wavelength, a set of elements will be generated in a radial pattern radiating from the emitter location into the plate until it intersects with a boundary.

i. H0: The size of the smallest elements to be created closest to the emitter location. In Kang's terminology: The first wavelength

ii. Q: the ratio of Hn/Hn-1: This is the ratio of each successive wavelength to the prior wavelength. This value **cannot** be less than nor equal to 1.

iii. 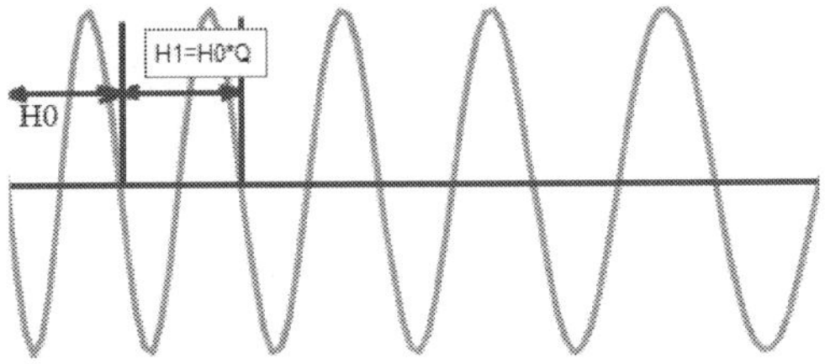

iv. An example of a regular mesh and a mesh using default emitters

v. 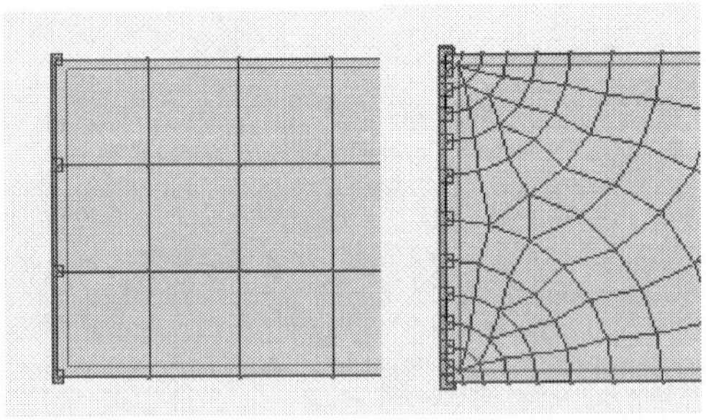

2- Mesh Generation controls: You can opt for an automatic determination of mesh size or specify mesh size here. "User" option will allow you to specify the number of divisions along edges 1 and 2 of the panel. Edges 1 and 2 are determined by creation order, but control of which edges are 1 and 2 can be managed by using **Analysis>Meshing>Base Mesh Points...** You can alternatively specify an element size. This is preferable, as you can tailor the mesh size to the element being meshed, taking into consideration size and shape of any irregularities. See Mesh Density Study below.

3- Advanced options gives you much greater control over mesh creation algorithms. Access advanced parameters by pressing this button. You can find information in the help files to better understand each option.

Mesh Density Analysis

There is no exact formula for mesh density which will guarantee you the proper mesh density to give you accurate results and the most optimum matrix size from a computational standpoint. In order to understand the effects of mesh density on your model, it is recommended to perform a mesh density analysis which will allow you to edge up to and watch for mesh convergence. The finer your mesh, the closer and closer each measured value will get to its theoretical max/min value.

Basic Steps:

1- Create a relatively coarse mesh to begin, element size roughly 1/5 of the element size and analyze the model.
2- Review result maps as isolines and get a sense of the discontinuity of the isolines. Take note of measured quantity of interest (a force or a stress or a displacement)
3- Refine mesh to be approximately ½ of the previous size, re-run analysis, review isolines (should be smoother) and check your measured quantity of interest. It will be slightly more accurate. Start a hand plot of this quantity.
4- Refine the mesh once more as above, re-run analysis and review isolines and measured quantity. You should see the measured quantity approaching a stable value in your hand plot of the measured quantity.
5- Continue this process until measured quantity is adequately converged.

Example: Simple plate in bending

N=3:

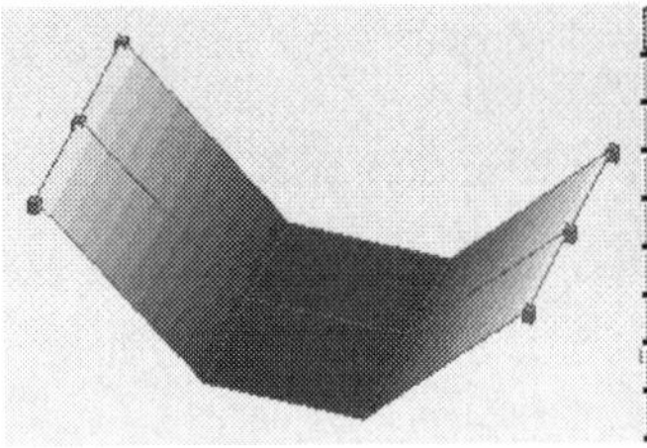

	UZ (in)
MAX	0.0
Node	1
Case	1
MIN	-0.0151
Node	7
Case	1

N=7

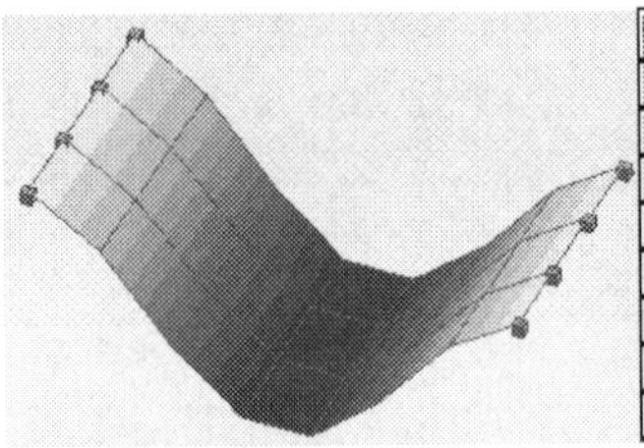

	UZ (in)
MAX	0.0
Node	1
Case	1
MIN	-0.0184
Node	18
Case	1

N=24

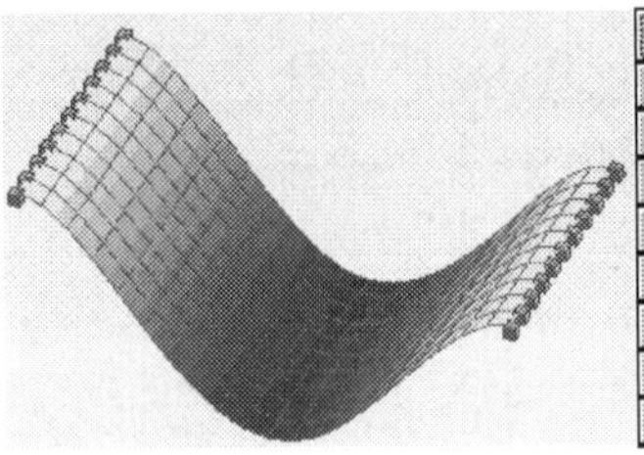

	UZ (in)
MAX	0.0
Node	1
Case	1
MIN	-0.0193
Node	23
Case	1

N=42

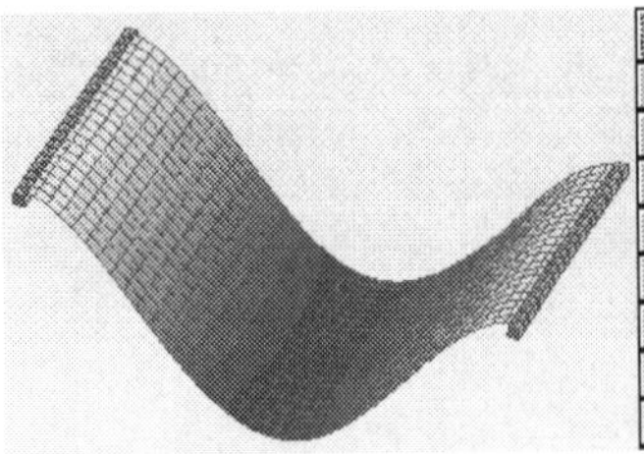

	UZ (in)
MAX	0.0
Node	1
Case	1
MIN	-0.0193
Node	43
Case	1

Plot of Max Displacement:

From the plot below you can quickly see that as the mesh is refined the value of maximum displacement converges to 0.0193" at 24 divisions along the long edge.

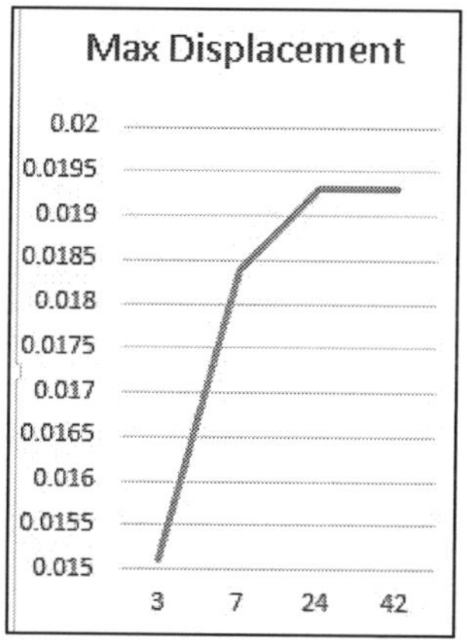

You can repeat this exercise for any measured quantity in the panel. It can also be useful to look at the isolines (see Exploring Results for Panels below) for various quantities. If your mesh is refined, you should see nice smooth isolines. If you see them breaking, or discontinuous across finite element boundaries you may need to further refine your mesh.

Exploring Results for Panels

Results presentation for panel elements is similar to diagrams on bars though only one quantity can be displayed as a map at one time. You can use the option to "open the result display in a new window" to view multiple quantities at once as you apply the map settings. Panel cuts are also a great way to take a look at what's going on at a slice through the panel.

Sign Conventions

Probably the most important key to understanding results presented by Robot is to understand the sign convention used. Here are sign conventions for both displacements/rotation and for forces/moments. These signs are always oriented with respect to the currently defined X direction for the panel. This may or may not be aligned with the panel local x-axis. Take a look at the direction option (#2) below to see what other options are available.

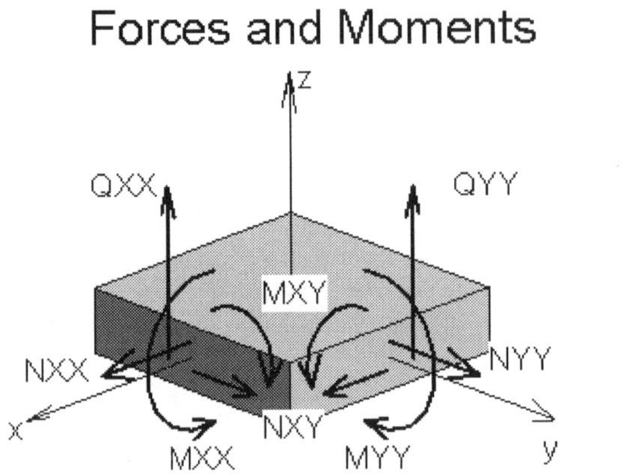

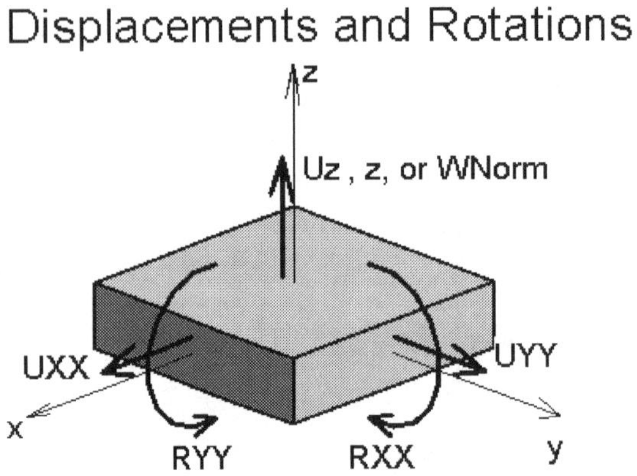

Maps

Once you have run an analysis involving panel elements, you can start investigating the results with the maps tool. **Results>Maps...** When the Maps dialog opens you will find several tabs across the top to offer different measured quantities which can be displayed as a map on the surface. You will also find options for displaying the active deformation which will show the deformed shape along with a map of the currently selected quantity to give you a better sense of what's happening in the model. The help files are pretty good at explaining the options, so we won't go into depth on each tab here. For more information on general functioning of this dialog and the tabs, please refer to Exploring Preliminary Results earlier in the text.

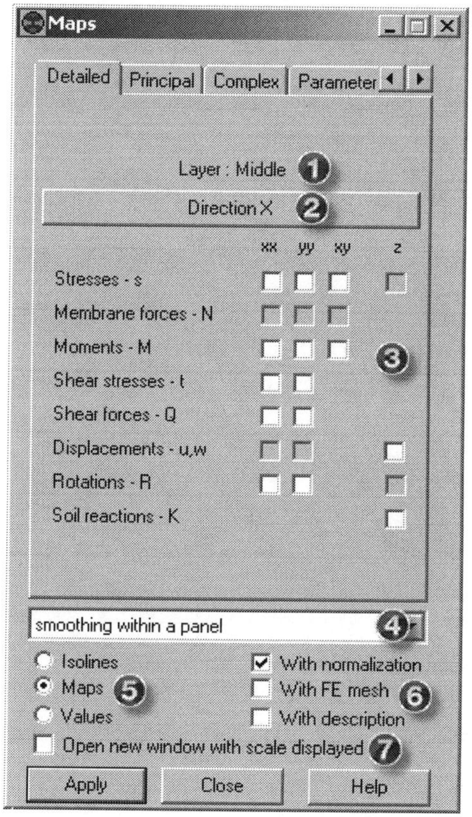

1- Layer: This reports the current layer set in the parameters tab. This will have no small effect on, for instance, the Stresses presented. For example, if you have a slab with pure bending, stress at the middle layer will be zero! Check the Parameters tab to change layer for which results are currently presented.

2- Direction: Access settings to control which direction is currently considered local X for the panel results presentation. Use Automatic option to respect the panel's local axes or choose Cartesian to specify a global axis or a vector direction. The panel normal will remain the same, but the x-axis for results will be aligned with the selected axis or the local panel coordinate system if Automatic is selected. If you would prefer to use the local axes ("Automatic" option), but they are not oriented correctly, you can adjust them using the **Geometry>Properties>Local Panel Direction...** tool.

3- Measured Quantity: Select the quantity for which you would like to display a map. Only one quantity can be selected at a time from any of the result tabs in this dialog.

4- Smoothing: Results are presented at the common nodes of adjoining elements. Due to the discontinuous nature of finite element analysis, the results at each node meeting the joint may vary slightly and cause the map or isolines to be disjointed (break or be discontinuous). Select a smoothing option to have Robot average the results at these locations and present a smoother map/isoline diagram. Use no smoothing to use isolines as an indicator of mesh refinement.

5- Display Style: You can display Color Maps of values, Isolines along contours of equal value, or the actual values themselves as text. (When you select text you have additional options for placement location).

6- Additional Display Attributes:

   a. With Normalization: Scale max and min are adjusted to the global maximum and minimum rather than max/min for the currently displayed panels.

   b. With FE mesh: Also display a graphical representation of the mesh on the surface

   c. With Descriptions: Also add text descriptions to the panel at finite element centers.

7- Open in a new window: As only one measured quantity can be displayed at one time, it can be useful to display the results by opening a new window with the map displayed so that you can open several maps at once to review the measured quantities as they vary over the surface.

Panel Cuts

Cuts can give you a diagram of a measured quantity as it varies along a defined line. Operation of the cuts dialog is not immediately discoverable, but it is a powerful tool for investigating quantities in the panel. Start the cuts tool from the results menu: **Results>Panel Cuts...** to open the Panel Cuts dialog.

Basic Workflow:

1- Name cut, Define cut geometry

2- Select Measured Quantity to display (Detailed, Complex, Principle, etc. tabs)

3- Press "Apply" or "Normalize".

4- To modify current cut, select it on the Cuts tab and adjust measured quantity or cut definition then press "Apply".

5- To Add a new cut, on the definition tab, press "New", Name the cut, define geometry, select measured quantity, then press "Apply" or "Normalize".

The Panel Cuts dialog:

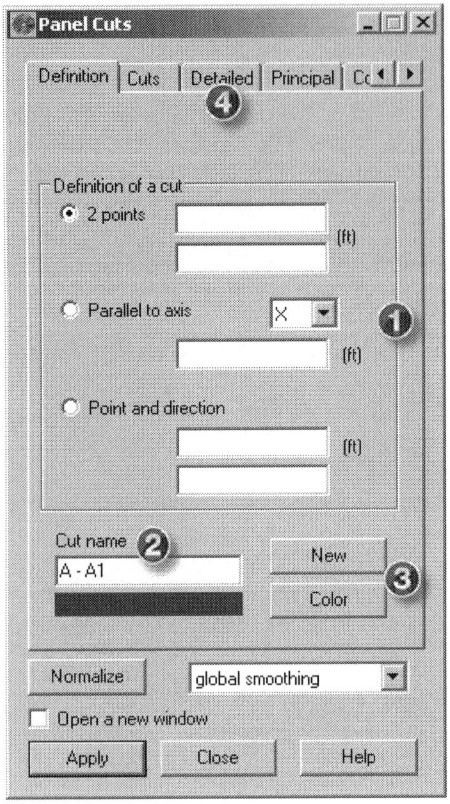

1- Definition of cut: Use one of the options to define a cut line, either by entering or choosing 2 points, selecting it parallel to an axis at a distance from the origin, or at a point with a direction vector. Experiment with these options to see how they work.

2- Give your cut a name that makes sense to you. As they are populated in the list on the cuts tab, they may become difficult to distinguish without a meaningful name.

3- New is only used to add a new cut to the list of cuts. If you do not yet have a cut defined, "New" will do nothing. It is only after one cut is defined that "New" becomes useful. Color is simply the color that will be used for the fence lines in the diagram.

4- After defining the geometry, name and color of the cut, use the detailed, principle, complex or other tabs to select a measured quantity to display as a diagram along the defined line (cut).

Chapter 12 - Basic Seismic Analysis

We will cover basic equivalent lateral force method as implemented in Robot to automatically generate and apply seismic loads to our structure. This does not require a modal analysis and is the most basic method for loading a structure seismically. Make sure that it is allowable to use the equivalent lateral force method in the area and seismic category for which you are designing. Robot provides several more advanced seismic analysis capabilities although they all require a modal analysis of the building. As modal analysis is a bit of an advanced topic, we will leave that to an advanced text.

IF YOU WOULD LIKE TO EXPERIMENT WITH MODAL ANALYSIS ON YOUR OWN, PLEASE KEEP IN MIND THAT THERE IS A SETTING CALLED "DISREGARD DENSITY" WHICH HAS A HUGE EFFECT ON THE MODAL ANALYSIS. IT WILL EITHER INCLUDE OR EXCLUDE MATERIAL MASS, REGARDLESS OF WHICH CASES YOU HAVE CONVERTED TO MASSES. UNCHECKING IT, WILL HAVE THE EFFECT OF DOUBLING THE MASS IF YOU HAVE ALREADY CONVERTED A DEAD LOAD CASE TO A MASS.

The methods of investigating and understanding the seismic analysis generated loading below are also applicable to more advanced seismic analysis types.

Seismic Equivalent Lateral Force Method

To generate seismic loads per the Equivalent Lateral Force method, first open the Analysis Types dialog from the Analysis Menu: **Analysis>Analysis Types...** or from the main toolbar with the Analysis Types Icon: 

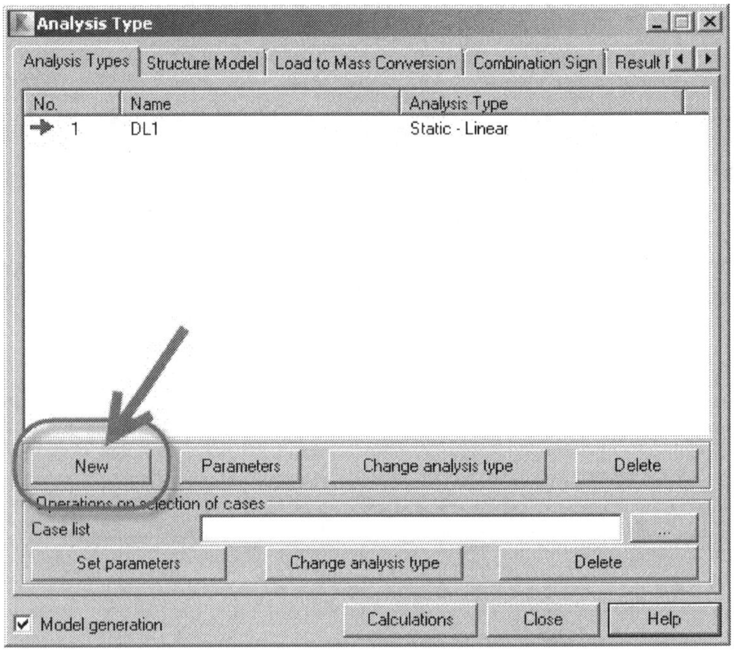

In the Analysis Types dialog press "New" to see a list of available analysis types:

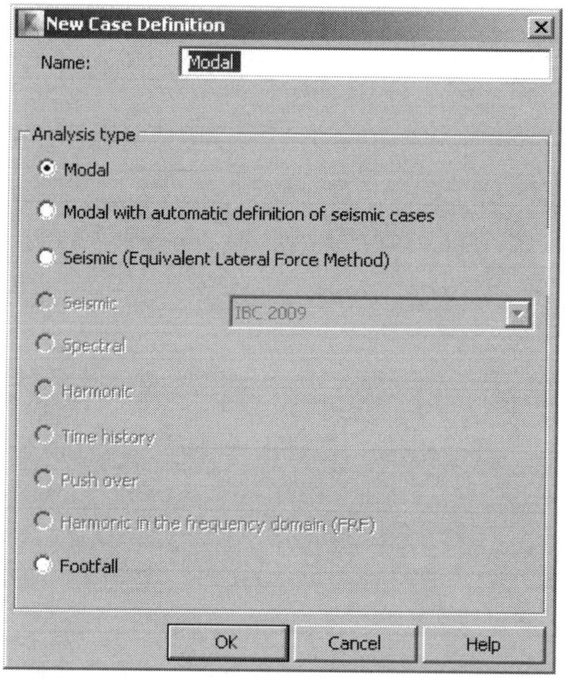

NOTE THAT ANY ANALYSIS TYPE WHICH REQUIRES A MODAL ANALYSIS IS GREYED OUT AND CANNOT BE CREATED. IF YOU NEED TO ADD A MORE ADVANCED ANALYSIS TYPE WHICH REQUIRES MODAL ANALYSIS, FIRST ADD A NEW MODAL ANALYSIS TYPE, THEN WHEN YOU PRESS "NEW" AGAIN, THE PREVIOUSLY GREYED OUT ANALYSIS TYPES WILL BE AVAILABLE FOR SELECTION. ROBOT USES THE RESULTS OF THE MODAL ANALYSIS CASE TO DRIVE THE OTHER TYPES OF ANALYSIS (AS AN INPUT, SO TO SPEAK, TO HARMONIC, DYNAMIC, TIME-HISTORY, ETC.)

In our case, we'll want to select "Seismic (Equivalent Lateral Force Method)" and press "OK". This will open the Seismic Analysis dialog where we can configure the parameters of our equivalent lateral force method. Once we have configured those parameters and started the analysis, Robot will automatically generate the forces to be applied to the structure and have them ready to be counted as a series of load cases for combination with other structure loads.

Select "Seismic (Equivalent Lateral Force Method)

This will launch the Seismic Analysis Dialog:

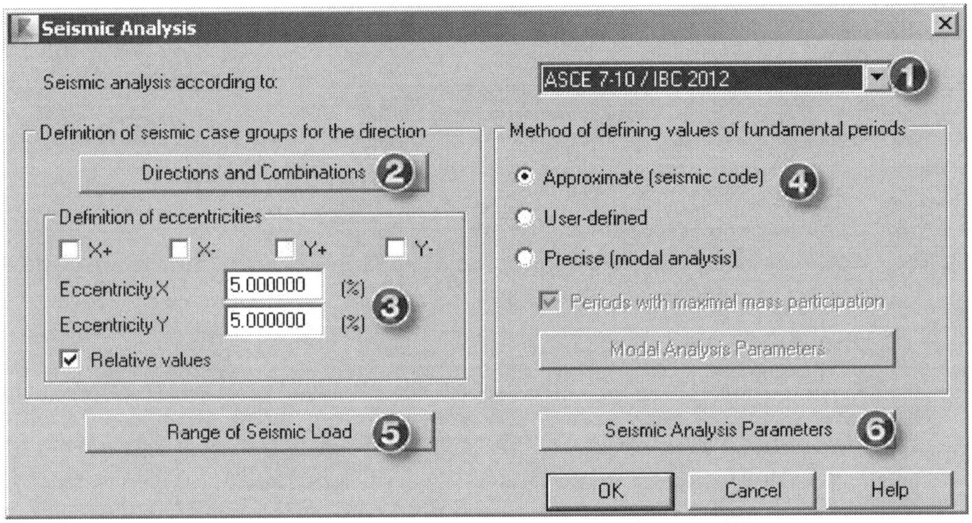

1- Code Selection: For the US, select ASCE7-10/IBC2009. Robot only implements a few equivalent lateral force codes.

2- Directions and Combinations: Robot will automatically have X and Y directions set for load generation so for equivalent lateral force method you will not typically need to access this dialog.

3- Definition of Eccentricities: In addition to the X and Y load cases, Robot can also automatically create eccentric load cases which can be combined into your loading combinations. Select which eccentricities you would like generated and whether you would like to specify the eccentricity as a percentage ("Relative value") or as a specific distance. (uncheck "Relative values").

4- Method of defining values of Fundamental Period: You can select "Approximate" to access a choice of building lateral systems which Robot will, in turn, use to apply the code estimated fundamental period, or you can specify your own fundamental period by checking "User-defined". Robot can, of course, perform a modal analysis of the structure to determine fundamental period, but this is a more advanced topic. Selection of the lateral system for automatic and specification of user defined period will be found in the Seismic Analysis Parameters dialog (#6)

5- Range of Seismic Load: Use this dialog to specify top and base levels in the structure. You can also leave this at the automatic setting.

a.

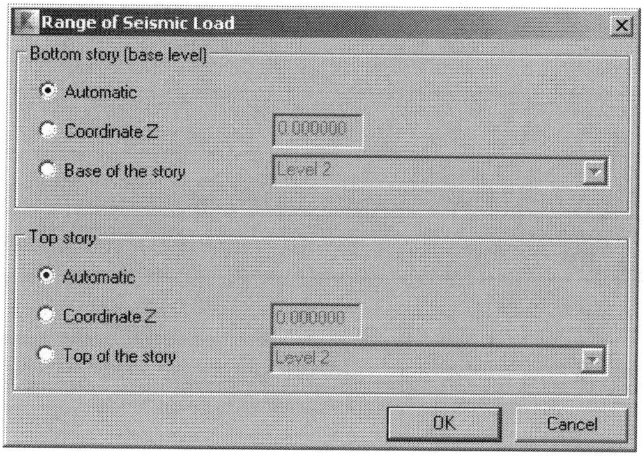

6- Seismic Analysis Parameters: Here are the typical settings for Equivalent Lateral Force Procedure. Fill in all parameters and select the site class which fits your design site. Fill in S1 and Ss, Robot will automatically select Fa and Fv based on selected site class and calculate the design spectral accelerations SD_s and SD_1.

a. 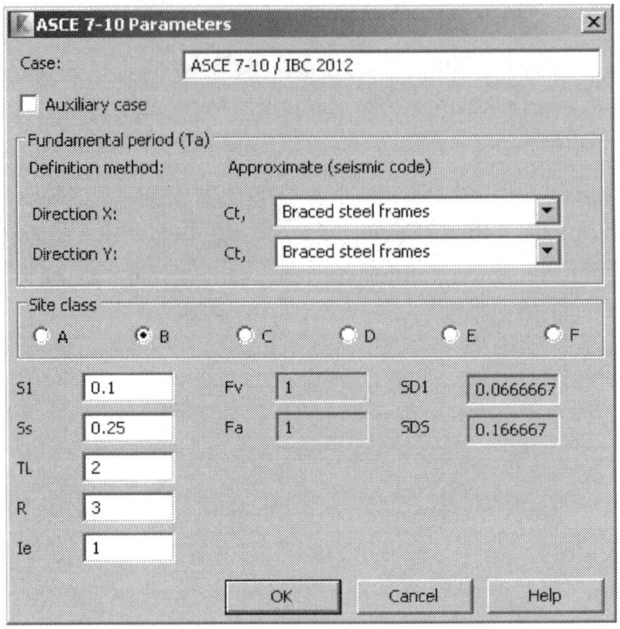

Press "OK" in the ASCE 7-10 Parameters dialog, then "OK" again in the Seismic Analysis dialog and Robot will create the specified load cases based on the code, the building mass and any desired eccentricities. Here is an example of the generated load cases:

No.	Name	Analysis Type
1	DL1	Static - Linear
2	ASCE 7-10 / IBC 2009 Direction_X	Static - Seismic
3	ASCE 7-10 / IBC 2009 Direction_Y	Static - Seismic
4	ASCE 7-10 / IBC 2009 Ecc X- Direction_X	Static - Seismic
5	ASCE 7-10 / IBC 2009 Ecc X- Direction_Y	Static - Seismic
6	ASCE 7-10 / IBC 2009 Ecc X+ Direction_X	Static - Seismic
7	ASCE 7-10 / IBC 2009 Ecc X+ Direction_Y	Static - Seismic
8	ASCE 7-10 / IBC 2009 Ecc Y- Direction_X	Static - Seismic
9	ASCE 7-10 / IBC 2009 Ecc Y- Direction_Y	Static - Seismic
10	ASCE 7-10 / IBC 2009 Ecc Y+ Direction_X	Static - Seismic
11	ASCE 7-10 / IBC 2009 Ecc Y+ Direction_Y	Static - Seismic

Seismic Weight (Mass)

In generating the seismic load per the equivalent lateral force method, Robot will automatically use the mass density of the elements (as defined in material properties and based on the physical geometry of the elements) in the structure as the seismic weight. To additionally consider cases of applied loads as masses in the structure, you can use the Load to Mass Conversion utility. Loads converted to masses here will be used in addition to the basic mass density of the physical structural elements in the model.

NOTE: IF YOU APPLY A SELF-WEIGHT CASE AND CONVERT IT TO A MASS, IN THE EQUIVALENT LATERAL FORCE METHOD, THE SEISMIC WEIGHT WILL BE TWO TIMES WHAT IT SHOULD BE. EQUIVALENT LATERAL FORCE METHOD IS ALREADY TAKING MASS DENSITY (SELF-WEIGHT) INTO CONSIDERATION.

Access Load to Mass conversion from the Analysis Types dialog: **Analysis>Analysis Types...**>Load to Mass Conversion tab...:

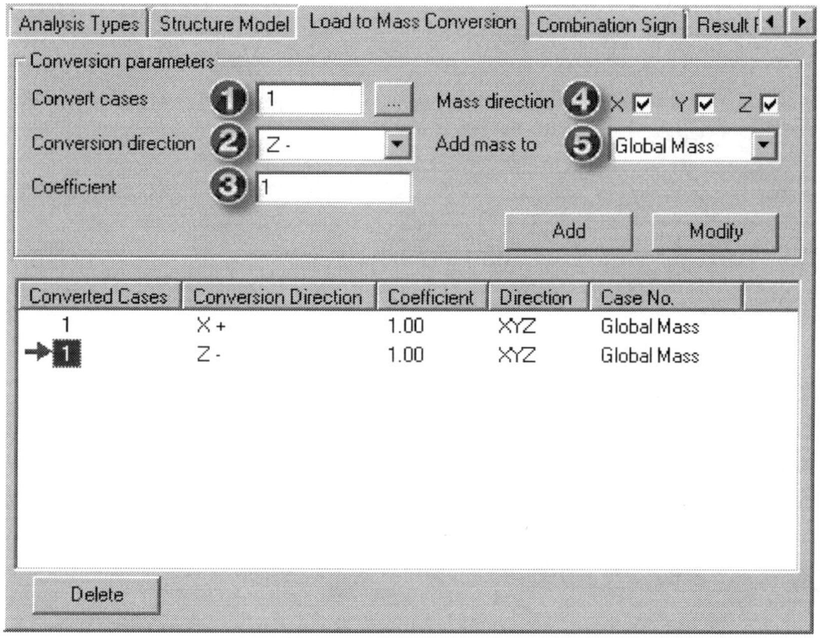

1- Convert Cases: Enter the case numbers for load cases which you want included as part of the structure mass. Use the ellipsis button to access the load case selector.

2- Conversion Direction: You must select the direction of load action which will be included as part of the mass. For instance, if you have one load case which includes vertical (gravity loads), but also includes lateral or traction loads, you can control which loads are converted with this selector. Only loads which act in the direction of the selected conversion direction will be converted to masses for the structure. If you have loads in one load case which act in different directions, you will need to add the conversion twice, once for each direction of load action in the load case. Otherwise, in the vast majority of cases, you will leave this set to Z- to account for gravity structure loads which you wish to have contribute to the structure mass.

3- Coefficient: You can adjust the contribution of the mass by using this coefficient. In cases where only a percentage of the load must be used in the seismic load calculations, you can factor its contribution here.

4- Mass Direction: Unless there are some really special circumstances, the mass will act in all directions (X,Y, and Z). Selecting only X or Y can have strange effects on the model. Make sure that if you have a case where only X or Y should be included that you carefully review the reduced building forces as shown below to make sure that you thoroughly understand what is happening in your model.

5- Add Mass to: For our purposes leave this set to Global. You can be selective about the analyses in which this mass acts, but they are advanced analyses that are not covered here. You can control for which cases conversion should be active by specifying whether the converted mass is:

a. Global – Mass is active for any load cases including static ones (like centrifugal or inertia loads)

b. Dynamic – Mass is only active for dynamic cases

c. Active for selected case – Only active for specified case except that if a modal analysis is specified here, this mass will also be active also for all successive spectral cases after it until the next modal analysis type. Because a modal analysis is an input for subsequent spectral cases. More about that just below:

Mass type and load to mass conversions are assigned to modal analysis types, not to spectral/seismic load cases. The modal load cases preceding the seismic analysis are used by subsequent spectral/dynamic cases. In other words, the order of the cases (analysis types) is important as the modal analysis which precedes a spectral case will be used to inform the spectral case (structure periods and mode shapes). You can add multiple modal analyses combined with subsequent spectral analyses to investigate different effects. The modal analysis will be used by subsequent spectral analyses until another modal analysis is encountered in the list of load cases (analysis types).

Investigating Seismic Analysis Loading

Now that we have added automatically generated seismic load to our structure, it is a good idea to review the loading and make sure we understand what is happening in our structure. The following sections are tools we can use to start understanding our model and the loading which has been generated for us.

CALCULATION NOTES:

First check the calculation notes for information on total seismic weight and base shear. **Analysis>Calculation Report>Full Note**. Here is an example of the calculation note for an equivalent lateral force analysis in the X direction. You should be able to pick out the approximate fundamental period selected, total seismic weight, the calculated base shear, and the vertical distribution of forces at each story (only one story in this case).

Case 2 : ASCE 7-10 / IBC 2009 Direction_X
Analysis type: Static - Seismic
Excitation direction:
X = 1.000
Y = 0.000
Z = 0.000
Data:
Soil : B
S_1 : 0.100
S_S : 0.250
Spectrum parameters:
F_a = 1.000 F_v = 1.000
S_{MS} = 0.250 S_{M1} = 0.100
S_{DS} = 0.167 S_{D1} = 0.067
T_o = 0.080 T_S = 0.400
T_L = 2.000
I = 1.000 R = 3.000
Fundamental period:
Approximated method T = 0.284 (s) <<---
Braced steel frames C_t = 0.03 (0.0731) x = 0.75
Structure range:
Top story Story 1
Bottom story Story 1
Effective height H_n = 20.00(ft)
Base shear
C_s = 0.056
$C_{s\,max}$ = 0.078
$C_{s\,min}$ = 0.010
Effective seismic weight W = 36.82(kip) <<-----------------------------------
Shear force V = 2.05(kip) <<---

Vertical distribution of seismic forces <<--

Story	Height (ft)	Weight (kip)	F(kip)	M(kip*ft)
Story 1	20.00	36.82	2.05	0.00

PSEUDOSTATIC REACTIONS

Next, we'll look at the loads which Robot has generated for us and automatically applied at the structure nodes. Open the Diagrams dialog from the results menu: **Results>Diagrams for Bars...** Switch to the Reactions tab and select Pseudostatic, select all forces and moments and also check "Descriptions".

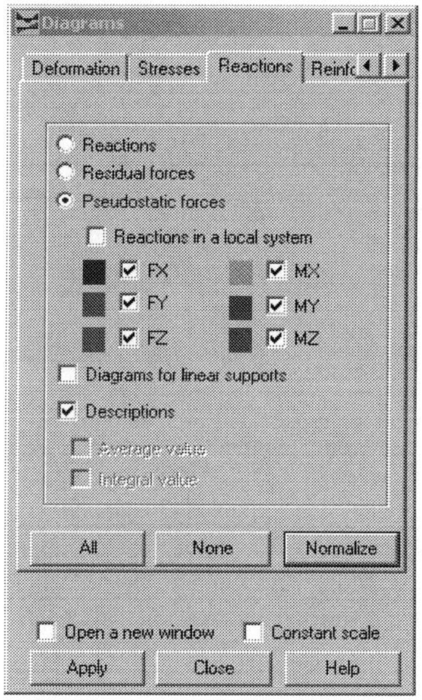

Every node which has a generated load applied will show the load, the direction, and the magnitude in the description tag as shown here for a one story structure:

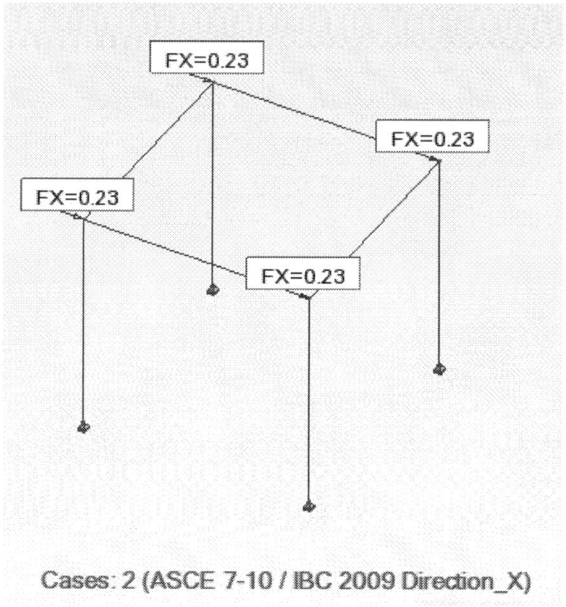

Now we can see exactly where Robot is applying the load and what the magnitudes of the load are. From the calculation notes, we find that for the X direction the shear is V = 0.93(kip) and distributed evenly (as we might expect for no eccentricity) at each node 0.93/4 = 0.23k.

You can take a look through each seismic loading case to see what loads Robot is applying for you and you can also easily verify them against the calculation note indicating shear at each level.

DIAGRAMS FOR BUILDINGS

As structures start getting larger, it is nice to be able to take a look at reduced forces at each story to get an overall picture. With the Building Diagrams tool, you can also look at things like the position of the center of rigidity and the center of mass. Once the analysis has been run, you can launch the Diagrams for Buildings tool from the Results menu: **Results>Diagrams for Buildings...** which launches the Diagrams for Buildings dialog.

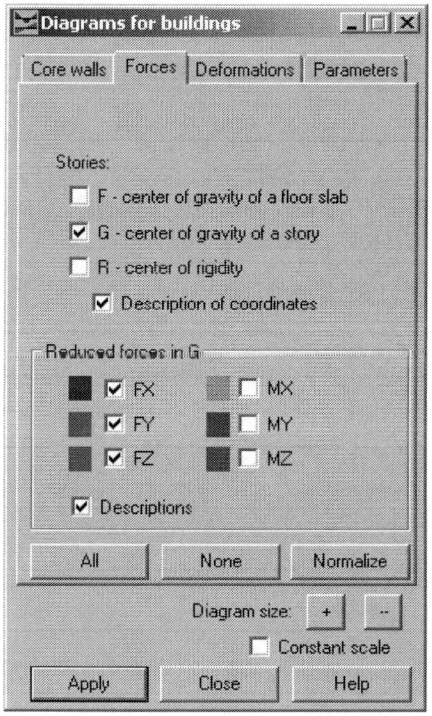

Switch to the Forces tab and select Center of Gravity of a story, select Fx, Fy, and Fz from the Reduced Forces in G, also check "Descriptions" for both. This will give you a summation of forces at that story for the selected load case as shown here, along with the position of the center of gravity and the coordinates.

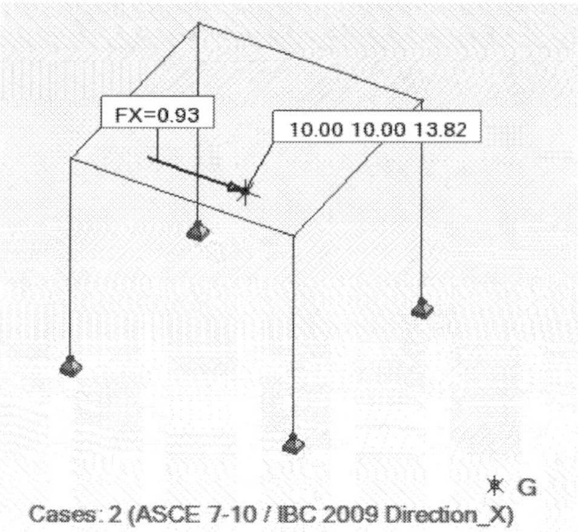

These three tools together should help you quickly review and understand what is being loaded and how it is being loaded when you are working with seismic analysis loading automatically generated by Robot.

STORIES TABLE

One final option for viewing lateral loads and displacements resulting from analysis is the Stories table. Also launched from the Results menu: **Results>Stories...** you can view tabular results for reduced forces at stories, overall displacements and drifts. Robot will also break down forces which have been taken by walls vs. forces which have been taken by columns. When you launch the Stories table, take a look along the bottom for additional information accessed via the table tabs:

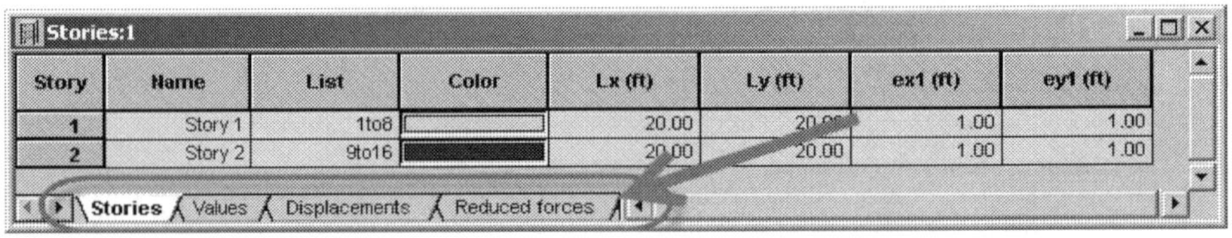

Stories:1

Story	Name	List	Color	Lx (ft)	Ly (ft)	ex1 (ft)	ey1 (ft)
1	Story 1	1to8		20.00	20.00	1.00	1.00
2	Story 2	9to16		20.00	20.00	1.00	1.00

Stories | Values | Displacements | Reduced forces

In the Displacements tab, you can find story displacements and drift ratios:

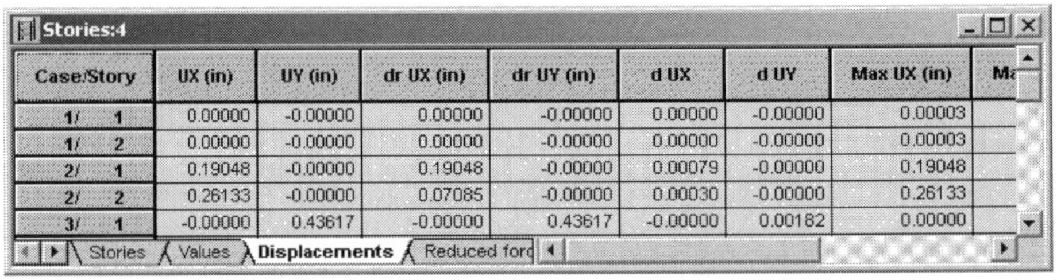

Stories:4

Case/Story	UX (in)	UY (in)	dr UX (in)	dr UY (in)	d UX	d UY	Max UX (in)	Ma
1/ 1	0.00000	-0.00000	0.00000	-0.00000	0.00000	-0.00000	0.00003	
1/ 2	0.00000	-0.00000	0.00000	-0.00000	0.00000	-0.00000	0.00003	
2/ 1	0.19048	-0.00000	0.19048	-0.00000	0.00079	-0.00000	0.19048	
2/ 2	0.26133	-0.00000	0.07085	-0.00000	0.00030	-0.00000	0.26133	
3/ 1	-0.00000	0.43617	-0.00000	0.43617	-0.00000	0.00182	0.00000	

Stories | Values | Displacements | Reduced forc

To save you some heartache, we should be very clear what these values represent:

1- UX, UY = Total average displacement of the story in the X or Y direction

2- drUX, drUY = Story drift for each story ($UX_i - UX_{i-1}$)

3- dUX, dUY = Inter-story drift ratio for each story ($drUX_i/h_i$ or $drUY_i/h_i$)

4- MaxUX, MaxUY = maximum point of displacement in that story. Remember UX and UY are average displacements for the entire story.

Chapter 13 - Beginning Concrete Design

Robot offers two different methods of concrete design: Required Reinforcing and Provided Reinforcing. Required reinforcing will calculate recommended bending and shear reinforcing for concrete members in your model and acts similarly to steel design. Provided reinforcing is a separate module and requires transferring elements into that module in order to perform design. Provided reinforcing module is by far the more powerful of the two options. We will look at beam and column reinforcing in the Required Reinforcing module and how you can use the Provided Reinforcing module to view capacity diagrams and generate rebar layouts.

On Concrete Elastic Displacements

It is worth noting that deflection calculations in the elastic analysis model are based on the gross moment of inertia of the concrete section. It is possible to do calculations based on a reduced moment of inertia (something closer to the cracked moment of inertia) by adding a reduction factor in the section properties of the beam.

In the section properties for a concrete type member, you can opt to have the moment of inertia reduced by a factor of your own choosing. Launch the sections dialog and configure a concrete section: **Geometry>Properties>Sections...** Choose "New Section" then select "RCBeam" or "RCColumn" in the member type selector at the bottom:

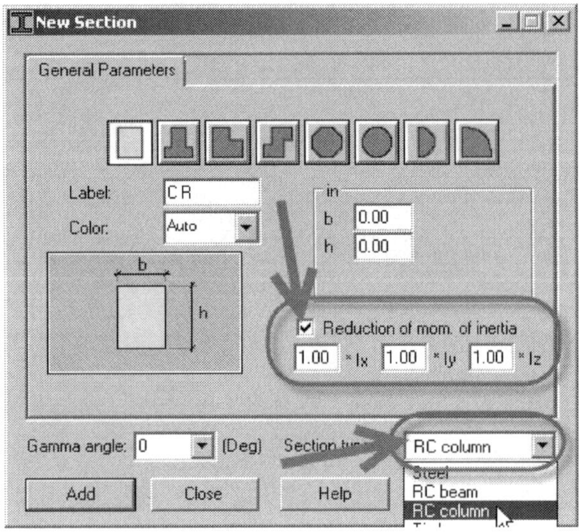

Notice that there is an option for "Reduction of moment of inertia". If you check this box, you will have the option to specify a reduction factor for the moment of inertia to be used for this member in structural analysis calculations. You can also set this in the Properties table for many elements at once. If you open the properties table **View>Tables...** and select "Properties", you can then right-click in the table and select "table columns..." and choose "Reduction of moments of inertia"

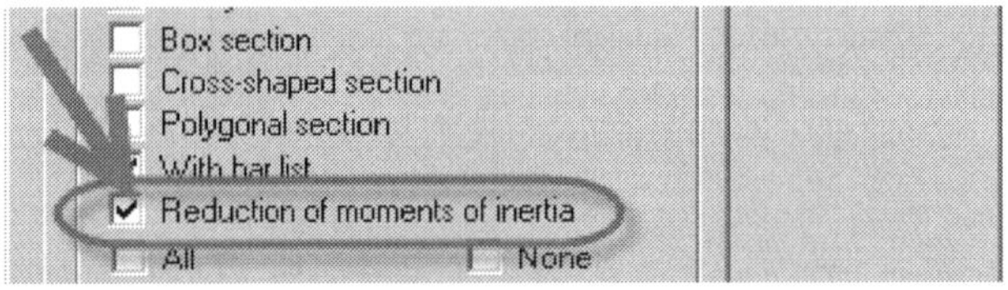

Then the properties table will show the reduction factors which you can review.

Properties

Section name	Bar list	AX (in2)	AY (in2)	AZ (in2	IX (in4)	IY (in4)	IZ (in4)	Red. Ix	Red. Iy	Red. Iz
B R18x30	1	0.000	0.0	0.0	36499.395	20250.000	14580.000	1.000	0.500	1.000

There are many theories on applying a reduction factor for elastic analysis of concrete structures. Some engineers believe that simply using the gross moment of inertia will yield sufficient results, and some codes actually require a reduction factor for elastic analysis.

NOTE: BEWARE THAT, IF YOU'RE USING REDUCTION IN CONCRETE MOMENT OF INERTIA IN CONJUNCTION WITH THE DIRECT ANALYSIS METHOD, ANY REDUCTION SPECIFIED IN THE DIRECT ANALYSIS PARAMETERS WILL BE APPLIED ON TOP OF THESE REDUCTIONS IN THE SECTION DEFINITION. IF YOU ALREADY APPLIED REDUCTIONS HERE, USE 1.0 FOR CONCRETE IN THE DIRECT ANLAYSIS METHOD.

Analysis Techniques for T-Beams

In the situation where you would like to use Robot to analyze and design concrete beams with concrete slab interaction we have a situation where we need to think about relative stiffness of the member in positive and negative bending. Obviously when the section is in positive bending (T-flange in compression) the moment of inertia will be larger in the cracked case than the same section in negative bending (T-flange in tension) in the cracked case. This variation of moment of inertia has no small effect on the outcome of an elastic analysis of the structure.

Several techniques for addressing this variation have been suggested. One method might be to use member offsets such that the beam is analytically positioned below the slab. The result of this is that we then, through finite element analysis, introduce additional tension and compression in the slab as well as the member which may not be desirable in terms of how you want to analyze the section in design.

The most accurate yet currently most time intensive method would be to not use offsets for the member and to break it up into several length-wise sub-members (recommended length of sub-member is 1-2 times the depth of the member) to which you can assign values of the cracked moment of inertia based on the moment condition in the beam. This would be an iterative process where each iteration would lead to successive refinement by manually updating the moment of inertia to coincide with the value of the moment demand at the section in question. This method of analysis would lead to a reasonably precise solution though is somewhat intensive in terms of the manual work required to effect such a precise analysis. (Perhaps a good candidate for Robot Dynamo to reset moments of inertia based on the analysis results?) It is important to note that this method of analysis is generally valid until the plastic moment of the beam has been reached.

Another method which proposes a simplified analysis is suggested by Artur Kosakowski where he proposed to use no member offsets and to artificially increase the moment of inertia of the beam to account for the effect of engaging the flange. He proposes multiplying the beam moment of inertia by the ratio of the T-section moment of inertia to the non-Tsection moment of intertia then running a straight elastic analysis. This method does, however, seem to overestimate the stiffness of the cracked T section in negative bending. The current ACI code leaves the method of analysis completely up to the user stating something to the effect of "any set of reasonable assumptions"

Required Reinforcing

The Required Reinforcing module is one part of the design process for selecting reinforcing for concrete members. There are some drawbacks but it does provide a nice exportable table from which you can develop your own more streamlined design routines. The required reinforcing workflow consists of creating and assigning member type labels for each member, then creating and assigning calculation option labels. Once member type and calculation options have been configured for each member you want to design, you will start calculations from the calculations dialog by selecting both which members you wish to design and which combinations you wish to apply. Robot will then calculate reinforcing requirements in terms of Area of steel (and resolve this into required number of bars) at each section along the length of the member. In reality, unless your concrete structure is very simple, this is basically a good starting point, but is not the end-game. The advantages of the Required Reinforcing module are quick, simple results which can give you a picture of steel quantity required for a structure along with a nice exportable table of results from which you can continue your design and check procedures manually in a spreadsheet program. On the downside, Required Reinforcing is not terribly verbose in terms of design, there are no easy-to-output individual member design/check reports available, not all quantities you might desire for hand verifying designs are available in the table, and last and probably least, member verification based on pre-configured reinforcing is not currently available.

There are some very important things to note that Robot does not check.

Things Robot does not check:

- member anchorage lengths
- minimum covers and diameters
- minimum and maximum bar spacings
- minimum and maximum geometrical elements values
- minimum reinforcing for temperature/shrinkage

Meaning Robot does not check if the bars fit in the beam, does not check to see if there is enough length to develop the bar, does not check minimum bar diameters, and does not consider spacing of members making up the required areas of steel.

Things Robot does check for ACI code in required reinforcing:

Beams

- Calculation of theoretical (required) reinforcement area as for ULS.
- Checking (correction) reinforcement for cracking using a simplified method that does not need SLS calculations.
- Calculation of theoretical (required) stirrup spacing for shear forces.
- Resolution of theoretical (required) shear reinforcing area to a number of bars.
- Resolution of theoretical (required) stirrups spacing to a modular spacing in a given number of equal sections.

Columns

- Calculation of theoretical (required) reinforcement area for ULS taking into consideration slenderness for sway and non- sway frames.
- Calculation of theoretical (required) stirrups spacing.

- Resolution of theoretical (required) stirrups spacing to a modular spacing at the length of a column.

Specifying Member Type Parameters

Open the concrete member types dialog from the design menu: **Design>Required Reinforcement of Beams/Columns - Options>Code parameters...** (Note: Use **Tools>Job Preferences**>Design Codes to choose concrete design codes.) This will open the R/C Member Types label dialog:

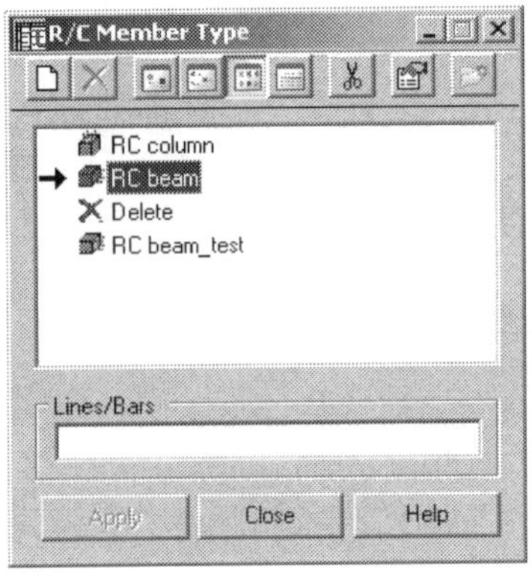

This dialog works exactly like all other label management dialogs. Labels are simply collections of parameter settings which are applied to members (bars). Double click an existing one to start or click the new button to create a new member type.

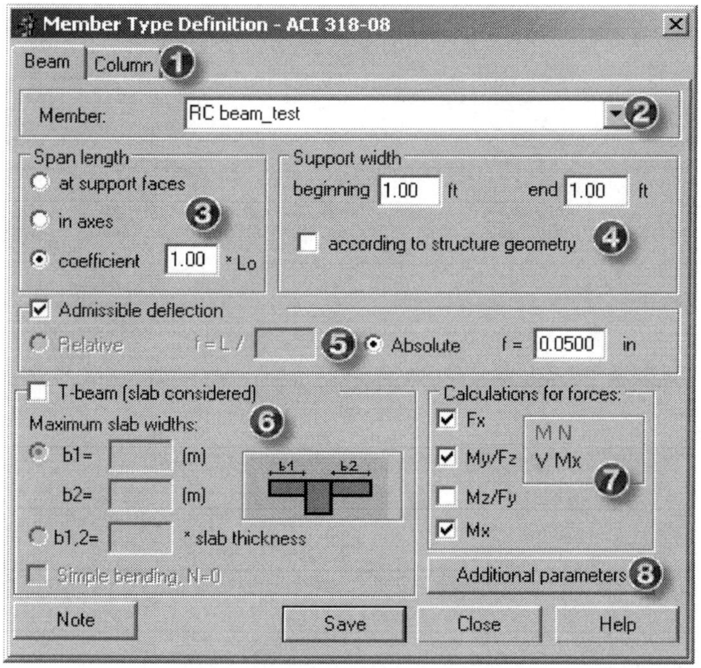

1- Beam types and column types can be configured in this dialog. It operates in a slightly different manner than other label parameter dialogs in that you can select between different existing labels using the drop-down control (or specify a new one by typing into the edit control of the dropdown). Configure column types on the column tab.

2- Member type label selector: Choose the member type label you want to modify or type in a new name here to create a new member type.

3- Span Length: This will affect the design sections taken for the member.

a. At Support Faces: Moments and shear sections for design will be taken starting at the support faces of the member. See #4 below for support definition options.

b. In axes: Use the actual length of the member, ignoring support widths.

c. Coefficient: Specify a percentage of the length to use as the effective member.

4- Support Width: Specify width of the support as measured from the end of the member along its length (e.g., if column is 2'-0" use 1'-0" for support width). Alternatively you can have Robot determine the support width automatically from the columns supporting the member in the model. To use this option, select "According to Structure Geometry".

5- Admissible deflection: Specify an absolute value for deflection of this member type in the local z direction. Relative values are not available. Check will be performed at calculation time and offer warnings which can be investigated in the calculation results dialog.

6- T-beam (Slab Considered): Note that you can have the slab with which the beam is coincident, considered in the design of the member. Specify effective widths or a number of slab thicknesses to be taken as the effective flange width. In order to accurately design these, you must use offsets for the members such that the panel representing the slab is at the top of the member. If there is no panel geometrically connected to the beam, the "T" portion will be ignored. It is also worth noting that the panel and beam will be double counted at the intersection of the beam and panel which will result in slightly stiffer members and smaller deflections. Read more about this in the Robot help files under "Calculations of RC T-beams considered with the slab". Simple Bending (N=0) will ignore axial compression in the member.

7- Calculations for forces: You can select which forces you want to have considered in the member design calculations. Minimum requirement is for My/Fz (major axis bending and shear), but you can also have minor axis bending/shear (Mz/Fy), axial force (Fx represented as "N" in the list), and/or torsion (Mx). Note that if you want to have deflections checked, the member cannot be checked for both My/Fz and Mz/Fy, the deflection criteria can only apply for My/Fz design forces. The admissible deflection will become unavailable if you opt to have Mz/Fy considered in design.

8- Additional Parameters: Here you can specify how you want Robot to handle the calculation of member rigidity. Robot can use the minimum value or an average value over the member. This button is only available when admissible deflection option has been checked.

Once you have configured different member types you will need to apply them to appropriate elements of the model. Use the display options (**View>Display...**) to turn on Bars>Member Types display for visual confirmation while you are working through applying member types.

Specifying Member Calculation Options

In the required reinforcing design workflow, calculation options are where you will configure reinforcing steel grades, preferred bar sizes and other options related to the calculation of As(req'd) for the members.

Open the calculation parameters dialog from the design menu: **Design>Required Reinforcement of Beams/Columns - Options>Calculation parameters...** (Note: Use **Tools>Job Preferences...**>Design Codes to choose concrete design codes) This will open the Calculation Parameters labels dialog:

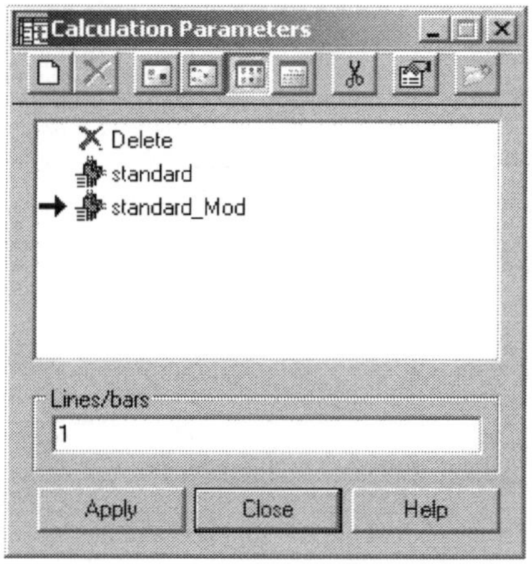

This dialog works exactly like all other label management dialogs. Labels are simply collections of parameter settings which are applied to members (bars). Double click an existing one to start or click the new button to create a new member type.

General Tab:

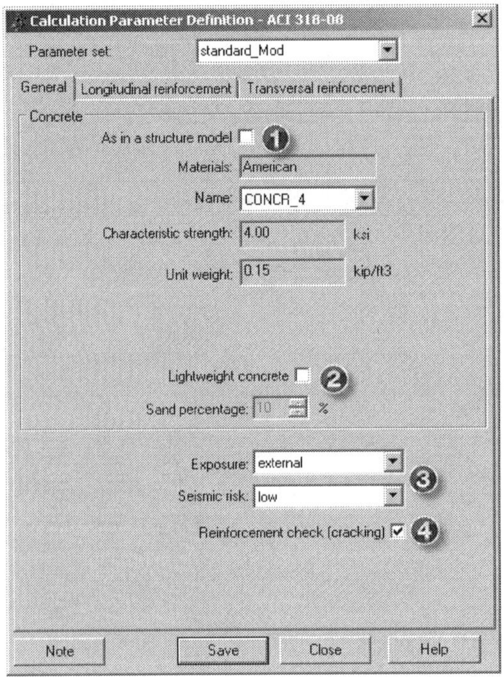

1- Select a material name to use for reinforcing design calculations or select "As in structure model" to use the material defined for the member in the model.

2- Select whether or not this will be designed as lightweight concrete

3- Select exposure and seismic risk

4- Reinforcement Check (cracking): This determines whether Robot will apply the ACI provisions for limiting crack width.

Longitudinal Tab

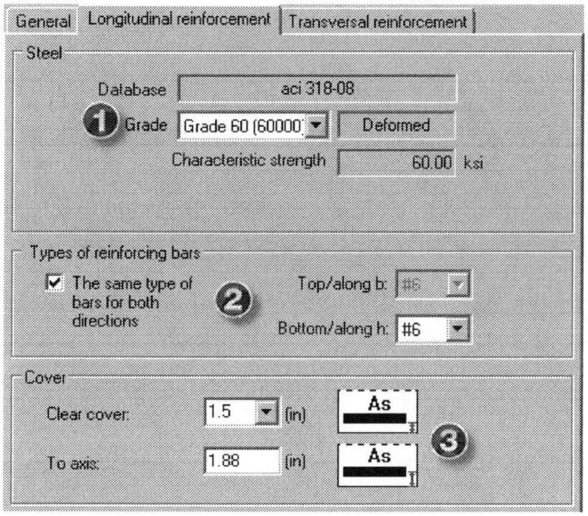

1- Select steel grade for reinforcing bars.

2- Select bar size which will be used for determining number of bars. Note that the bar you select is the bar type which will be used. Robot does not have any capability in the Required Reinforcing module to select bars, it can only take the As(req'd) and calculate a number of bars (of the type you specify here) that can be used to achieve the required area of steel. You can look at this as a sanity check by specifying #9 bars here (approximately $1in^2$ for each bar) to see if the rebar is about what you expect for each member). You can create many calculation option labels with different bar sizes, but only the bar size selected in the calculation option assigned to the member will be used.

3- Specify cover distance. Specify clear cover and Robot will calculate the distance to axis for you. Or you can specify the cover distance to the centroid of the steel bar. (set the bar type first then set these settings so that they update correctly)

The Transverse Tab

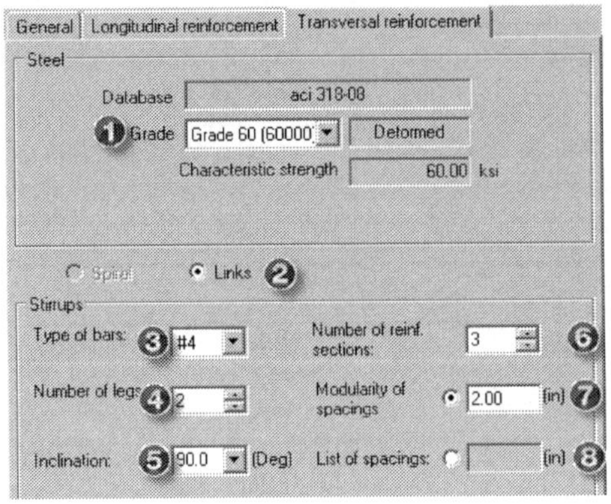

1- Grade: Specify rebar grade to be used for transverse (shear) reinforcing

2- Links is the only option (this may say stirrups as well).

3- Select the type of bars to be used for this calculation model. Only this size bar will be used to determine spacing of shear reinforcement.

4- Number of legs: Specify number of legs considered for the stirrup diameter specified. (How many times does the transverse reinforcement cross the shear zone. A "U" stirrup has 2 legs because it crosses the shear zone twice.)

5- Inclination: If you intend to slope the stirrups to cross the shear tension (diagonal tension) zone more closely to perpendicular to the shear induced diagonal tension. Only fixed options are available.

6- Number of reinforcing sections: Specify the number of spacing zones you want to have calculated. (The number of segments along the beam/column length of equal stirrup spacing.) Robot will take the worst shear at each section and generate a uniform spacing of bars for that section. Section length is defined as effective length divided by the number of sections.

7- Modularity of spacings: Increments of inches or mm which Robot will select for spacing numbers. Setting 5" here will result in spacings which are multiples of 5".

8- List of Spacings: You can specify a list of modularities for Robot to use. Spacing will be selected to nearest increment within the list you specify. Use commas to separate values.

Once you have configured your desired calculation options, you can assign them to members. Use **View>Display...** and Bars>Code Groups option to display calculation options for each member as a visual check.

Running Calculations

Once you have both member types configured and assigned as well as calculation options configured and assigned to your members, you are ready for running calculations of member required reinforcing. From the design menu select **Design>Required Reinforcement of RC Beams/Columns...** Your interface will be rearranged into the required reinforcing layout and you will not be able to close any of the dialogs opened until you switch your layout back to the structure layout.

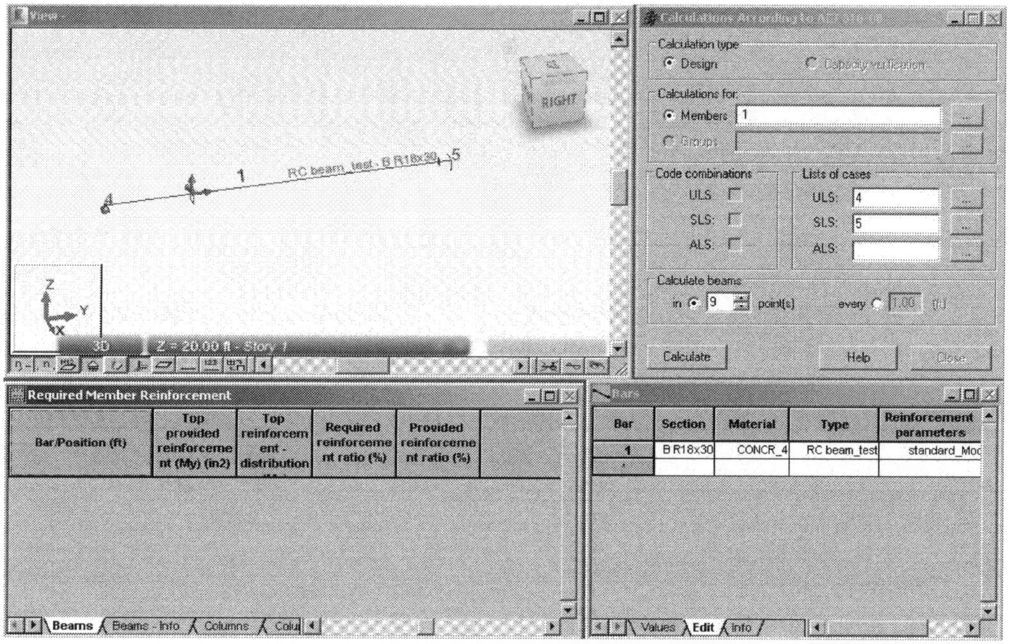

You will have the calculations dialog, the required member reinforcement table, and the bars table with table columns reduced to just section, material type, member type and calculation option assignment. You can always use right-click>table columns… to adjust the columns shown in this dialog if you need additional information shown.

THE CALCULATIONS DIALOG:

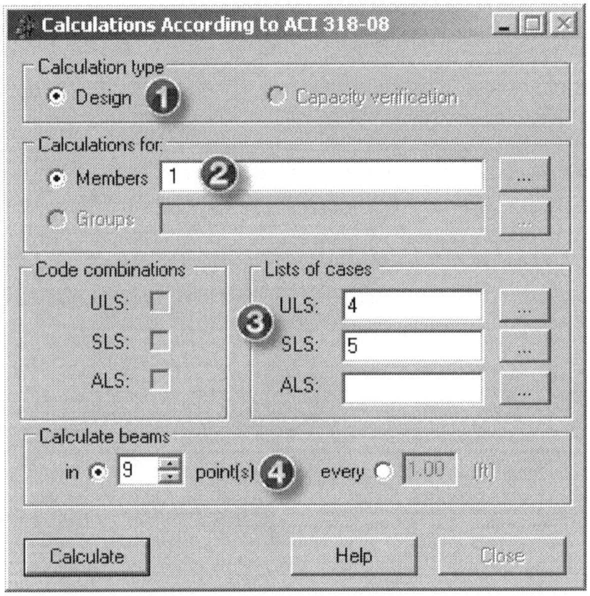

1- Calculation Type: The only option available currently is "Design".

2- Calculations for: (Groups is not available) Select the members for which you want to run design calculations. Only the members listed here will be designed. You can use the ellipsis button to the right to access the member selection dialog.

3- List of cases: Only the ULS and SLS cases listed here will be used in member design. Be sure that you have carefully checked which cases/combinations you want to have considered. Robot will ignore any cases/combinations not listed here. Use the ellipsis button to access the cases/combinations selection dialog

4- Calculate beams: Determine the number of design sections you want to have considered in your beam/column. Be warned that only the forces at the sections specified here will be checked and designed. It is highly recommended to use a larger number of design sections or a smaller increment of "every x feet". Use your engineering judgment to determine the proper number of design sections to be considered based on the variation of member forces along the length of the member.

After pressing Calculate in this dialog, Robot will perform the member design calculations to determine required area of reinforcing steel. Then will present the RC Member Calculations Report indicating the results:

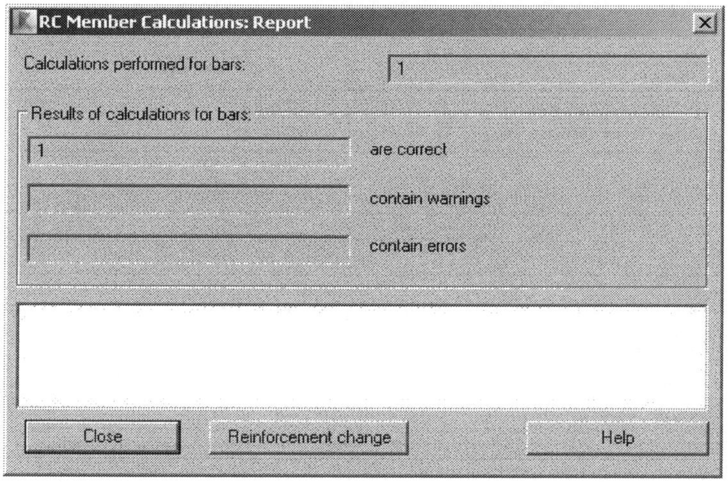

This dialog lists the bars designed, the members which are designed successfully with no issues (are correct), the members which have some warnings associated (usually deflection check failed), or members which contain errors of some kind. You will use the remarks column in the reinforcement results table to understand more about the errors and warnings.

NOTE: IF YOU HAVE WARNINGS AND THE REMARKS SEEM TO SAY "NO PROBLEM" THEN CHECK "REINFORCEMENT CHANGE" TO REVIEW MEMBERS WHICH HAVE FAILED FOR ADMISSIBLE DEFLECTION.

THE REQUIRED REINFORCING RESULTS TABLE

After reviewing the calculation results in the RC members calculation report dialog, the results table will be populated with the calculation results for each member at each design section along its length as shown here:

Required Member Reinforcement

Bar/Position (ft)	Design moment My (kip-ft)	Bottom required reinforcement (My) (in2)	Required reinforcement ratio (%)	Bottom provided reinforcement (My) (in2)	Provided reinforcement ratio (%)	Bottom reinforcement - distribution (My)	Min. reinforcement area (in2)	Remarks
1								
1/ 0.0	0.0	0.0	0.0	0.0	0.0	-	1.68	Calculations OK
1/ 5.00	255.00	2.11	0.42	3.00	0.60	3#9	1.68	Calculations OK
1/ 10.00	340.00	2.85	0.57	3.00	0.60	3#9	1.68	Calculations OK
1/ 15.00	255.00	2.11	0.42	3.00	0.60	3#9	1.68	Calculations OK
1/ 20.00	0.0	0.0	0.0	0.0	0.0	-	1.68	Calculations OK

Beams | Beams - Info | Columns | Columns - Info | General

This is just like any other results table in Robot with exception of the tabs available at the bottom of the dialog. You can use right-click>Table Columns...or **View>Table Columns...** to adjust the information shown to you in the table, you can use filtering as well. The Remarks column is particularly important in terms of catching any messages Robot has for you concerning the design checks of the member.

Recognizing the limitations of the required reinforcing design routine, you can take this farther by exporting the results table to a spreadsheet (right-click>Convert to CSV format), then apply additional checks for reinforcement configuration and other necessary checks to finish your design and verify bar anchorages and spacings.

REINFORCEMENT CHANGE

In the case that a member fails the deflection criteria, you can use the reinforcement change dialog to adjust the amount of reinforcing in the member to reduce the deflection of the member. Your other option, of course, is to go back and manually change the member size to allow a larger moment of inertia thereby reducing the overall deflection. If you receive warnings on members, you can select the "Reinforcement Change" button to open the following dialog to see the failing members and also adjust the reinforcement for individual members:

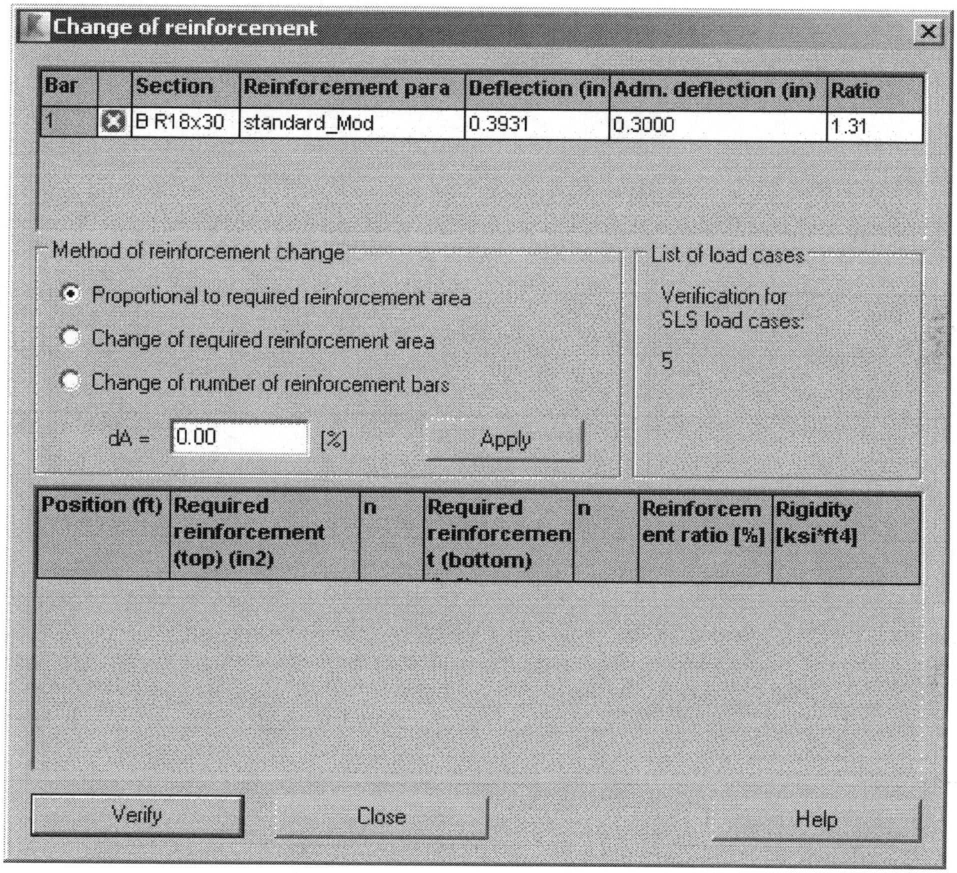

You'll notice that the B R18x30 has a small deflection problem where it is currently exceeding its specified allowable deflection of 0.3". Selecting the bar number will show a list of design sections in the table below:

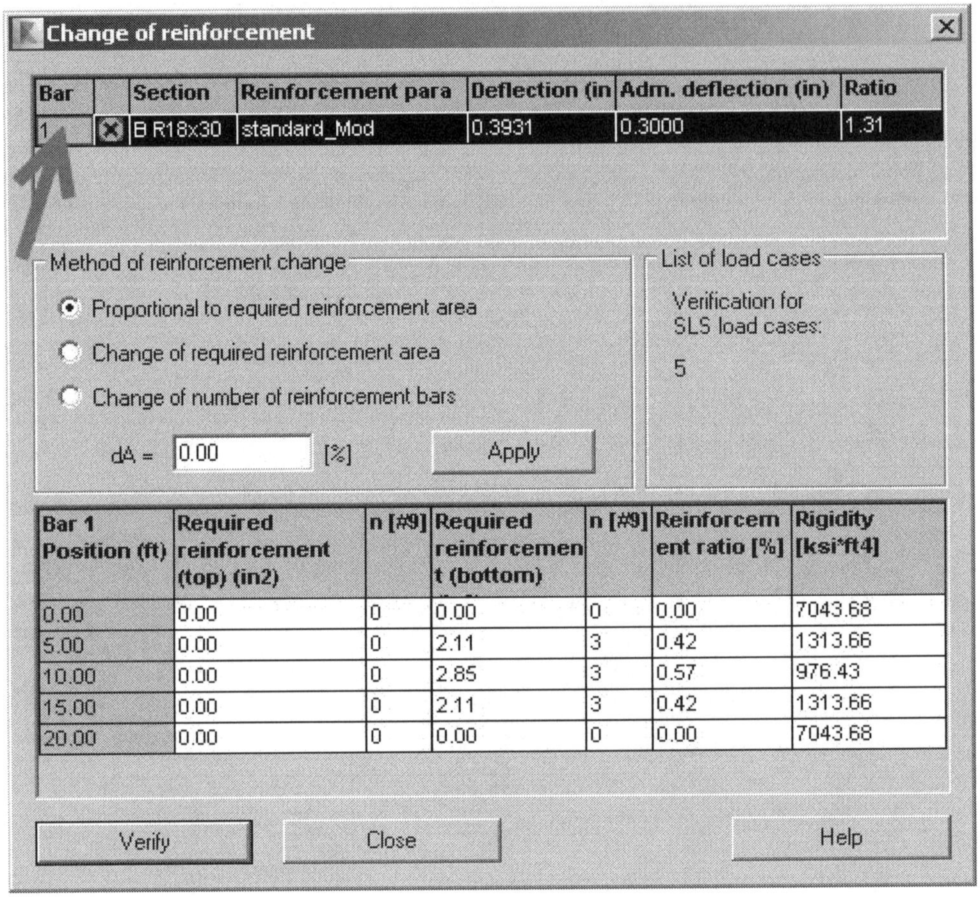

Once you have selected the member for which you want to adjust the reinforcement, you can specify the amount of change in the "Method of reinforcement change" section. Then use the "Apply" button to see the effect in the table below.

-Proportional to required reinforcement area: Add a percentage of the already calculated reinforcement area. If you specify 25% then 25% of the current area will be added to the total area at each design section.

-Change of required reinforcement area: Add a specific area of steel to the current area at each section.

-Change number of reinforcement bars: Specify instead a number of bars to add to the total number of bars already calculated.

Use the "Verify" button at the bottom to see the effect of additional reinforcement on the calculated deflection. Once finished with reinforcement changes, close this dialog and close the calculation report, then continue with your design and documentation.

Provided Reinforcing

The Provided Reinforcing module is a powerful feature for individual concrete member design where specific rebar configurations, spacings, anchorage, construction bars, and bar cutoffs are all considered and calculated. Additionally, member diagrams showing plots of the required capacity vs. the provided capacity are available in this module as well as full 3D visualization of the member reinforcement, member reinforcement drawings, and member calculation notes.

Organizationally, Provided Reinforcing is a separate module which originally served as an independent tool to perform individual member design based on user settings for the member sizes and spans as well as user input loads. The Provided Reinforcement module can also take members directly from a Robot model thereby accessing member forces, load combinations, span length, and member size all directly from the model. The fact that the module is separate does make it feel slightly strange to work with in the sense that, while "linked" you do have to provide some management of that link and occasionally manually update something if the original model changes. The member with which you are working in Provided Reinforcement is effectively a copy of the element in the model and lives separately from it.

I believe that the most difficult part of this module is understanding the process of moving elements into the module. The topic of Provided Reinforcing and all of the settings are a large topic and will be left for a more advanced text. We will cover the process of moving elements into the module from the model and methods of interacting with the module and let you explore on your own from there.

Transferring elements into the provided reinforcing module

Once you have modeled and analyzed a concrete structure, and maybe even done a few runs of theoretical reinforcing to get an idea of whether the member sizes will work, you can push the members (along with their results) over into the Provided Reinforcing module. In the example below, we will select member 11 for transfer.

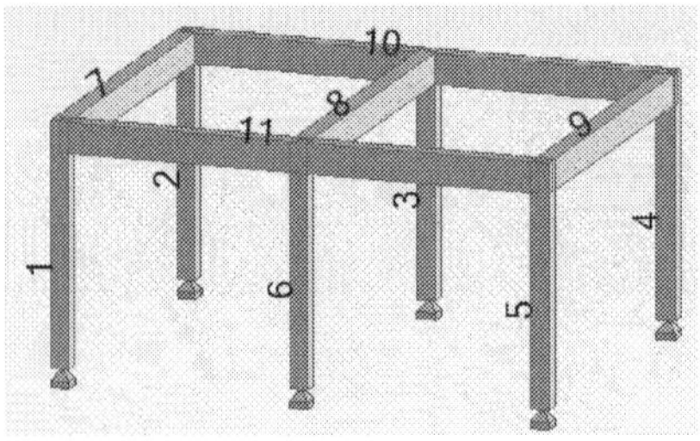

Once member 11 is selected, press the Provided Reinforcement button 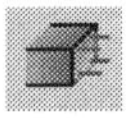 on the modeling toolbar along the right hand side of the screen. You may also select the Provided Reinforcing layout or from the menu selected **Design>Provided Reinforcing of RC Elements**. Any of these methods will start the process of transferring this member to the Provided Reinforcing module. You may select multiple elements at the same time or one at a time.

Once the transfer has started, you will be presented with a dialog where you can configure how you want the member to be treated by the Provided Reinforcing module. The dialog below is specific to beams as member 11 is a beam type element. Other versions of this dialog will be shown depending on which element or set of elements you have selected for transfer to the Provided Reinforcement module. You will be setting the parameters for each type of element (beam, column, footing, etc.) being brought in, not for each individual element.

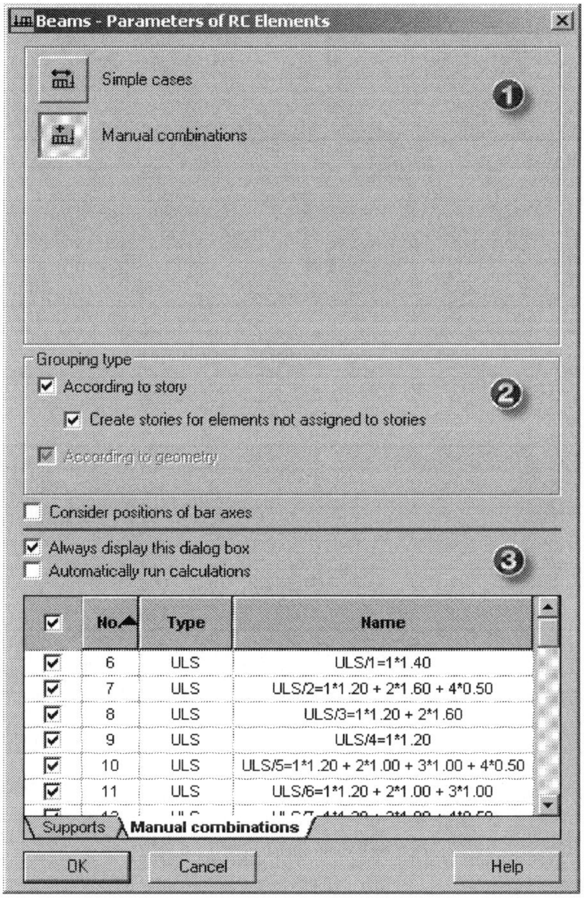

1- Select which method of calculating member forces to use. Simple cases will bring in load cases (end moments and loads distributed along the member) and will apply load combinations to determine design forces. Manual combinations will allow you to select from your previously created manual load combinations to determine design forces for the member.

2- Grouping type: Decide if you would like elements grouped in the object browser (see below)

a. For beams:

i. According to story: Beams will be designed individually and organized by their assigned story in the browser

b. For columns:

i. According to story: Columns will be designed individually and organized by their assigned story in the browser

ii. According to geometry: Columns with identical geometry and supports will be grouped together in the browser **and designed together!** The columns will all be designed for the worst case design forces among them and have identical reinforcing.

iii. Column chain: Columns will be designed individually and yet organized in chains of vertical columns in the browser.

c. Consider positions of Bar Axes: This function will preserve the relative position of bar objects (considering their offsets). Here is an example with the same set of two beams where one is

i. Considering position:

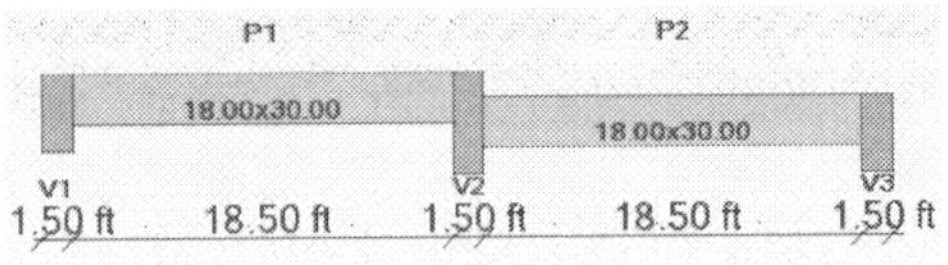

ii. Not Considering position:

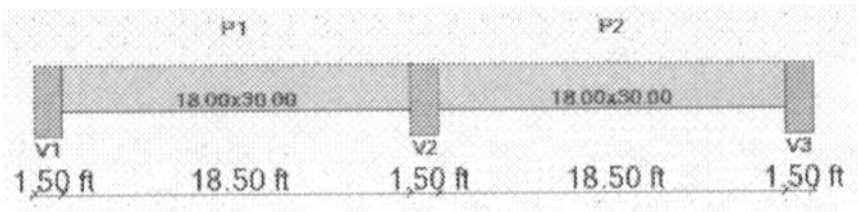

3- In this portion of the dialog you can select which load combinations will be used in design of this type of member. In the supports tab for beams, you will have the ability to disregard certain supports. The support has the effect of causing Robot to use design forces at the face of the support. Removing a support will cause Robot to use the forces at the center of the support, but will lead to issues with rebar anchorage where Robot will believe that there is not enough length to develop the bar between the end of the member and the location of the maximum design load.

The Provided Reinforcing Interface

As we begin to look at the Provided Reinforcing module interface, remember that this module is intended to design individual elements. The method of interacting with the module is distinctly modulated by that fact. Once you have completed the transfer process of elements into the module, your interface may look something like this:

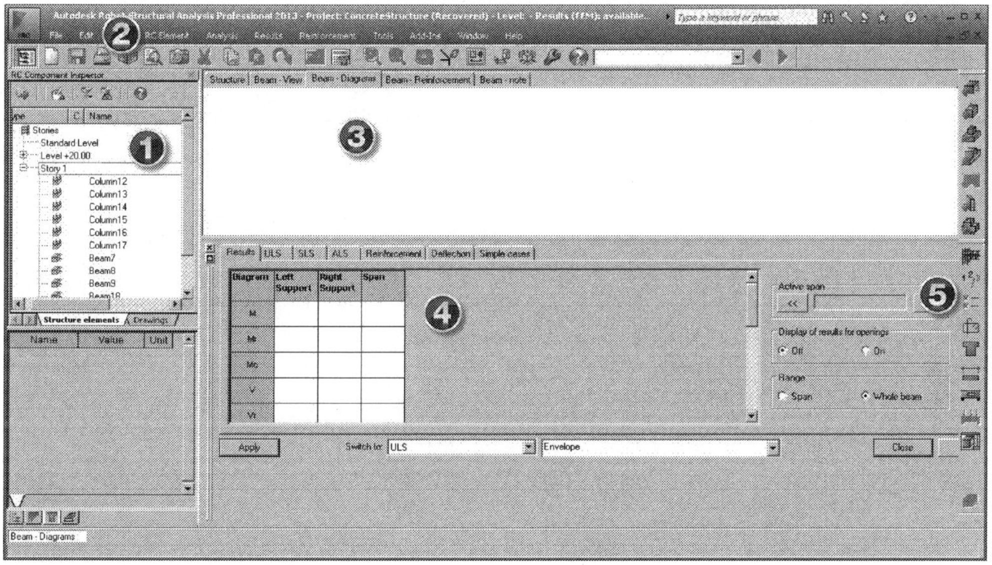

Parts of the interface:

1- The element browser (RC Component Inspector)

a. 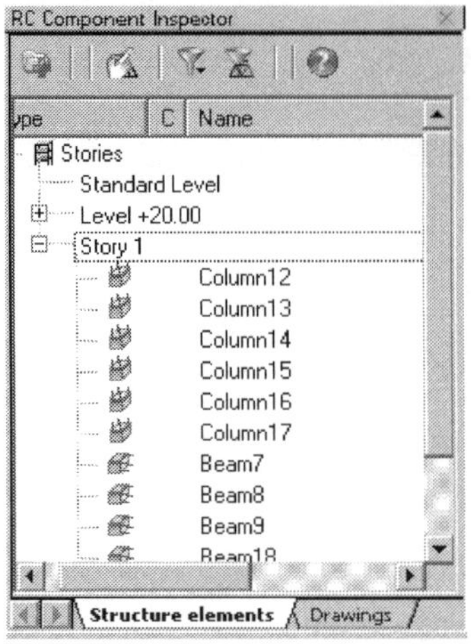

b. Each element transferred into the Provided Reinforcing module will be listed here in the groupings you specified at the time of transfer. Once you have created drawings for elements, the drawings will populate in the drawings tab at the bottom of the dialog.

2- The menu bar: (Note that the menu items are distinctly different from the standard robot menu items. All items on this menu are particular to RC Element design:

a. 

3- Display area: In this area there are several tabs across the top of the display area to allow you access to different options of viewing the current member:

a. Structure: Gives you a view back into the Robot model. The menus will change back. You can select additional members here for import to the module or review moment and shear diagrams etc.

b. Element View: A view dedicated to the geometry of the current element. (The current element is the one selected in the RC Component Inspector on the left hand side of the screen. Double-click on an element to select it.)

c. Results View: A view dedicated to either interaction diagrams or moment and shear diagrams with provided capacity plots.

d. Reinforcement View: Once calculations have been run, reinforcement will be shown in 3D in this view. You can select individual bars to highlight them in the corresponding table for further investigation.

e. Note View: Calculation notes for the currently selected element. (Available once calculations have been run.)

4- Supporting information or settings area: This area of the interface will change to show information pertaining to the currently selected view tab at the top.

5- Toolbar access to configuration settings: All buttons here are simply quick access to options available in the menus. A note about some of the buttons on this toolbar (and their counterparts elsewhere in the interface)

a. Buttons at the top: 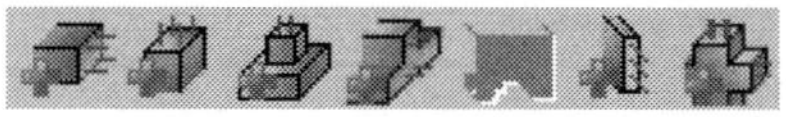 are intended for adding new stand-alone members to the Provided Reinforcing project. You will not need to use these if you have transferred elements directly from the Robot model. You can add members in this way which are custom and not related to the Robot model but I believe there is an issue in the interface where the span geometry is fixed and so it is not actually possible to add custom members in this way. If you want to work with individual element designs separate from a Robot model, start a new Provided Reinforcing project and add all necessary members there.

b. The Typical Reinforcing Button: 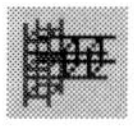 is a separate option for member reinforcing. You can use this to put a standard (you configure) reinforcing pattern into the element. Once you have done this, you can use the Provided Reinforcing module to check the rebar you have provided for the element, view capacity diagrams, view the reinforcing in 3D and also generate calculation notes and drawings. You will either use typical reinforcing or let Robot calculate it for you, but not both. (This is, other than the situation where you want to try some typical reinforcing first and see if it meets capacity and if not, you can, of course, allow Robot to redesign the reinforcing for you.) Once you have used the typical reinforcing option, you will want to freeze the reinforcing while running calculations on the element.

c. Of the remaining buttons, the following buttons are particular to each element and control the calculation and configuration of reinforcing in the element: 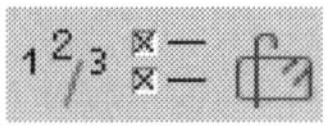 You will configure each of these options for each member prior to running calculations.

Using the Provided Reinforcing Module

Likely the best way to illustrate how to use this module in a general sense is to go through an example demonstrating the functionality.

Beam Example:

The following bent exists in the Robot model and will be transferred into the provided reinforcing module:

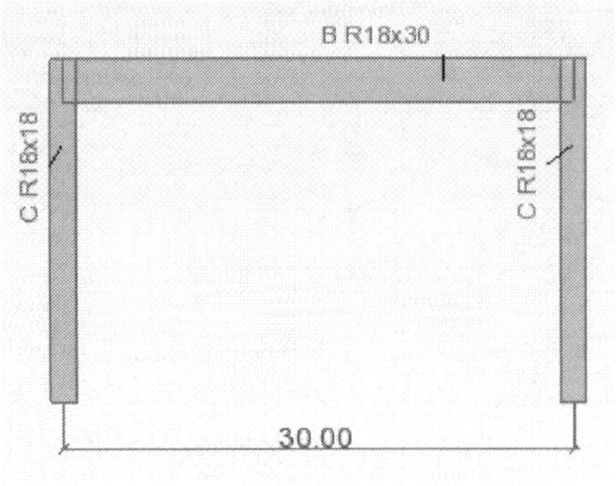

Once the transfer process is initiated (elements selected and Provided Reinforcing button 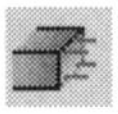 is pressed) we will first configure import options for columns:

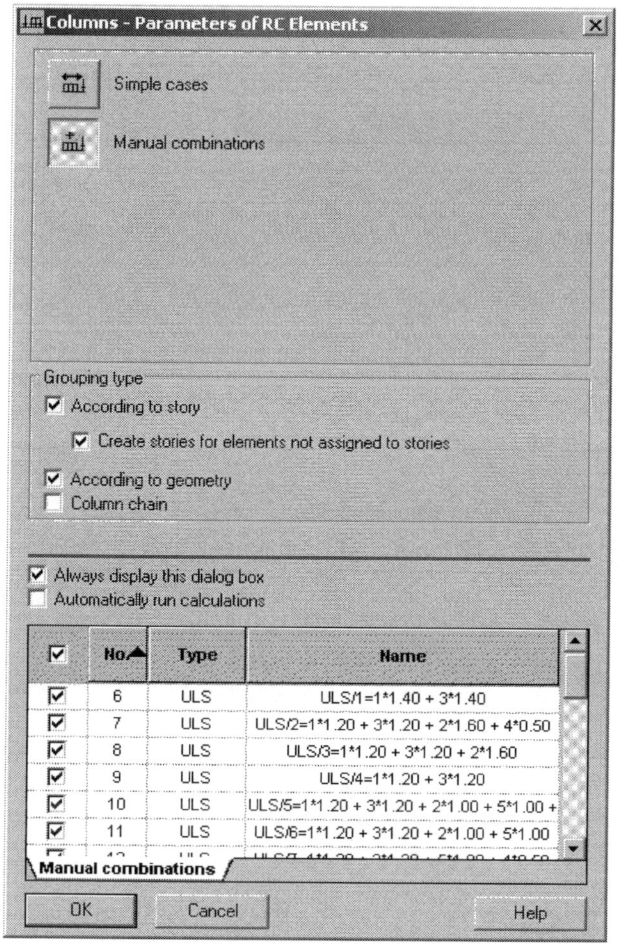

Selecting Manual Combinations, Grouping by Story and Grouping by Geometry as well as making sure that all of our relevant combinations (generated in Robot) are selected. Pressing "OK" will let Robot import the columns and then show you the Beam properties dialog. Similar settings in this dialog will import the beam:

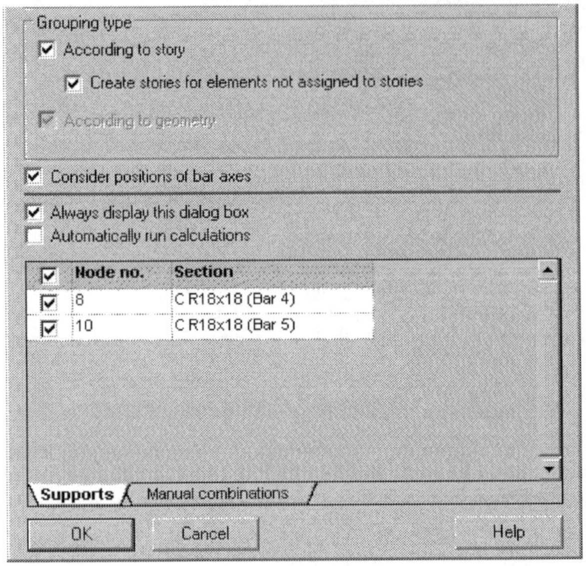

Checking not only the Manual Combinations, but also the Supports tab to ensure that the supports are properly selected. Pressing "OK" will bring you into the Provided Reinforcing module interface: (If the beam is not currently selected on entering the Provided Reinforcing module, find it on the RC Component Inspector browser on the left hand side and double-click it)

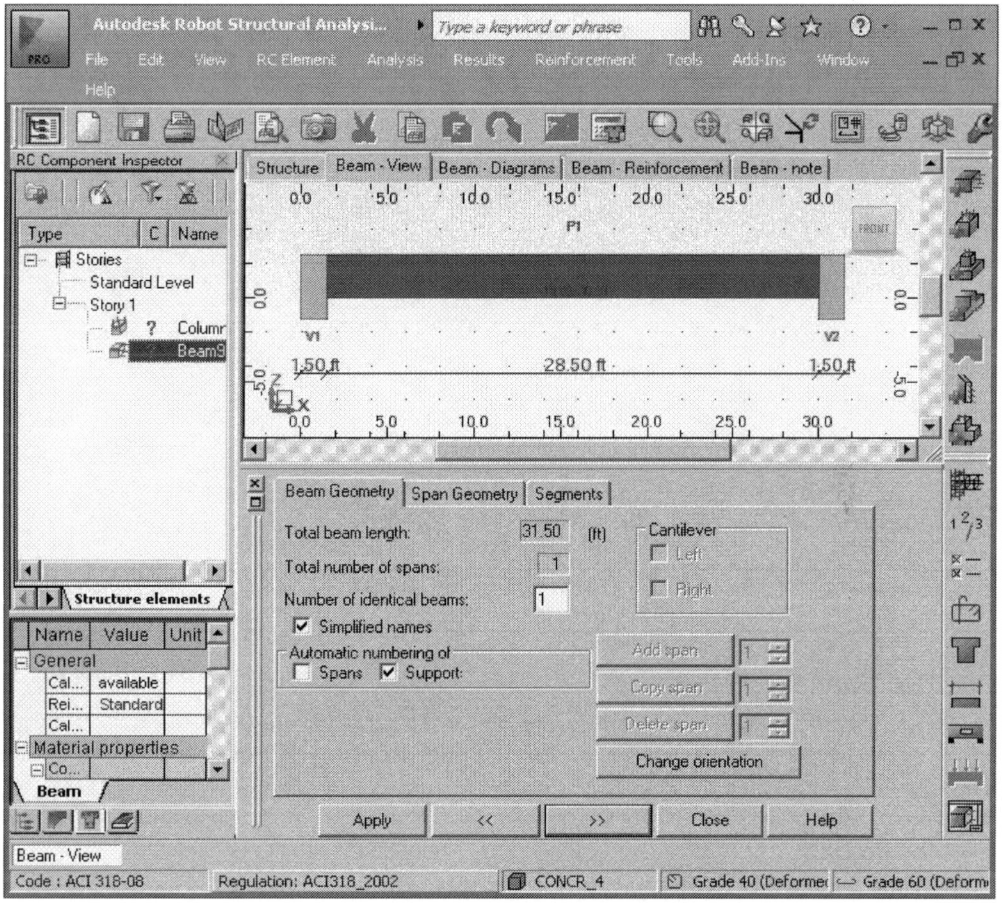

In the Beam View, you can review the span information though you cannot modify it as this member has been imported from and is linked with the element in the Robot model.

The first thing we will do is configure the settings for member calculation options and reinforcing parameters. **Selecting Analysis>Calculation Options...** we are presented with the Calculation Options dialog for ACI 318-08:

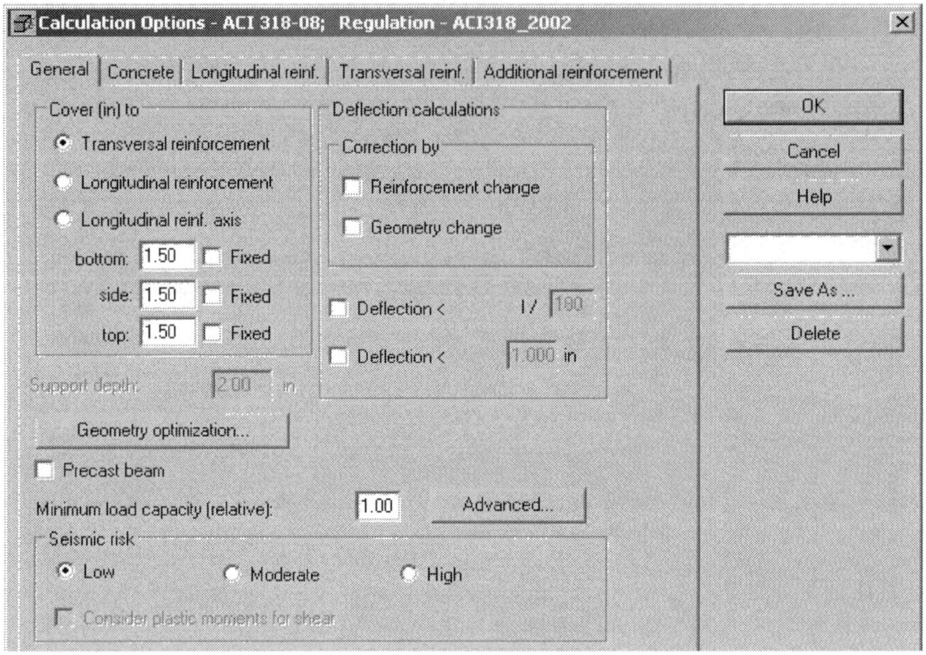

We will not be considering deflection in this example. The Concrete tab will be used to specify concrete strength, the remaining tabs will be used to specify reinforcing grades and allowable bar sizes.

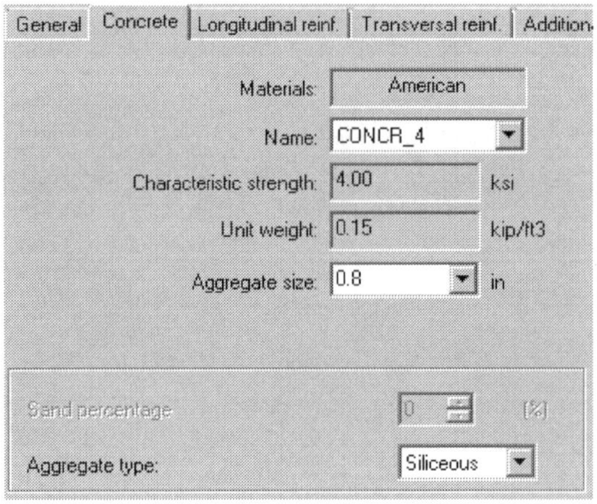

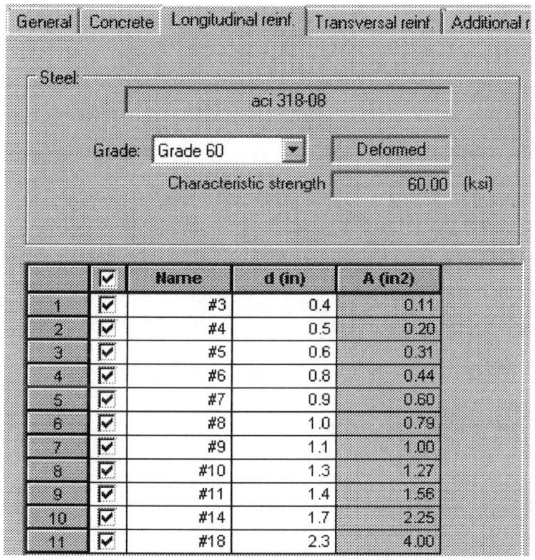

These options will only apply to the current member under design. You can use the template option on the right hand side of this dialog to save the current settings as a template:

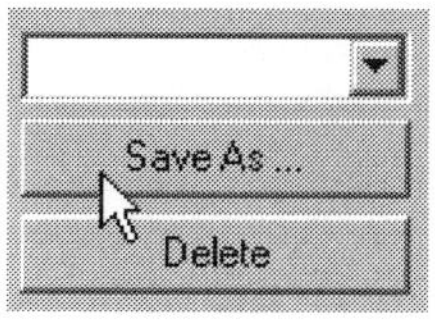

Just press Save As… and enter a new name in the subsequent dialog.

Next we will configure the reinforcing pattern options from the menu **Analysis>Reinforcement Pattern...**

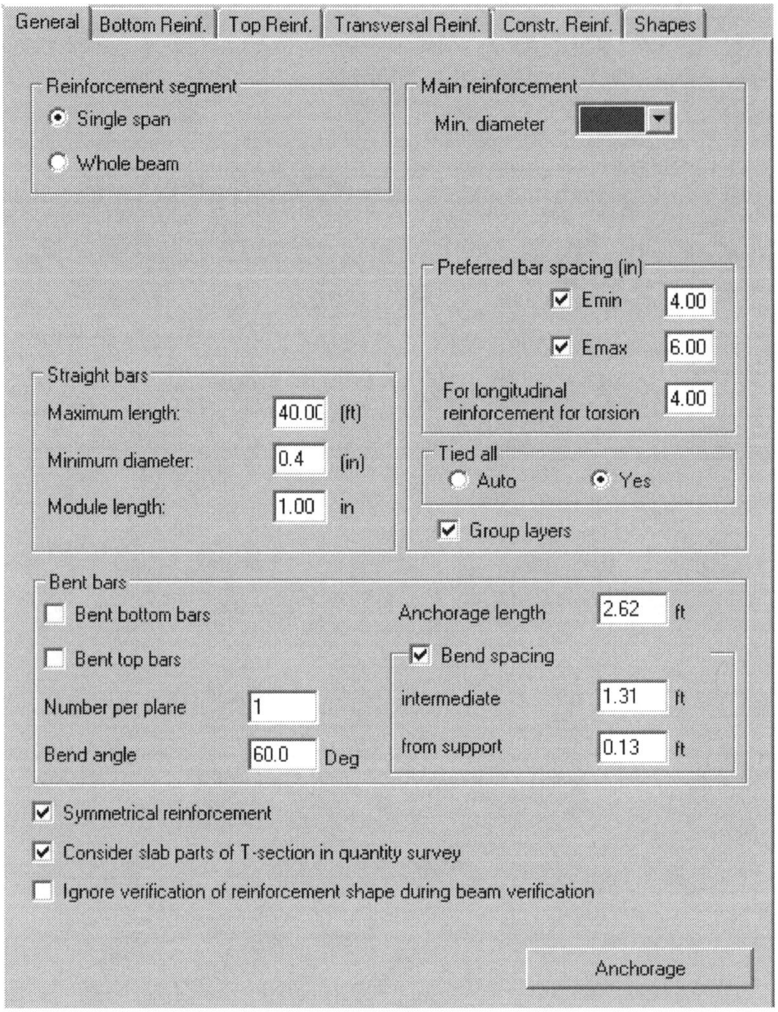

As you can quickly see, there are many, many options to control how Robot will configure the reinforcing for this member. On the Bottom and Top Reinforcing tabs, we will only allow one layer of reinforcing:

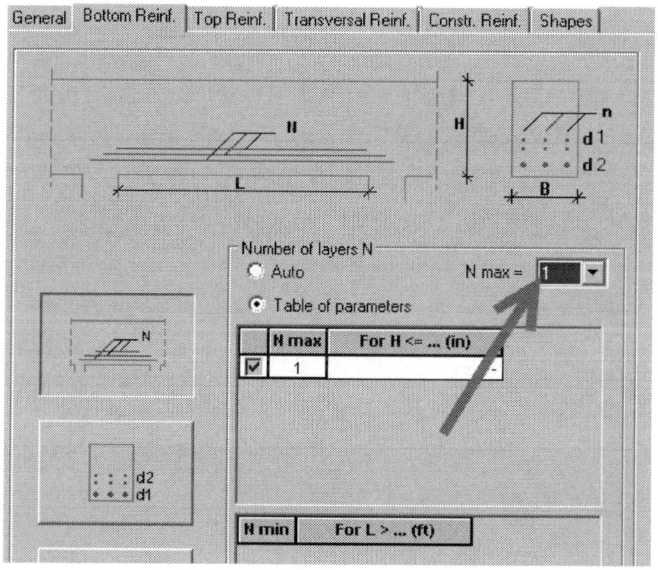

Then on the Transverse Reinforcing tab we will use a constant segment spacing for stirrups:

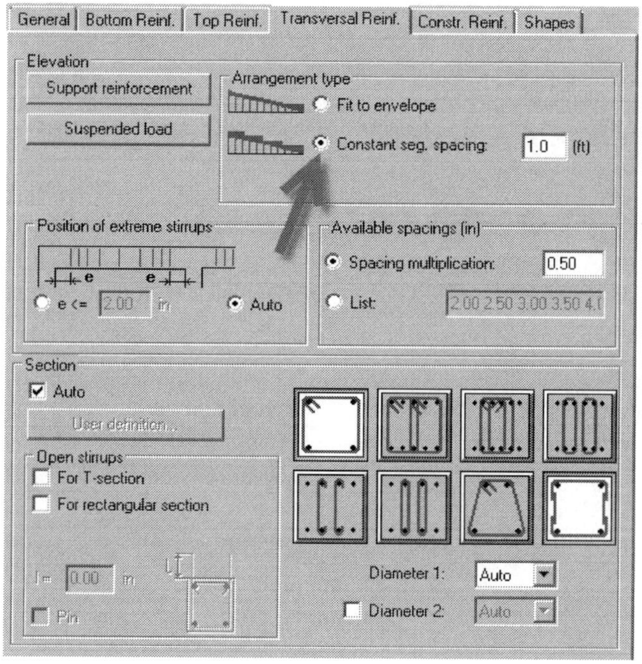

Once we have finished this configuration we can run calculations either from the menu **Analysis>Calculations...** or by pressing the calculations button 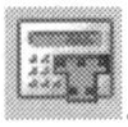. This will launch the calculations dialog where you can opt to simply analyze the existing reinforcing by selecting "Keep existing reinforcement" and also opt to generate the drawing for the element by selecting "Generate drawings for calculated element". This will only make the calculation process last longer and probably is not warranted at this point as we still need to check the member and reinforcing that Robot produces.

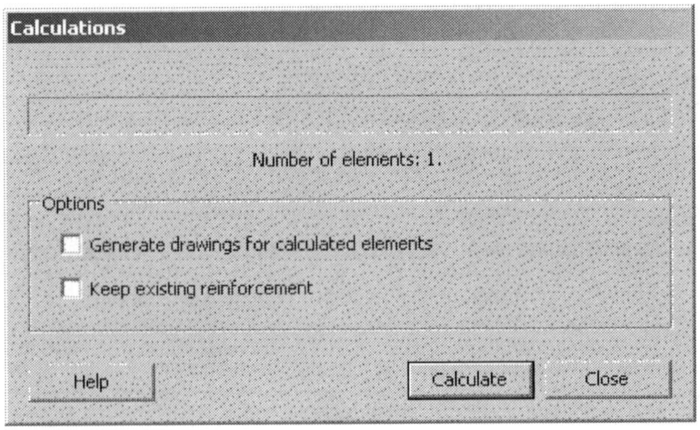

Pressing "Calculate" will begin the process. Once complete, select the Beam-Diagrams View:

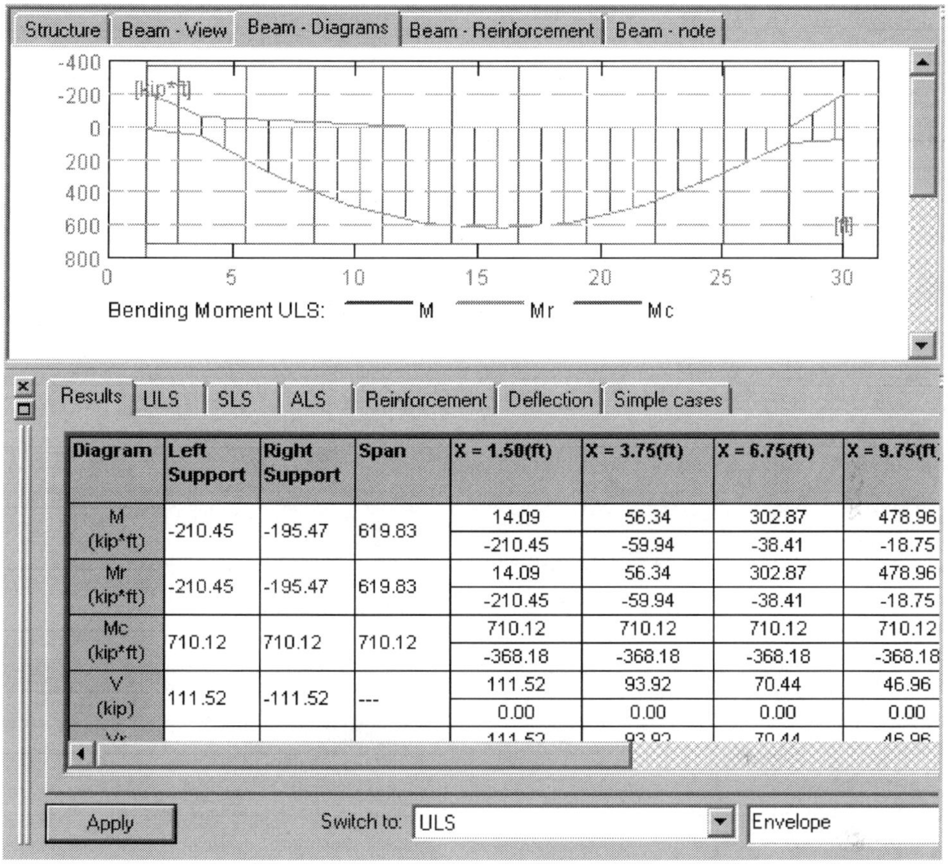

Diagram	Left Support	Right Support	Span	X = 1.50(ft)	X = 3.75(ft)	X = 6.75(ft)	X = 9.75(ft
M (kip*ft)	-210.45	-195.47	619.83	14.09	56.34	302.87	478.96
				-210.45	-59.94	-38.41	-18.75
Mr (kip*ft)	-210.45	-195.47	619.83	14.09	56.34	302.87	478.96
				-210.45	-59.94	-38.41	-18.75
Mc (kip*ft)	710.12	710.12	710.12	710.12	710.12	710.12	710.12
				-368.18	-368.18	-368.18	-368.18
V (kip)	111.52	-111.52	---	111.52	93.92	70.44	46.96
				0.00	0.00	0.00	0.00

Here you can fully investigate the design moment envelope in both diagram and table form. Hovering over the diagram with your mouse will also give you point by point resolution on the required and provided capacities:

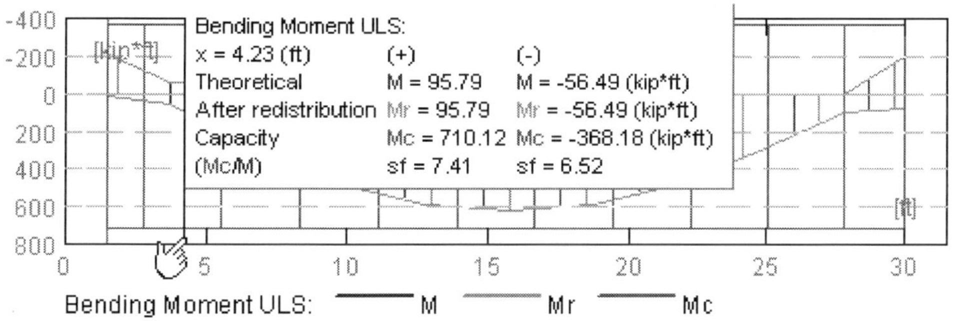

As you can see, the reinforcing Robot has selected completely encompasses the required moment diagram. You can also review shear and required minimum reinforcing diagrams. Use the tabs at the bottom to reveal additional options for diagram display:

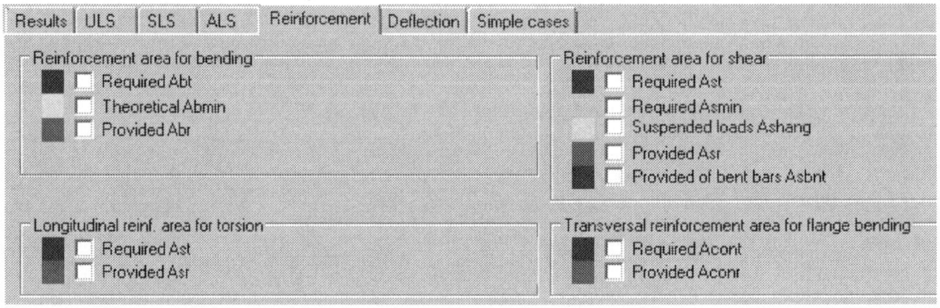

After taking a look at the beam diagrams, let's take a look at the actual reinforcing selected by Robot on the Beam-Reinforcing tab:

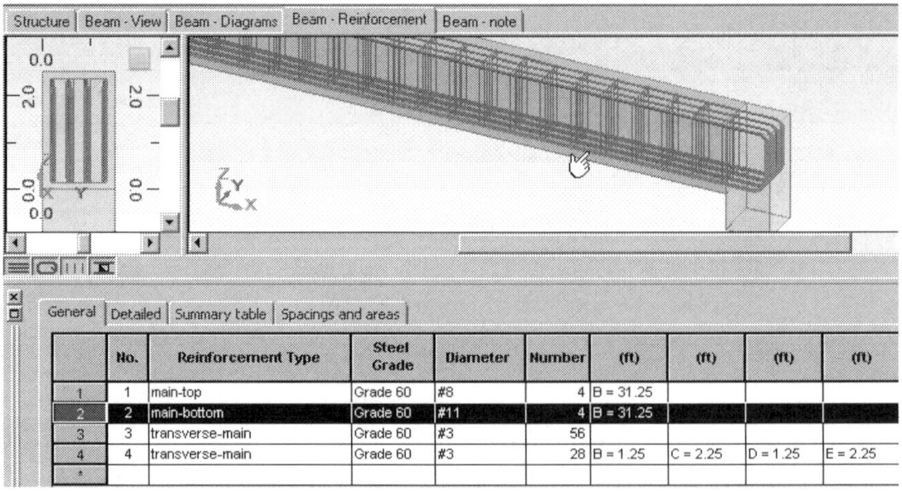

	No.	Reinforcement Type	Steel Grade	Diameter	Number	(ft)	(ft)	(ft)	(ft)
1	1	main-top	Grade 60	#8	4	B = 31.25			
2	2	main-bottom	Grade 60	#11	4	B = 31.25			
3	3	transverse-main	Grade 60	#3	56				
4	4	transverse-main	Grade 60	#3	28	B = 1.25	C = 2.25	D = 1.25	E = 2.25
*									

You get two great views of the reinforcing in 3D which you can orbit, pan, etc. Selecting any bar will highlight the corresponding row in the reinforcing table. Select the "Summary table" tab to view stirrup spacings.

Select the Beam-note view to view a summary of the design including material information, all beam diagrams, and member forces at design sections.

Once you are satisfied with the configuration of reinforcing you can select the drawings option to generate an annotated drawing of this beam and its reinforcing.

Select from the menu **Results>Drawings...** or press the drawings button 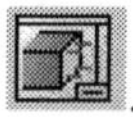.

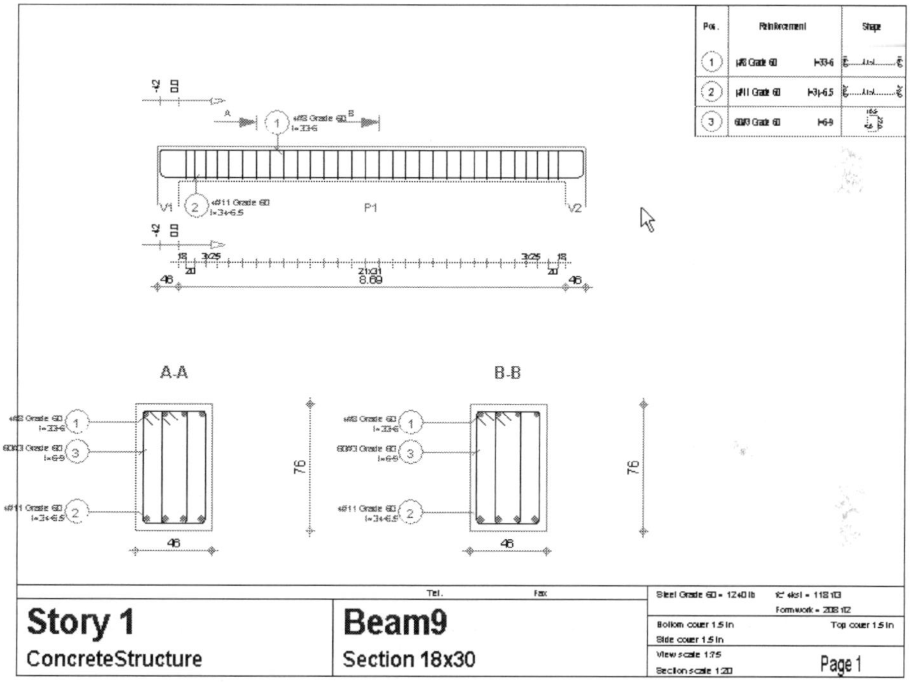

This is a great presentation of the reinforcing and a huge time-saver. These drawings cannot be directly included in your printout composition but can be printed to include with your project documentation.

Wrapping up Provided Reinforcing

As you can see, the Provided Reinforcing module is a very powerful and robust tool for designing reinforcing for individual members. The preceding was, admittedly, an incredibly brief introduction to the Provided Reinforcing module. The number of settings and configuration options to control how Robot handles designing the reinforcing as well as options to redesign the section to accommodate service conditions are numerous and requires an in depth discussion which is currently beyond the scope of this book. I hope that this gives you enough of an introduction to the module and the basic workflow that you will continue exploring on your own with the various options and testing Robot's results against beams you may have previously calculated to get a better feel for what Robot is doing and how it handles the different settings.

Chapter 14 - Trouble-shooting

In the course of using Robot and running analyses, you may run into issues which vex you. Here are some of the techniques you can use to try to get things running smoothly again.

Robot interface issues:

Robot can sometimes get lost and it helps to remove any user configuration files and let Robot rebuild them from scratch. This is especially useful if you are in a situation where you are unable to enter the Provided Reinforcing module. In the case you think you may have an interface issue, go to the following locations and delete these folders where personalization settings are located (you can make backups before deleting them). If this fixes the issue and you need your customizations you can move them selectively until you find the issue.

C:\Users\<username>\AppData\Roaming\Autodesk\Autodesk Robot Structural Analysis 2016\CfgUsrv

C:\Users\<username>\AppData\Roaming\Autodesk\Structural\Common Data\

Verification errors

Calculation errors can be some of the most frustrating to deal with. Sometimes they will pop up as Robot is assembling the calculation model forcing you to go searching for offending members. Sometimes they will pop up during the actual solution of the matrices in the form of instabilities.

Locating offending members in error list:

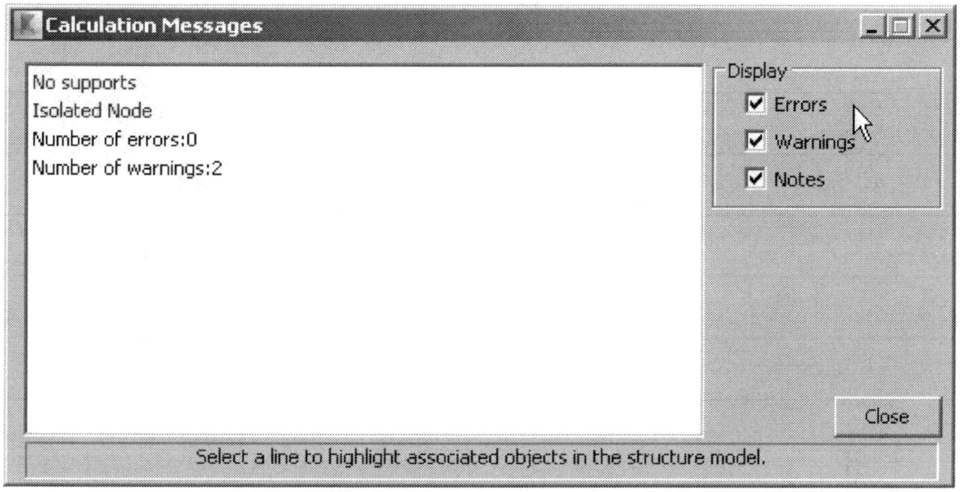

Selecting a row with an error or warning will select the elements in the Robot environment. Once that happens you can use an awesome tool called "Edit in New Window" to get a view with only the selected elements shown. This will give you the ability to immediately zero in on where the elements are in your model and let you get a leg up on resolving any issues with those elements.

"Edit in New Window" can be accessed from the Edit menu **Edit>Substructure Modification>Edit in New Window** or by pressing the "Edit in New Window" button in the selection toolbar: 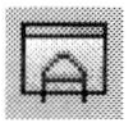

Resolving Instabilities:

Unfortunately resolving instabilities is a deep and winding topic and there is no simple answer to locating nor resolving instability issues in a model. It takes time and patience, but it is possible to resolve them.

BEWARE: Robot will present you a dialog which will allow you to continue the solution. That dialog looks like this:

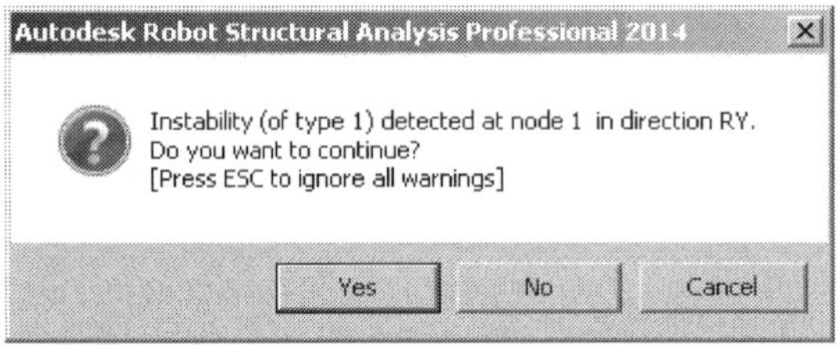

Pressing "Yes" or "Esc" will cause Robot to introduce small stiffnesses to allow the model to be solved. In other words, Robot found a zero on the diagonal and introduced a slight stiffness which negates the zero; however, the model will very likely be incorrect in this state. Which degrees of freedom Robot chooses and why it chooses them is not immediately clear. **The first place to begin looking at your model is in the displacements to see if there are any drastically unexpected displacements.** Also be watchful for unexpected force diagrams and investigate anything that doesn't seem right. These are indications that there is something wrong in your model. Helpfully, when instabilities are detected and ignored during calculations, the Calculation Messages dialog will show, letting you know what issues occurred during calculations so that you can go back and track them down:

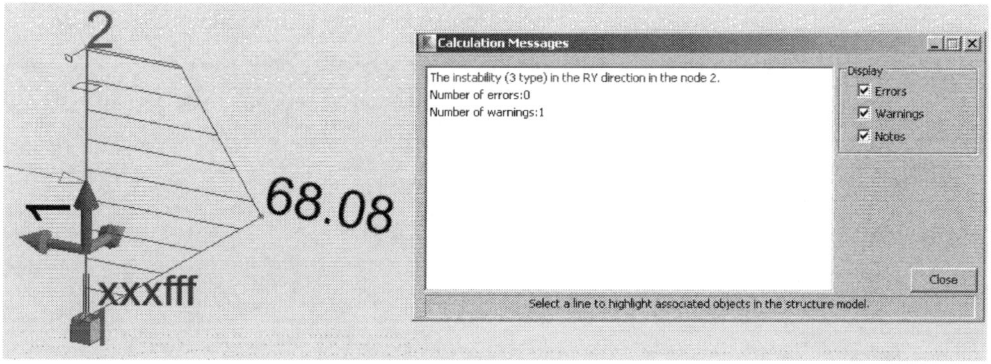

In this model above I've released too many degrees of freedom (Note that the column is rotationally free at the fixed support and the top has no restraint in the direction of loading). The model solved when I ignored the warnings and it looks like I have a real moment result along the beam; however, take a look at the displacements for node 2 and you'll see immediately where the problem is:

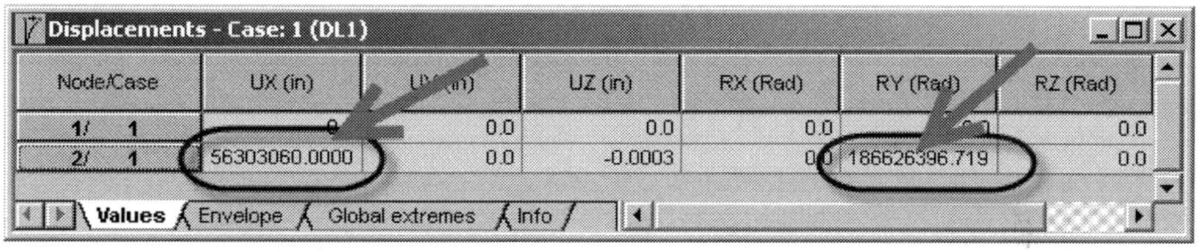

The type of instability may differ, the degree of freedom that Robot identifies as the culprit may vary but you should always do a gut-check of the displacements to be sure that the model is behaving as you expect.

By all means, it is best to work to resolve any instability issues with the model so that it performs the way you expect it to.

Instabilities of Type 1 and 2:

These instabilities are typically the result of "mechanisms" in your model. This typically indicates issues where members are improperly restrained or the supports are insufficient to restrain the elements. You can think of the following as a visualization of the connection of elements to each other or to support locations. In this graphic the end of the member releases act at a point between the actual end of the member and the supports. This is a visualization ONLY. There is no geometric distinction between where member end releases are applied and the member end supports (if any). I just want to give you a method to visualize what is happening with member releases. In this example, the major bending moment has been released at the end of a member supported by a pinned support. Here is what happens mathematically, demonstrated visually:

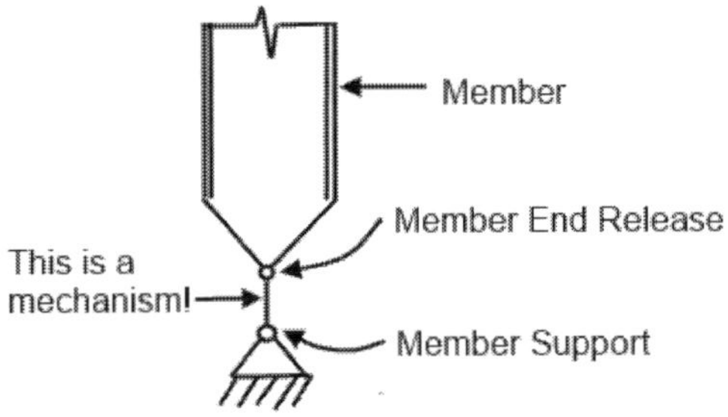

This can be the case for any of the member releases combined with lack of support fixity. One area which is particularly an issue is in attempting to release bending moments in beams or columns. If you release the rotation about x (along the member axis) at both ends you will have a mathematical instability. The visualization of that is something like a straw on a wire where the straw can spin freely around the wire. Effectively this is what happens when you release both end rotations about X (Rx).

Instabilities of Type 3:

This type of instability is typically the result of large differences in the element stiffnesses in the matrix being solved though they can sometimes be thrown for mechanical instabilities of the model as well.

When you are experiencing issues with the stability of a model, it can help to understand the range of member stiffness in the matrix which Robot is attempting to solve. Vast differences in member stiffness can result in serious issues in attempting to solve the system of equations due to machine precision inherent in all computer based matrix solvers. To better understand the matrix that Robot is attempting to solve, take a look at the calculation notes and find the stiffness range:

Stiffness matrix diagonal elements
Min/Max after decomposition: 5.960464e-008 9.481197e+008

If the difference between these numbers is greater than 1e11 or 1e12 Robot will report an instability of Type 3. Sometimes this vast difference in stiffnesses can come from incredibly short members (stiffness is partially a function of member length). You can use the intersect technique below under "Find Wicked Short Members". Eliminating the causes of the short members (usually nodes too close) will often resolve this instability.

If you have looked for and cannot find any particularly short members, the next step in diagnosing this issue is to prove that the assigned member sections are not at fault. Take a uniform section size and assign it to all members of the model (eliminates member cross section as the source of the issue). If you find that this resolves the issue then you likely have an issue with the assigned sections though this is far less likely than an issue of with the relative lengths of the elements (finding wicked short members).

Find wicked short members

Very, very short members can sometimes cause issues with relative stiffnesses of model elements. (Recall that member stiffness is often a function of the member length). Locating these very short members and resolving the conditions which gave rise to them, can help resolve some instability issues.

To locate short members, working with a copy of the model, use the Intersect tool to divide the model at every node. (Select everything in the model and select **Edit>Intersect**). Once this has been accomplished, you can view the bars table (**View>Tables...** and check "Bars") and order by member length by double-clicking the length heading in the values tab. Any members which are unusually short can be the source of issues. Wherever you find one, select the member/nodes and use the

"Edit in New Window" function to select and locate it in the model. You will likely find that there are several nodes or members meeting at single location very, very close to each other and are in need of some correction.

Merge nodes:

You may also run into issues where you aren't even aware that members aren't connected. Using tools like **Edit>Correct...** detailed earlier in this text. Will locate nodes which are very close (within a tolerance you specify) and merge them together. As you go through this exercise, you will want to exercise care that the model isn't getting worse. Merging nodes will result in either one or the other of two nodes being selected as the final node, or a point halfway between the nodes selected as the final location. For a simple model this may be fine, but in a complex model you may find that this causes issues where panels are no longer flat and connection issues in the model grow worse. In cases like this, you may opt for tools like "Detailed Correct" which will offer more precise options for pushing a model into shape and making sure that things line up nicely.

Repair Mode

In the case of models exhibiting issues which are more obscure, Robot's "Repair Mode" can sometimes resolve pesky issues. In the Robot Open File dialog, notice that off to the right is the "Repair Mode" checkbox shown here:

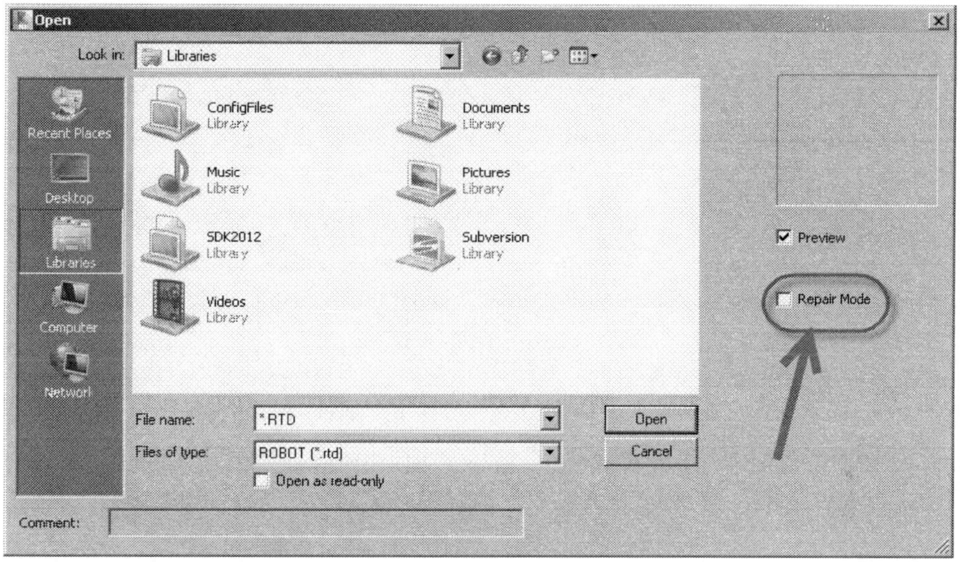

Repair mode (in general):

-Is similar in principle to the "Recover" command in AutoCAD®

-Checks, audits, and attempts to correct the internal consistency of the model.

-Compacts the data file (renders externally saved results unusable, you will have to recalculate the model).

-Is intended as a file recovery option in cases where nothing else seems to work.

Some of the types of problems Repair Mode can help with are the following, gleaned from Robot forums:

Problems like not being able to see nodes, empty element lists in tables indicating which finite elements a load was applied to, issues with graphical representation of panels resulting from overlapped panels, meshing problems, situations where you cannot change the design code, etc.

Meshing Problems:

If you are faced with a situation where panels won't mesh properly, you may also be in a situation where small inaccuracies in modeling have resulted in elements not being co-linear or co-planar. If Repair Mode does not resolve your problem, the next course of action to take is a careful review of the positions of adjoining elements to verify their accuracy.

1. Increase the display precision of calculation nodes in the tables

2. In areas where things are not lining up, use the "Detailed Correct" functionality to accurately line up elements. Be careful of simply merging nodes as this could move nodes *OUT* of alignment with other elements you intend them to be meshed together with. See section on Drawing Correction in Structure Modification – Chapter 10

About the Author

Ken is a former Structural Engineer and Autodesk® Revit® Quality Assurance Analyst. Ken has recently joined PTAC Edge for Revit (www.edgeforrevit.com) as a Senior Software Developer working on Revit API based solutions for the Precast Industry.

While working as a structural engineer, Ken was always looking for a better way to coordinate production drawings. After searching for a tool to do the job and finding nothing suitable, he started coding his own tools built on AutoCAD®. After a couple years of solo development effort he found Revit Structure 2.0. Even at this early stage, the product was so exciting that he jumped at the opportunity to join Autodesk and help make Autodesk® Revit® Structure an even more amazing and awesome product for structural engineers. Ken worked closely with Revit Structure over a 7 year tenure as a Revit Structure Quality Assurance Analyst. Probing the depths of Revit looking for bugs and participating in design teams in the role of subject matter expert, developing awesome new features and functionality for Revit.

Ken continues working to enhance the profession of structural engineering through developing Revit API based solutions for the precast industry as part of PTAC Edge for Revit. Ken truly enjoys working to create smoother workflows, accelerate operations, enhance coordination and make Revit an awesome product for structural engineers everywhere.

A champion of Revit for structural engineers, Ken hopes that our industry can eventually realize a fully integrated engineering analysis, design, modeling, and documentation workflow which is possible at the intersection of Revit and Robot Structural Analysis.

The future is bright.

Further Reading

Logan, Daryl L. (2011) *A First Course in the Finite Element Method* Cengage Learning, Stamford, CT

Kassimali, Aslam (2012) *Matrix Analysis of Structures* Cengage Learning, Stamford, CT

Bathe, Klaus-Jürgen (1996) *Finite Element Procedures* PHI Private Learning, New Delhi

Index

Made in United States
Orlando, FL
04 November 2026

90200173R00295